New Atlantis Revisited

New Atlantis Revisited

AKADEMGORODOK,

THE SIBERIAN CITY OF

SCIENCE

PAUL R. JOSEPHSON

PRINCETON UNIVERSITY PRESS

PRINCETON, NEW JERSEY

Library of Congress Cataloging-in-Publication Data

Josephson, Paul R.
New Atlantis revisited : Akademgorodok, the Siberian city of science /
Paul R. Josephson
p.   cm.
Includes bibliographical references (p.) and index.
ISBN 0-691-04454-6 (cl : alk. paper)
1. Science—Russia (Federation)—Akademgorodok (Novosibirsk)—History.
2. Russia (Federation)—Politics and government. I. Title.
Q127.R9J67   1997
338.947′06—dc21   96-45577

Portions of this book previously appeared in somewhat different form in " 'Projects of
the Century' in Soviet History: Large-scale Technologies from Lenin to Gorbachev,"
Technology and Culture, vol. 36, no. 3 (July 1995): 519–559, copyright © 1995
by the Society for the History of Technology, reprinted with permission from
University of Chicago Press; and "New Atlantis Revisited: Akademgorodok, Siberian
City of Science," in Stephen Kotkin and David Wolff, eds., *Rediscovering Russia in
Asia: Siberia and the Russian Far East* (Armonk, N.Y.: M. E. Sharpe, 1995),
pp. 89–107, reprinted with permission from M. E. Sharpe, Inc.

CONTENTS

ACKNOWLEDGMENTS

IN THE FALL of 1989 I first visited Akademgorodok. At that point I was gathering research materials for a political and cultural history of postwar Soviet atomic energy programs. From the first minutes of my arrival I could feel differences in lifestyle and work ethic that were to distinguish Akademgorodok from other Soviet cities. People were more helpful and open in all personal contacts. In contrast to hectic Moscow, the pace of life was slow and no crowd crushed you in the streets, although buses were often packed. But then, you could walk everywhere on beautifully wooded paths. The variety of food, clothing, and other goods in stores was smaller than in Moscow, but the lines rarely as long and the people always more polite. With great joy, in one of those stores I bought my son cross-country skis for less than a dollar that he used for two winters in New Hampshire before outgrowing them. Akademgorodok impressed me in all respects as a fine place to live and work. On this first occasion I finished preliminary research after an all too short ten-day trip, concluded a four-month stay in the USSR, and returned home.

A few months later Peter Dougherty, my editor at Princeton University Press, called me to find out if I knew anyone who might write a history of Akademgorodok. After a few months' discussion, I decided I was that person. I am grateful to Peter for suggesting the topic of this book and seeing it come to fruition.

In the fall of 1991 I returned to Akademgorodok. I had the good fortune to work in the archives of the Siberian Division of the Academy of Sciences; the Communist Party Archive of the Novosibirsk Region; and the archives of several institutes that figure prominently in this story. They are the Institute of Nuclear Physics; the Institute of Cytology and Genetics; the Institute of Economics and the Organization of Industrial Production; and the Computer Center, including the personal papers of computer scientist Andrei Ershov. I would like to express my thanks to the director of the Siberian Division, Valentin Koptiug, and to the directors of those institutes, Vladimir Shumnyi (Cytology and Genetics) and Anatoli Alekseev (Computer Center) for granting open access to their institutes. Andrei Trofimuk, director emeritus of the Institute of Geology and Geophysics, kindly granted access to his personal papers concerning Lake Baikal. Boris Elepov, director of the public library in Novosibirsk, put the resources of his staff and the library's collection at my disposal. I was also able to read the institutional newspaper at Sibakadem-stroi, the organization responsible for building Akademgorodok. Tatiana Khodzher helped me find my way at the Limnological Institute in Irkutsk; Raissa Astafieva made my visit to the Computer Center in Akademgorodok run smoothly.

I especially wish to thank my colleagues among the physicists. Aleksandr Skrinskii, director of the Institute of Nuclear Physics, endorsed any request I had for access to other facilities with remarkable speed; Stanislav Popov, its academic secretary, made it possible for me to work in the institute's archive (and to play basketball on Tuesday evenings); and Andrei Prokopenko, of the foreign department, facilitated my living and travel plans. Without their help I could never have accomplished what I did in such a short time.

Even more important was the assistance of several close friends in Akademgorodok who made life away from my family pass quickly. Experimental physicist Vadim Dudnikov and his wife, Galia, made me feel that I had a home away from home. Margarita Ruitova, a theoretician and expert on solar physics, and her husband, Dmitrii Ruitov, a plasma specialist, made certain that I met with their colleagues and friends who could tell me something about the history of Akademgorodok. Iurii Eidelman, an experimentalist, and his wife Violeta, shared their warm home and knowledge of Siberia with me on many cold nights. Theoretical physicist Viktor Fadin ensured that my early trip to Akademgorodok led to fruitful later visits. Biologist Sasha Kerkis and his family often invited me to share in the good Siberian life. They welcomed me into their homes, took me to the market, on drives in the country, even into the sauna. Natalia Pritvits and Zamira Ibragimova shared with me their great knowledge of the history of Akademgorodok and Siberia in their written work and in warm discussions. Evgenii Vodichev always tried to find fresh beer and introduced me to other historians who shared our love of Akademgorodok.

A number of individuals read early drafts of chapters, helping to make the final product better: Julia Thompson, Vadim Dudnikov, Igor Golovin, and Evgenii Shunko, the physics chapter; Deborah Fitzgerald, the biology chapter; Sarah Allen, Greg Crowe, and Vadim Kotov, the computer science chapter; Ray Clarke and Mike Bressler, the chapter on the environment; Regina Arnold, Frank Roosevelt, Susan Linz, and Cindy Buckley, sections of the economics and sociology chapter; Linda Lubrano, the chapter on the Communist Party; and Evgenii Vodichev, who commented on the introductory chapters. Zora Essman and Lydia Kesich polished and annotated my translations of poems. Boris Mordukhovskii, one-time photographer for Sibakademstroi, and Rashid Akhmerov, the official photographer of the Siberian Division of the Academy of Sciences, contributed photographs to this history of Akademgorodok.

The National Endowment for the Humanities, National Council for Soviet and East European Research, the International Research and Exchanges Board, the Fulbright-Hays Faculty Research Program, the Russian Research Center at Harvard University, and the Dibner Institute all contributed financial and/or institutional support to this book. Charles Constanza assisted me with the research effort early on.

Most important, Cathy Frierson, my colleague, friend, critical reader, and wife, and Isaac Josephson, my sports companion and son, let me leave them in New Hampshire while I went off to Siberia, and then gave me the room to write this book upon my return. Without my family I could not have experienced the great joy of having written the story of a Baconian city of science in the USSR.

I dedicate this book to my Mother, Lili Josephson, and to the memory of my Father, Jules Josephson. They paid for my piano lessons, only suspecting that I was instead sneaking off to Forbes Field to watch the Pittsburgh Pirates play.

Paul R. Josephson
Durham, New Hampshire

NOTE ON TRANSLITERATION

I HAVE USED a modified version of the Library of Congress system of transliteration throughout this book. In the sources I remain faithful to the system but in the text have dropped virtually all soft and hard signs and diacritical marks, especially for names, for ease in reading—hence "Lavrentev," not "Lavrent'ev"; "Vasiliev," not "Vasil'ev"; and so on.

IN DECEMBER 1956, the mathematician Mikhail Alekseevich Lavrentev journeyed to the Siberian industrial center of Novosibirsk where he met with T. F. Gorbachev, a geologist and chairman of the West Siberian branch of the Soviet Academy of Sciences. They walked through the Golden Valley, a hilly, forested region, twenty-five miles south of the city and twenty-five hundred miles east of Moscow on the western shore of the Ob River. They agreed it would be a fine place to build Akademgorodok, the Siberian city of science. The scientific institutes of Akademgorodok would be located somewhat to the north of this site. The Golden Valley itself would be preserved for a handful of cottages for the leading scientists, several orchards connected with biological research, and trails used for hiking, cross-country skiing, and gathering mushrooms. In 1958, in the middle of that site, construction on Akademgorodok began. A mere seven years later, fifteen scientific research institutes had opened. Tens of thousands of scientists and administrators, accompanied by their families, began their life and work anew.

Akademgorodok is noteworthy in world and Soviet history. Visionaries have long imagined the creation of a utopia where a community of scientists could toil away in the pursuit of knowledge for the sake of knowledge. Divorced from such pressures as public accountability, the need to beg government and industry for financial support, or concerns about how the fruits of their research might be applied, the scientists would somehow discover the "objective truth" about nature. Most modern scientific societies have been modeled to varying degrees on this view of the appropriate relationship between scientists, their work, and the broader political, financial, and moral questions that have an impact on research. The question for Mikhail Lavrentev was whether Akademgorodok could ever approach the utopian dream of science, free from political, financial, and moral constraints.

From the point of view of the Soviet experience, Akademgorodok was significant as a symbol of de-Stalinization, the postwar democratization and decentralization of scientific forces, and the development of rich Siberian resources. Under Stalin, science in Moscow and Leningrad had prospered at the expense of the periphery. Indeed, in 1934 Stalin required the transfer of the Academy of Sciences from Leningrad to Moscow, close to his suspicious eyes. Whole areas of research were dominated by individuals or their institutes. This thwarted the growth of entire fields and led in such cases as genetics to pseudoscience. The overly centralized, highly bureaucratized Soviet economic system presented its share of obstacles to the smooth conduct of research. The founders of Akademgorodok intended to create world-class scientific research institutes free from these forms of political, ideological, and economic pressures. Akademgorodok was also prominent in the history of planned cities in Russia, whose tradition dates to the construction of St. Petersburg by Peter the Great in the early 1700s, an event accompanied by the founding of the Imperial Russian Academy of Sciences in that city in 1725.

Much of Akademgorodok's great promise lay in the faith of the independent-minded researchers who gathered there in the power of rational science to conquer any problem, so long as they could avoid political and economic uncertainty. Many of the first Akademgorodok scientists were young recruits, the so-called children of the twentieth party congress at which Khrushchev had set in motion the de-Stalinization thaw. In Akademgorodok a kind of glasnost, or openness, reigned that presaged that which developed under Gorbachev decades later. The roots of this openness were the city's geographical and psychological distance from Moscow and the culture of informal exchange of ideas explicitly intended by Akademgorodok's founders to foster the creative impulse among researchers. Openness extended down the corridors of scientific research institutes to newly founded social clubs and cafés whose very existence was largely unheard of elsewhere in the Soviet empire. Here the scientists talked about their research fields at the cutting edge, and also read poetry, viewed art exhibitions, and listened to traveling bards. This openness was facilitated by Akademgorodok's location, far from Moscow, the central party apparatus, and strict ideological control. The city of science had its own party organization that was not nearly as vigilant as Moscow intended, especially at the level of the primary party organizations in the institutes.

The Siberian scientists accepted intellectual freedom as a given, exploring fields previously closed to them. Previous restrictions may have been the result of philosophical prohibitions owing to the alleged idealism of a particular field or of political centralization that precluded new investigations. Scientists in Akademgorodok opened physics, biology, sociology, economics, and computer science to new investigations. Physicist Gersh Budker employed new techniques for colliding beam accelerators to achieve unheard of luminosities in high energy physics. Other physicists produced dramatic results in alternative fusion research. Scientists used studies in animal husbandry and plant hybridization to commence research in genetics even before the condemnation of T. D. Lysenko in 1965. Tatiana Zaslavskaia freed Soviet sociology from increasingly metaphysical Marxian concepts of class and society to study labor migration and worker (dis)satisfaction, studies which revealed that all was not well in the Soviet countryside. In economics, Abel Aganbegian accelerated the rebirth of interest in linear programming and scientific management of labor. In computer science and technology, future Academy of Sciences president Gurii Marchuk and Aleksei Liapunov, among others, struggled to overcome resistance to cybernetics in the USSR. Owing in part to a relatively open atmosphere, they achieved significant results in short order in these fields. Both the pace and scope of the developments in these disciplines quickly became distinct from the other branches of the Soviet scientific establishment.

Yet the utopian designs of these scientists and Communist Party leaders to build a city of science fell prey to the same handicaps endemic to Soviet science in general. Ideological constructs, political desiderata, and economic uncertainties dogged the efforts of Akademgorodok's founders to create a unique scientific community in the west Siberian forest. This community of scientists never fully escaped the constraints imposed by a centralized command economy and a party apparatus that grew increasingly conservative and ideologically vigilant in the Brezhnev years. As in other societies, the power of the purse always shapes the face of research, and the USSR was no different as the state pressured scientists to abandon basic science in the name of accountability to Siberian economic development.

Local and regional party officials came to fear the autonomy of the scientists, indeed of any interest group. They mistrusted western science and its attendant ideological precepts that seemed to be gaining a foothold in some fields. In these circumstances, Akademgorodok remained a scientific utopia only until 1968 when, in response to the affair of the *podpisanty* (signatories), a festival of bards, and the invasion of Czechoslovakia, the Communist Party brought its wrath down on the scientific community. The signatories had extended Akademgorodok's openness beyond the limits tolerable to the party apparatus when they signed a letter protesting the regime's violation of international standards of human rights in the persecution of dissidents. The festival of bards, popular among young and old alike, encouraged naked satire of Soviet political institutions. The invasion of Czechoslovakia was accompanied by a nationwide crackdown on dissent in the USSR.

Still, a uniquely Siberian science emerged in Akademgorodok. It was Siberian in terms of personnel, for many of the young recruits came from Siberia and eventually became laboratory directors. It was Siberian also in terms of novel research approaches. In addition to the exploration of previously closed fields of study, Akademgorodok scientists were directly involved in the study of Siberian natural resources. This led them in some instances publicly to oppose policies emanating from Moscow concerning economic development and to become involved in a nascent Soviet environmental movement to preserve Lake Baikal and derail the costly, irrational project to divert the flow of Siberian rivers to Central Asia. Finally, Akademgorodok was Siberian even in terms of its novel architecture and city planning.

The Soviet New Atlantis

In 1608, during the early years of the Scientific Revolution, Sir Francis Bacon, a British statesman, scientist, and philosopher, and father of the modern inductive method of reasoning, composed his *New Atlantis*. *New Atlantis*, one of the major utopias of western literature, described the activities of Salomon's House, a learned society and city of science that served as the model for the British Royal Society. In Salomon's House, scholars searched for knowledge of and power over nature, using laboratories with modern equipment, instrumentation, and libraries to facilitate research. The purpose of this scientific center was "the knowledge of causes and the secret motions of things; and the enlarging of the bounds of human empire, to the effecting of all things possible."[1] For Bacon, as for his contemporaries and many social commentators since, "science" meant the acquisition of objective knowledge of nature to improve the human condition. Akademgorodok, too, was intended to subject Siberian natural resources to economic development. Many of its main goals involved the cataloguing and study of indigenous flora and fauna and the charting and exploitation of minerals, timber, and water and other natural resources.

Akademgorodok was not only the product of the Baconian scientific tradition but also the logical outcome of decades of Soviet political, economic, and ideological developments. On the eve of the Russian Revolution, biogeophysicist V. I. Vernadskii and playwright Maxim Gorky, among others, advanced ideas of something akin to Salomon's House in order to raise the cultural level of Russia and to bring the backward nation into the twentieth century as an industrial and technological power. Throughout Soviet history leading party representatives, economic planners, and members of the scientific intelligentsia hoped to build socialism on the basis of science and technology. Scientists would uncover the secrets of nature; the Soviet state would master them and, in the process, build a modern industrial power. As in Salomon's House, Akademgorodok's success in facilitating these ends would be secured through a series of architectural, organizational, and scientific innovations. Its architecture would include styles that meshed with the beauty of Siberian pine forests, its spaciousness encouraging openness in personal and scientific exchanges. Organizational innovations would foster a symbiotic relationship between science, education, and industry, something that had failed to materialize elsewhere in the postwar USSR. The scientific innovations concerned entirely new fields and methodological approaches.

The conditions were propitious for the endeavor. First, there was broad support nationwide for the expansion of science as part of a "cult of science." The cult was initiated by postwar successes in nuclear weapons research and development, was cemented by achievements in the peaceful uses of atomic energy and space exploration, and was part of the general environment of de-Stalinization. It provided fertile soil for embracing science and technology as a panacea for Soviet economic, political, and social problems. Although ideology and politics eventually hindered Akademgorodok's normal operations, they were crucial to its success during its planning stages. Officials saw modern science and technology as the key to constructing communism and surpassing the economic production of the capitalist world.

The city of science was crucial from a strategic standpoint as well. The great human and capital costs of the Nazi invasion and occupation during World War II underlined the importance of developing Siberian wealth far from Russia's European borders, east of a natural geographic barrier, the Ural Mountains. Akademgorodok would achieve the slogan promulgated in the 1920s, "Science to the Provinces!" In order to become more than a slogan, however, forces of localism, openness, and freedom would have to counter those of centralization in the Soviet economic and political system.

Third, officials believed that Akademgorodok would contribute to the reinvigoration of Soviet science. The scientific enterprise had made remarkable strides since the Revolution in terms of numbers of researchers, institutes, publications, and other quantitative indexes of performance. During Khrushchev's rule the rate of growth of each of these categories increased rapidly. And there were always successes in space or nuclear power to consider. Unfortunately, from a qualitative perspective Soviet science lagged. Whether by such indexes as Nobel prizes and scientific citations, or by more subjective evaluations such as those offered by the western peers of Soviet scholars, Soviet science did not fare as well. In many areas of biology, chemistry, the new field of computers, in fields both fundamental and applied, Soviet science had stumbled, and in spite of Soviet pronouncements, the promise of communism seemed a long way off. Furthermore, the Soviet economy only grudgingly introduced the achievements of scientific research and development into the production process. The question was how best to wed the advantages of the Soviet system with the power of science, to accelerate economic development, promote social welfare, and compete with capitalism. Scientists and officials alike believed that Akademgorodok, like Sputnik and nuclear power before it, would serve as a symbol of the innate advantages of the Soviet social system over the capitalist system, and could not fail to contribute to the improved performance of Soviet science.

The project for a city of science logically found great response in this environment. Scientific and political authorities carried out a large press campaign, announcing the plan to build Akademgorodok and trumpeting every success that followed. Leading officials in the USSR Academy of Sciences shouted their approval. Those scientists who accompanied Lavrentev to Siberia recognized Akademgorodok's uncertain future but dreamed of its limitless promise, and they gave their scientific and organizational talents to the city. Young communists signed on in droves to assist in the task of construction. Promising students flocked to the newly opened institutes to commence scientific careers. Indeed, Akademgorodok had the crucial endorsement of the general secretary of the Communist Party himself, Nikita Sergeevich Khrushchev, who ordered that Lavrentev be given the full support of Novosibirsk economic planners and policy makers.

Ultimately, however, the freedom the scientists enjoyed came under assault from the Communist Party. The central party apparatus grew increasingly conservative once Brezhnev replaced Khrushchev. As part of a general ideological clampdown on all sorts of literary, artistic, and political freedoms achieved during the de-Stalinization thaw, the party campaigned against the autonomy of the scientists in matters of research. It demanded devotion to the ideological precepts of the Brezhnev era. Regarding science, these commandments included warnings against becoming "tainted" by western scientific or philosophical thought, and the requirement that applied science assume preeminence over basic research. The party instructed Akademgorodok scientists to concentrate on the big science of Siberian economic development. Officials in the local and regional party apparatus used these shifts in Moscow to force changes on the local level. These officials had long envied the favor Lavrentev enjoyed at the top of the party, including his personal acquaintance with Khrushchev and his individual pass to the Kremlin. They also resented Akademgorodok's unusual access to goods and services, and the scientists' intellectual freedom. Faceless bureaucrats were enlisted to "administer" science; before, they had "supported" its development. Long-term problems of equipment supply and shortfalls of financial support increasingly handicapped efforts to undertake new programs. The party apparatus vigilantly rooted out perceived ideological deviations after the Soviet invasion of Czechoslovakia in 1968. Natural aging of the young, enthusiastic, and talented scientific population had begun to run its course. Party officials eventually shut down the social clubs, symbols of Akademgorodok's openness. Its entire atmosphere was redrawn in gray, monotonous tones, devoid of the originality that had characterized scientific thought for the first decade of the city's existence. And the glory days of rapid construction, scientific achievement, and intellectual freedom came to an end.

PERSONALITIES, INSTITUTIONS, AND POLITICS

The history of Akademgorodok is the history of the interaction of personalities, institutions, and politics. Creating the city involved assembling a scientific diaspora in Siberia. Akademgorodok's visionary founders gathered other visionaries to fill the new institutes and the university. Many were scientific renegades. Others, because of outspoken personalities, simply did not fit in among the scientific status quo in Moscow and Leningrad. These scientists rapidly built physics, chemistry, mathematics, and biology research centers in Akademgorodok. They began to train talented, if very young Siberian students to do research with them.

Three interrelated tensions hampered the efforts of these individuals and their institutions to carry out their research programs. The first concerns the ironic tension between the view of science as a supremely rational endeavor that holds a central place in Soviet science and the insistence of many administrators, economic planners, and scientists alike that science is inherently political since it must reflect broader social goals and cultural aspirations. Whether in Lenin's writings on the development of capitalism or his fascination with Taylorism, Bukharin's deterministic version of historical materialism, Stalin's urgings for the worker to master science and his infatuation with large-scale technologies, or Khrushchev's embrace of scientists and a cult of science, the application of objective knowledge was intended to accelerate the construction of communism.

Many of the visionaries who gathered in Akademgorodok shared a view of science as something supremely rational. They hoped to see science applied to the solution of a wide range of economic and political tasks. Armed with the achievements of modern science and technology, they, and Soviet ideologists, proclaimed there were no fortresses the communists could not conquer. The assumption was that science involved discovery of natural laws, a value-neutral endeavor that unfailingly served those who embraced it in their pursuit of the control of nature. This explains the central place in Akademgorodok of the mathematical, physical, and technological sciences—physics, geography, geophysics, computer science, and mechanics. At the same time, however, science could not be separated from the culture and polity in which it developed. There was tension, in a word, between the rational, utopian ideal of modern science and the Soviet reality of a postwar society undergoing rapid economic and political change.

A second major issue concerned the importance of Akademgorodok science as a symbol of the decentralization of the scientific research and development apparatus in the postwar USSR. By the time of the city's founding, most leading scientists recognized the importance of "decentralizing" science away from Moscow and Leningrad. This provided an important impetus to the creation of Akademgorodok. But this impetus could not last long in a country whose institutions were so highly centralized. In time, the government bureaucracies and political organizations that were instrumental in the city's construction exerted pressure on Akademgorodok to conform to Soviet norms. These norms included the widely accepted view of science as a branch of the economy and a strategic military force. As such, Siberian science had to compete with all other branches of the economy to secure funding, manpower, and machinery and equipment. After initially garnering sufficient political favor and resources to begin its operations, Akademgorodok came to be seen by Moscow policy makers as no different than any other scientific organization. Like other scientific centers, it was intended to produce big science and applied science in service of the economy. Shortfalls of rubles, hard currency, machinery and equipment, and building supplies plagued construction from the start. Akademgorodok eventually became beholden to central economic and political organizations. All these phenomena sharply curtailed the development of a special, Siberian science.

The third tension, and perhaps the most crucial in determining Akademgorodok's fate, involved growing pressures from national and local economic and party organizations to conduct big science projects at the expense of little science, and applied science at the expense of basic research. In the case of high energy physics, the pressure originated in the physics community itself since the scientists realized they would soon fall far behind their western counterparts if they did not build larger and larger accelerators. But on the whole, the pressure for big science came from Brezhnev's central party apparatus, an outgrowth of the competition with other superpowers for prestige, economic might, and military strength. Paradigmatic of Brezhnevite big science were grandiose projects signaling the glory of Soviet rule, such as a new trans-Siberian railroad called "BAM" and the "Siberia" economic development program, both of which are discussed in this book.

I am not so concerned with evaluating the successes and failures of Akademgorodok according to world scientific standards. Soviet scholars achieved a great deal in many fields while being isolated from the international scientific community. They faced obstacles to doing research uncommon in the West. For example, the physicists had to move their first particle accelerator from Siberia to Moscow and back, all the while trying to beat Stanford physicists to the first results. The biologists had to labor under Lysenkoist prohibitions against genetics research until 1965. The computer specialists had to work out their own operating programs without any support from the computer industry. I am more interested in understanding the performance of Akademgorodok scientists from the perspective of Soviet standards. More important, in my view, is the fact that perhaps only in post-Stalinist Russia could such an undertaking as the construction of a city of fully equipped research institutes in the middle of the Siberian forest have taken place in the face of political, financial, and logistical problems that scientists in other societies may have found too daunting.

I have chosen to organize this book around an investigation of these themes or tensions in each of several disciplines, devoting most of my attention to the glory years of the 1960s. Chapter 1 explores the genesis of the idea of Akademgorodok, its organization and construction as conceived by the mathematician Mikhail Lavrentev and others. I describe the problems encountered in building and staffing fourteen scientific research institutes (now twenty-one) in the middle of Siberia. Next I turn to the community of scientists who gathered in Akademgorodok, whose history I explore both through their institutions and disciplines. The theme of the gathering of a scientific diaspora will appear in other chapters as well.

Chapter 2 considers the work of leading specialists in high energy physics and thermonuclear synthesis (fusion). Under the direction of the iconoclast Gersh Budker, Soviet physicists succeeded against all odds in building a colliding beam accelerator in competition with physicists at a number of western research facilities. Ultimately, Budker's desire to build larger accelerators conflicted with Lavrentev's vision of Akademgorodok as a city of moderately sized institutes. So Budker's next innovation was to have his institute produce industrial accelerators for sale at home and abroad to generate money to expand fundamental research.

Chapter 3 focuses on the painful, premature effort of Soviet biologists to escape Lysenkoism in Siberia. This chapter provides a fascinating case study of the relationship between politics and science. Such biologists as Nikolai Dubinin and Dmitrii Beliaev attempted to defend their nascent discipline, while Khrushchev, still a supporter of Lysenko, sought to close their Institute of Cytology and Genetics. After Lysenko was finally disgraced, geneticists labored to overcome nearly three decades of underground existence, but with insufficient financial support from the Brezhnev government. The result was an emphasis on applied programs in plant selection and animal husbandry rather than fundamental research, an emphasis that handicaps programs involving recombinant DNA research to this day.

Chapter 4 considers the development of computer science and technology in Novosibirsk. While subject to significant systemic impediments elsewhere in the Soviet Union, computer science found a healthy environment for development in Akademgorodok. Andrei Ershov, a visionary in artificial intelligence and programming, almost single-handedly raised the level of achievements to world heights. Aided by the fact that both Lavrentev and his successor as chairman of the Siberian division, Gurii Marchuk (later president of the Soviet Academy of Sciences), were specialists in mathematical methods and modeling and propagandized their application in other fields, computer science found response in all institutes of Akademgorodok: sociology, economics, physics, chemistry, and biology.

Chapter 5 continues the examination of big science, in this case its impact on environmentalism and the ecological sciences in Siberia. Such grandiose visions of the future proposed by planners and engineers as the Siberian river diversion project and the construction of a paper mill on Lake Baikal led scientists in and around Akademgorodok to unite in opposing them. A surprising union of such traditional, party member scientists as geophysicist Andrei Trofimuk and the independent Grigorii Galazii, a talented limnologist, ultimately succeeded in derailing the far-fetched projects of the so-called nature-creators.

Akademgorodok served as a center for the development of long dormant approaches in economics, sociology, and survey research, which chapter 6 explores. Quantitative economic analysis and modeling of regional development, the study of migration patterns and labor reserves, and surveys of job satisfaction and family income provided a foundation to question the long-held assumptions of Soviet leaders about the social achievements of socialism. Economist Abel Aganbegian and sociologist Tatiana Zaslavskaia found academic freedom in Akademgorodok to challenge both the dogma of leading scholars in their fields and to create new disciplines.

Chapter 7, the final chapter, brings the reader full circle, using rare archival documents to evaluate the relationship between science, politics, and ideology in the USSR. Here I examine the impact of the Communist Party's call for scientists to be accountable to the political, economic, and philosophical tenets of communism. I argue that ideological considerations were crucial in determining Akademgorodok's fate. For on the basis of these considerations the openness that scientists had known was ended. This was not simply a matter of tying scientists' research more directly to the party's economic programs. The crackdown on academic freedom that followed the affair of the signatories in 1968 made the Soviet New Atlantis a short-lived dream. Still, in a number of fields, from high energy physics to economics and sociology, alternative views persisted that were important both to those fields of science themselves and to undercurrents of criticism concerning how the Soviet Union had evolved and where it was going. An increasingly well-educated Soviet "middle class" began to question fundamental assumptions about Soviet politics and society. Ultimately this element in Akademgorodok would contribute actively to Mikhail Gorbachev's policy of perestroika.

A brief epilogue indicates the impact on Akademgorodok of the ongoing political and economic changes in Russia. Here I suggest that nascent market mechanisms and political reform are no more a panacea for modern scientific research than were Soviet centralized planning and ideological control. Notwithstanding the political and economic crises enveloping Russia, Akademgorodok remains a center of scientific excellence with great promise for the twenty-first century.

New Atlantis Revisited

CHAPTER ONE

From Moscow, Leningrad, and Ukraine to the Golden Valley

"From the integral to metal!"

—*A founding slogan for Akademgorodok*

AFTER HE HAD gained government and Communist Party approval for the construction of Akademgorodok, Mikhail Lavrentev faced an unending series of obstacles in realizing his vision for the creation of the Siberian city of science. The construction effort was a source of constant disappointment. The builders encountered shortages of materials and equipment. A capricious climate interfered with schedules. Cost overruns led to the abandonment of more expensive bricks for less attractive prefabricated concrete forms. Poorly skilled workers had trouble even laying the huge concrete slabs that were used in apartments and research institute buildings, almost interchangeably, as floors, walls, and ceilings. In the effort to cut costs, the plans gave low priority to social services such as day care facilities, hospitals, and stores. Roads, plumbing, natural gas and electrical services, and telecommunications all lagged behind Akademgorodok's needs. Still, drawing on Khrushchev's direct support, on the environment of reformism of fundamental research policy within the Soviet Academy of Sciences, and on his own tremendous energy, Lavrentev was able to bring Akademgorodok to fruition by the mid-1960s.

Lavrentev seized on two innovations that were crucial to his success. One centered on the physical plant itself, where expansive architectural styles unique to the postwar Soviet Union were applied in a setting of natural beauty—a Siberian forest. This physical innovation encouraged a relaxed atmosphere for scientific research where free discussion reigned and where junior and senior scholars had far more contact than in most Soviet scientific settings. Lavrentev and his colleagues were convinced that Akademgorodok's success depended to a large degree on maintaining the right mix between the natural geographical setting and the architectural design.

The second innovation concerned the administration of the research institutes, which was to be more decentralized, if not more democratic, than that of the scientific status quo of the USSR Academy of Sciences in Moscow. Lavrentev wanted institute directors to talk freely and informally among themselves, and he made himself accessible as well. By example and request, he encouraged them to be more open to their staffs than was the norm. In reality, of course, administrative style differed from institute to institute. Some, like the biologist Dmitrii Beliaev of the Institute of Cytology and Genetics, were imperious. But many, like the physicist Gersh Budker and Lavrentev himself, truly were more democratic. They welcomed discussion of issues scientific and social, and occasionally even art and politics.

This administrative innovation was closely connected to three principles, often referred to as Lavrentev's "triangle." The first side of the triangle was an emphasis on fundamental research based on an interdisciplinary approach. The goal was to ensure healthy feedback between the various branches of the sciences through the broad application of mathematical approaches, methodology, and languages. After all, cybernetics, genetics, computer science, and mathematics shared common features and approaches. The healthy interdisciplinary environment that dominated Akademgorodok was facilitated by the fact that many of the first scientists to arrive in Siberia had already adopted this approach in their consideration of pressing scientific issues. The fact that they had warm personal relationships with many individuals in institutes throughout the city of science contributed to the spread of this approach. The second side of the triangle, albeit a smaller one in practice, was the creation of a strong tie between science and production. Although this was always called for in the USSR, in Akademgorodok it was to be achieved by innovations in structure and management, including the formation of industrial research institutes and design bureaus located nearby. It was hoped that personnel from industrial institutes and from institutes of the Academy of Sciences would spend time in each other's facilities, learn from one another, and thus create this "tie." The third side was the training of "cadres," from which scientists and engineers of high quality would emerge as the product of close contact between scholars in the research institutes and students at the newly founded Novosibirsk State University. Training was crucial since so few scientists inhabited Siberia, far fewer than needed in Akademgorodok.

Akademgorodok scientists were continually frustrated in their efforts to achieve their ideals in geography, aesthetics, and administration. The building of the city required nearly superhuman effort. The Novosibirsk *sovnarkhoz* (the regional economic organization responsible for allocating most of the machinery, equipment, and building supplies) had to create a construction industry from scratch, before ever turning to the erection of institutes, apartment complexes, and stores. It had to lay roads, piping, sewage, and electric wiring. Supplies and equipment were shipped by trains that grew to be several kilometers long and always arrived later than needed. When buildings were finished, the laboratories and institutes that would fill them, often still in Moscow or Leningrad, were also loaded wholesale onto trains for the long journey to Siberia, and then they too had difficulties acquiring modern equipment. Construction brigades materialized slowly out of the taiga and often consisted of young, untested volunteers and conscripts. The question yet to be faced was how to find individuals willing to move to the uncertain promise of Akademgorodok. In spite of support from party, scientific, and economic organizations, the tasks proved to be daunting. Lavrentev, armed with administrative experience, a well-earned scientific reputation, and a vision of the future, hoped he was up to the effort.

Lavrentev Meets Novosibirsk

Mikhail Alekseevich Lavrentev,[1] the inspiration for Akademgorodok, was born in 1900 in Kazan into a fairly well-to-do family that had pulled itself up from modest origins. The intellectual environment in the household was westernized, embracing enlightenment views of science and rationality. His father was a well-known mathematician, and such leading scholars as Nikolai Luzin often met at his home. Luzin himself was ultimately attacked in the Soviet press in the mid-1930s for giving western science priority and praise over Soviet science. Lavrentev fell in love with mathematics in high school. He attended Moscow University where he studied with Luzin and specialized in differential equations, the theory of functions, and variational calculus. In 1926 Lavrentev defended his candidate of science dissertation. In 1927 he was chosen as a member of the prestigious Moscow Mathematical Society and was sent to Paris for half a year of study. This was a brief period in Soviet history when scientists were able to travel and study abroad for extended periods, often with foreign funding from such organizations as the Rockefeller Foundation's International Educational Board. Foreign contacts were sharply restricted in the 1930s when the xenophobic Soviet leadership under Stalin attempted to build "socialism in one country" and kept scientists and engineers from accepting these important research opportunities.

For a brief time Lavrentev worked under the mathematician Sergei Chaplygin in the theoretical department of the Central Aviation Institute on problems of hydrodynamics, turbulence, and aerodynamics. The institute was important in the creation of the Soviet navy and air force. Through this experience Lavrentev became interested in the theory of explosions and their application for economic purposes. Lavrentev received two doctoral degrees in the early 1930s, one in the technological sciences, the other in the physical-mathematical sciences. At the aviation institute Lavrentev also became acquainted with the young Mstislav Vsevolodovich Keldysh. Keldysh became president of the USSR Academy of Sciences during the crucial early years of Akademgorodok's history. Keldysh seems to have worried that Lavrentev's Siberian venture would eat into his own power and authority as Academy president, but, remembering their early association, he did not hesitate to help Lavrentev at crucial junctures. Lavrentev then worked in the Steklov Mathematics Institute in Moscow.

In 1939 the president of the Ukrainian Academy of Sciences, A. A. Bogomolets, invited Lavrentev to Kiev to become director of its Institute of Mathematics. This was a great opportunity for career advancement since he would have to compete with so many fine mathematicians in Moscow at the Steklov Institute. Lavrentev was elected academician of the Ukrainian Academy and was soon elected vice president. Bogomolets's own policies convinced Lavrentev that personnel, not organization, were central to the success of science, and he carried this knowledge with him to Akademgorodok. Lavrentev's contacts with Ukrainian scientists were vital when he sought out personnel for Akademgorodok. Lavrentev's own research in Ukraine was interrupted by World War II, but he and his colleagues worked on, carrying out artillery calculations. During this time Lavrentev also became acquainted with Khrushchev, first secretary of the Communist Party of Ukraine.

After World War II Lavrentev's interests in mathematical methods, education, and new approaches to the organization of science came together. As institute director and Academy vice president he focused on rebuilding Ukrainian mathematics. He invited the Muscovite Sergei Alekseevich Lebedev to begin work on modern computers, far away from the gossip in Moscow that cybernetics was a "pseudoscience." (Lavrentev was instrumental in saving both cybernetics in the late 1940s and genetics in the late 1950s from permanent stultification.) Lavrentev, now a leading mathematician, joined with Keldysh, the mathematician Sergei Lvovich Sobolev, a future Akademgorodok cosponsor, and many others, to create the Moscow Physical Technical Institute, a Soviet equivalent of the Massachusetts Institute of Technology, which was set up to train specialists for the burgeoning nuclear weapons effort.

The road signs to Akademgorodok were now in place. Lavrentev's westernized upbringing and training gave him faith in the power of modern science. His experience with Ukrainian Academy president Bogomolets and at the Moscow Physical Technical Institute convinced him that there ought to be a scientific setting in the USSR that included a university whose "life's blood" of teaching, research, and science supplied young scholars with the tools necessary to undertake world-class science. His training as a mathematician convinced him of the efficacy of its application in other fields—chemistry, biology, and physics—and of the vitality of an interdisciplinary approach. His early interest in computers would find response in the creation of the Computer Center in the Institute of Mathematics, and in the centrality of cybernetics in the research programs of all Akademgorodok institutes. His fascination with explosions for geological engineering and high-pressure water cannons for mining and excavation served as a central focus of his Institute of Hydrodynamics—the first Akademgorodok institute to open. And his work with other scientists persuaded him of the need to bring whole "schools," research groups or entire laboratories, not just individual scholars, to serve as the kernel of future institutes.

In the 1950s, and especially after Stalin's death, like other leading scientists Lavrentev lobbied behind the scenes to bring about administrative reforms in Soviet science. It was a science overly bureaucratized and dominated in many fields by a few leading institutes or individuals whose eminence often derived as much from political favor as from scientific excellence. Ideology and politics impeded the smooth conduct of research, including the overvaluation of applied science at the expense of basic research and the energetic rooting out of all perceived ideological deviations. Lavrentev believed that the decentralization of science policy and reemphasis of the great Russian tradition of fundamental research would go a long way toward rectifying the situation. Of course Lavrentev hoped to benefit personally by increasing his own political power and scientific authority through any adopted reforms.

At Akademgorodok Lavrentev's vision of the "triangle of science" would take shape. He believed that a city of science could be designed to take advantage of the relationship that existed between science and technology in the postwar world, a world that had emerged from the nineteenth century when thirty years might pass before a scientific advance found its way into the productive process to a time when that discovery found practical application in only a matter of years or even months. To take full advantage of that accelerated tempo, Lavrentev intended to organize Akademgorodok so that its research institutes had design bureaus, factory laboratories, and industry located nearby. This would result in an "innovation beltway" surrounding the city. Unfortunately the beltway never performed as intended owing to the fact that the ministries who paid for research were impatient for a return on their investment while scientists were unable to play the role of engineer or entrepreneur.

A second rationale for Akademgorodok was the increasingly interdisciplinary nature of modern science. For decades Soviet science had been organized along narrow themes or disciplines, with one institute in charge of a particular problem. Individuals ran entire fields like fiefdoms, parceling out assignments as they saw fit. With the advent of "big science" which required the input of specialists from a number of fields—space exploration and atomic physics, for example—Lavrentev argued that institutes could no longer be permitted to operate in this way. He wanted to design Akademgorodok so that its research institutes were geographically proximate, allowing for the natural development of close personal contacts between the scientists. He also hoped to avoid the drawbacks of "big science," namely, massive institutes requiring huge sums for each project. Akademgorodok would have moderately sized institutes united by a common philosophy. Computer science and mathematics would provide the methodological language. The third aspect of Lavrentev's "triangle," the training of young scientists, is explored in detail later in this chapter.

Akademgorodok would never have lived up to its goals had Lavrentev not provided able leadership. He wielded great power but did not abuse it. All indications are that although he could be inflexible and dismissive, usually he was a fair and judicious man, a man with vision and energy rare in this world. Those who remember him refer to his occasionally sharp criticism and sarcasm. They recall his tense disagreement with Budker over the size of the physics institute when Budker pressed to build ever larger particle accelerators. To

Lavrentev, the modern accelerators had become an affront to the human scale of Akademgorodok. But they praise Lavrentev for not escalating this conflict into sharp personal attacks so often the standard in Soviet science. Whatever Lavrentev's personal qualities, he would never have been able to achieve any of his goals had he not had the support of Nikita Sergeevich Khrushchev.

KHRUSHCHEV AND AKADEMGORODOK

For three years after Stalin's death in 1953 a power struggle unfolded in the Kremlin between Politburo members Khrushchev, Malenkov, Molotov, and Kaganovich. When Khrushchev finally emerged as first among equals in 1957, he unleashed the forces of de-Stalinization to secure his position. De-Stalinization was set in motion by Khrushchev's so-called secret speech at the twentieth Communist Party congress in 1956. It consisted of a kind of reformism and internationalism not seen since the 1920s. On the domestic front, although some coercive aspects of party rule—arrest, intimidation, media control, and so on—were retained, an increasingly well-educated stratum of society—writers, poets, technologists, engineers, and scientists—were permitted to participate in the policy process. Indeed, the input of specialists was required after the war to rebuild the Soviet economy, rejuvenate agriculture, build housing, and manufacture consumer goods. In the international arena, the thaw yielded "peaceful coexistence" with the West, whereas previously "hostile capitalist encirclement" and the inevitability of war held sway in Soviet ideological pronouncements.

The de-Stalinization thaw was more a series of haphazard reforms than a full-scale relaxation of restrictions on public and private life. Khrushchev permitted publication of Aleksandr Solzhenitsyn's *One Day in the Life of Ivan Denisovich* (1962), but sanctioned the repression of Boris Pasternak's *Doctor Zhivago* (1957) and refused to allow Pasternak to accept the Nobel prize in 1958. Soviet literary journals and academic forums were filled with debate unheard of months earlier, yet Trofim Lysenko was allowed to ravage genetics until 1965. Still, it was a time of vibrant intellectual and cultural activity during which constructivist visions of the communist future were resurrected. Scientists were expected to play a major role in building this future. Those young people whose first encounters with Soviet politics centered on the thaw, and who derived great hope from reforms expected to follow, are often called the "children of the twentieth Communist Party congress." Many of these "children" found their way to Akademgorodok.

Khrushchev's triumph in the post-Stalin succession struggle was critical to the emergence of Akademgorodok as he proved to be a major patron of science. Khrushchev used the achievements of Soviet science and technology to solidify his position as party leader.[2] He had come from a peasant family and made his career as a party boss in agriculturally rich Ukraine. Now, in addition to the modernization of Soviet agriculture, Khrushchev would show himself to be a twentieth-century man whose visions extended beyond the farm to the city and laboratory. He promoted nuclear power, recognizing its value both for the Soviet economy and for his own purposes. His personal interest in the "peaceful atom" led him to accompany I. V. Kurchatov, head of the Soviet atomic bomb project, to Harwell, England, in April 1956, to visit the major British nuclear research facility and participate in the first Geneva conference on the peaceful uses of atomic energy.

The most important scientific achievements of the Khrushchev years, of course, occurred in space exploration. These successes were the capstone of physicists' increasing power and prestige. A series of "firsts"—the first satellite, man in space, two-manned shot, woman in space, space walk, soft landing, and so on—convinced the Soviet populace, if not the majority of party officials, of the superiority of Soviet science. Khrushchev and other officials knew of the technological failings, if not backwardness, of their program and knew that the United States would respond to Soviet achievements with redoubled efforts in space. Still, they embraced Sputnik as confirmation that a policy which gave scientists increased autonomy in the design and administration of R and D was in the country's best interests.[3]

Khrushchev's trip to the United States in 1959 was important in cementing his ideas about the political utility, ideological significance, and social value of big science and technology. Khrushchev the internationalist rejected many of the xenophobic policies of Stalinism. When he visited the West, it was often to see for his own eyes what ought to be copied, emulated, or adapted in the Soviet context. To say he was a westernizer like Peter the Great centuries before would be an exaggeration, yet he saw in the vitality of the IBM corporation the promise of the computer. On Iowa farms he saw technological solutions to the USSR's annual harvest problems. In America's universities he saw what he took to be cities of science. Khrushchev was convinced that these cities had secured America's technological leadership and were the epitome of scientific achievement. He thus hailed the creation of research cities and ordered the construction or expansion of a series of scientific cities around Moscow—the physics centers at Troitsk and Dubna, biology in Pushchino, and, in Siberia, Akademgorodok where he placed his personal stamp of approval, twice visiting it during its construction.

In another section of his secret speech Khrushchev put Siberian economic development at the center of party attention. He referred to Siberia's great mineral wealth—boron, germanium, aluminium, gold—and its fossil fuels, timber, and so on. At the "extraordinary" twenty-first party congress held in 1959, just months into Akademgorodok's existence, he and other party leaders alluded to the role Siberian science would play in solving problems of controlled thermonuclear synthesis, new polymers, automation, geophysics, and new agricultural products. He had no doubt that science could be put in the service of "the creation of the material-technological bases of communism." Indeed, in the third party program of the Communist Party, promulgated in 1961, Khrushchev guaranteed the construction of communism by 1980, with science playing a major role in the process.

Soviet scientists took Khrushchev's personal interest in their enterprise as an endorsement to expand their autonomy far beyond the limits imposed by Stalin. Having gained authority from their scientific successes in space and nuclear technology, scientists pressed for organizational reforms. At the beginning of 1956 the central press carried several articles that questioned the concentration of scientific resources in Moscow and Leningrad, including articles in *Pravda* by Lavrentev and others. The Academy of Science corridors in Moscow were abuzz with rumors of various organizational reforms. Grabbing onto the coattails of the twentieth party congress, and embracing the newfound political power and prestige they commanded in the postwar years, Soviet scientists turned to the expansion of the scientific enterprise. Proposals to build a series of new scientific research institutes and cities of research quickly gained support. Soviet visionaries had long dreamed of the rapid construction of urban utopias throughout the nation. The most ambitious proposal in the scientific sphere was that of Lavrentev to create a Baconian city of science in Siberia. Such leading figures as chemist and Nobel prize winner N. N. Semenov, president of the Academy of Sciences A. N. Nesmianov, father of Soviet physics A. F. Ioffe, and I. V. Kurchatov defended the position of fundamental science within the Soviet system of organization. They urged the reform of the administration of science and the transfer of the technological sciences from the Academy of Sciences to the jurisdiction of the industrial ministries, a reorganization that occurred in 1961. The national political, ideological, and scientific preconditions for Lavrentev's Akademgorodok were now in place.

A City of Science in Siberia

At a series of closed meetings throughout the first half of 1957, Lavrentev worked through the bureau of the Central Committee of the Communist Party of Russia to get party, government, and scientific organizations to support the Akademgorodok project. Then things moved quickly. On May 10, 1957, after hearing Lavrentev's report on the proposed city, the presidium of the Academy of Sciences ordered a committee to prepare a project proposal for the Council of Ministers. The committee, under Lavrentev's direction, consisted of a specialist in various regions of mechanics, Sergei Khristianovich (deputy director), mathematician Sergei Sobolev, physicist Lev Artsimovich, geologist Andrei Trofimuk, and several others. Already on May 18 the Council of Ministers approved the following proposal:

> to create in Siberia a powerful scientific center; to organize the Siberian division of
> the Academy of Sciences of the USSR and build for it a city of science not far from
> Novosibirsk [with] premises for scientific institutes and well-designed housing for

the staff . . .; [and] to consider [as] the fundamental problem of the Siberian division the broad development of theoretical and experimental research in the area of physio-technological, natural scientific, and economic sciences, directed toward the solution of the most important scientific problems and tasks that will permit the successful development of the productive forces of Siberia and the Far East.[4]

Within a week Lavrentev returned to the Golden Valley with the special commission. They met in a pine forest surrounded by fields of berries, fruit trees, and medicinal herbs and grasses, two thousand meters from the shore of the Ob River reservoir and not far from the nearly completed Novosibirsk hydroelectric power station that provided nearly five hundred megawatts of electricity to the region. Here Lavrentev, Sobolev, Khristianovich, and others, approved the site, near to a soon-to-be-built commuter train (the "elektrichka"), and a newly constructed highway, though quite primitive by western standards.[5] Other sites were considered but rejected for one reason or another.

Novosibirsk, a massive industrial center and the sixth largest city in the USSR at 1.3 million inhabitants, was a logical choice for the site. In addition to being an industrial center, it was the gateway to western Siberia through which Tsar Nicholas's trans-Siberian railroad passed.[6] The city of Tomsk, some 150 miles to the northeast of Novosibirsk, was the center of Siberian academic and scientific endeavors until World War II, when a large number of important machine building, metallurgy, heavy equipment, electrical equipment, and chemical enterprises from the western parts of the country were relocated to Novosibirsk, barely in advance of Nazi armies from the west.

There was a small scientific tradition in Siberia on which to build Akademgorodok. The Russian Geological Society and various organizations concerned with local history, folklore, and nature studies had active memberships in Tomsk and Novosibirsk. Before the Bolshevik Revolution, the Imperial Russian Academy of Sciences limited its involvement in Siberia to the Commission for the Study of the Natural Productive Forces (Kommissiia po izucheniiu estestvennykh proizvoditel'nykh sil, or KEPS). KEPS, which was founded in 1915, was primarily concerned with cataloging flora and fauna, with some exploration of mineral wealth. KEPS, later called SOPS (the Council on Productive Forces), and later still supplanted by a number of state committees, was perhaps the central scientific organization in Siberia before the creation of Akademgorodok.

After the Bolsheviks seized power in 1917, Glavnauka, the Scientific Administration of the Commissariat of Enlightenment, provided quite modest funding for scientific research institutes and laboratories in Siberia. It devoted most of its resources to Moscow and Leningrad institutes.[7] But KEPS managed to expand its Siberian activities in the 1920s through Glavnauka subsidies to the Academy of Sciences, and a number of engineering organizations completed preliminary studies for the construction of hydropower stations on Siberian rivers. At a special 1932 Academy of Sciences session held in Novosibirsk specialists discussed fossil fuel resources, industrial research and development, and agriculture in Siberia, and called for the organization of a branch of the Academy in Novosibirsk.[8] None of these projects was realized until much later, however, as the government had more pressing needs for investment in science and industry in the European USSR.

The war with the Nazis, and the invasion and occupation of the most productive farm and industrial land of the USSR, convinced scientists and policy makers alike of the need to develop resources east of the Ural Mountains. In 1942 a commission commenced study on the mobilization of natural resources of the Urals, West Siberia, and Kazakhstan for military purposes. Dozens of Siberian scholars joined scientists, who had been evacuated from cities in the west currently under Nazi seige or occupation, in the commission. By October 1943 it was decided to create a Siberian branch of the Academy of Sciences with four institutes and a total staff of eighty persons whose profiles reflected government interest in the development of natural resources—mining-geological, transportation-energy, chemical-metallurgical, and medical-biological. Until the creation of Akademgorodok, this modest branch—with its geophysical and geological orientation—was the main center of scientific activity in western Siberia.[9] Yet at the time of the creation of Akademgorodok, there were less than a thousand scientific personnel, loosely defined, in all academic institutes of Siberia and the Far East, a huge region of some 5 million square miles.

Having selected the site for Akademgorodok, Lavrentev moved quickly to establish directions of research and approve construction plans. By June 1957 the first ten Akademgorodok institutes had been named, their directors selected, and scientists' internal visas authorized. By the end of the summer all of the governing and academic bodies had been appointed. Thus the Siberian division (Sibirskoe otdelenie) of the Academy of Sciences had come into existence on paper. As an indication of the importance of the division, the Academy designated Lavrentev one of its handful of vice presidents. At the same time, the division received the right to publish a series of new but as yet unestablished journals. Limited runs of mimeographed reports were allowed with minimal prepublication restrictions, unlike in the rest of the USSR. (In the Brezhnev years it was nearly impossible to get a new journal off the ground, so concerned about freedom of expression was the Communist Party. Further, the foreign department of each institute, where a KGB employee sat, had to approve the sending of reprints abroad. But in the first decade of Akademgorodok, publication and reprints were approved relatively simply by Soviet standards.)

In May 1958 the builders met with scientists to chart the construction of the future institutes. The workers organized the meeting in modest barracks, where the scientists received bouquets of the first spring flowers from young girls who lived in the area. In response to this gesture, Lavrentev promised to open a university promptly in Akademgorodok. Walking through the woods

TABLE 1.1
Scientific Research Institutes in Akademgorodok, 1970

Automation and Electrometry	Hydrodynamics
Biology	Inorganic Chemistry
Bioorganic Chemistry	Mathematics
Catalysis	Mining
Chemical Kinetics and Combustion	Nuclear Physics
Computer Center	Organic Chemistry
Cytology and Genetics	Theoretical and Applied Mechanics
Economics and Industrial Production	Semiconductor Physics
Geology and Geophysics	Soil Science and Agrochemistry
History, Philosophy, and Philology	Thermal Physics

not far from the Novosibirsk hydropower station, I. N. Vekua, soon-to-be rector of the university, exclaimed, "Here's where the new university will be!" In Akademgorodok, one year scientists picked mushrooms in the forests and the next year they lectured in a university building, albeit still under construction, in that very forest.

Initially ten and then fourteen institutes were planned; today there are twenty, one university, several libraries, botanical gardens, and other facilities (see Table 1.1). Not surprisingly, physics, mathematics, and computers have a central place in Akademgorodok.[10] Of the first scientists to be appointed to work there, 40 percent were physicists, mathematicians, or engineers working in related fields; 22 percent, geologists or geophysicists; 11 percent, chemists; 18 percent, biologists; and 9 percent, economists and sociologists. To this day Akademgorodok maintains the traditional Soviet orientation toward the exact and technological sciences.

However, the best laid plans of academicians and party officials often go astray. Lavrentev, Sobolev, and Khristianovich convinced their colleagues in the Academy and the Central Committee of the Communist Party that within five years of a government resolution an independent, world-class scientific research center would exist where before there was only forest. As communist visionaries, Lavrentev and the other founders of Akademgorodok may have truly believed they could keep Soviet reality from intruding on their city. Through a combination of central planning, government resolutions, and exhortation of workers, and armed with the doctrines of Soviet Marxism, how could they fail? Unfortunately, problems of finance, material and equipment supply, construction delays, low labor productivity, cost overruns, and the vagaries of the Soviet planning system combined with the harsh Siberian winter to waylay construction at every step. Construction started somewhat later than anticipated, moved along slowly, rapidly depleted the budget, and, more important, failed to attract the necessary equipment or manpower until the late 1960s. As a result, sacrifices were made in the architectural designs that lowered the aesthetic quality of Akademgorodok's physical plant.

THE ERA OF THE LOWLY BRICK HAS ENDED!

Financing for Akademgorodok came from the Novosibirsk sovnarkhoz, the construction budget of the Siberian division of the Academy of Sciences, and several other sources. The government allocated 200 million rubles for construction in 1958. By the summer of 1958, one source claims, construction on housing, institutes, parks, streets, and so on, reached 1 million rubles daily,[11] although only six miles of roads, forty miles of communications, and roughly six hundred apartments were completed in that year. By 1961, 160 million rubles had been expended, 110 million for construction. The nuclear weapons ministry, Minsredmash (the Ministry of Middle Machine Building), contributed funds to the project, most likely through the construction trust charged with building Akademgorodok.

The rapid pace of construction masked significant problems. The Ministry of Finance (MinFin) criticized project management. Its officials contended that construction was poorly organized and that Akademgorodok officials regularly violated finance rules by shifting amounts between line items without informing the authorities or by underspending and had illegally accumulated reserves of materials well above established norms. The deputy minister of MinFin of the Russian Republic (RSFSR) claimed that capital investment, staff hiring, and educational plans were all underfulfilled at the same time that budget requests were inflated. Lavrentev was told that his budget would be cut by at least 2 million rubles, after he had just requested 1 million rubles for two dozen automobiles. Transportation problems and the absence of telephones hindered scientific communication between Novosibirsk, where most of the scientists still worked in "institutes" housed temporarily in two or three rooms of a government building at 20 Soviet Street, and Akademgorodok, where they regularly traveled for business; there were only thirteen cars for hundreds of employees. It hardly mattered, since the muddy road from Novosibirsk to Akademgorodok was impassable after a rain shower.[12]

One of the sources of these difficulties was that initially no single body was responsible for the design and construction of Akademgorodok. The design institute Novosibproekt first handled architectural responsibilities. Novosibgesstroi-2, the firm responsible for building the Novosibirsk hydropower station, joined the effort in September 1957. Sibakademproekt then replaced Novosibproekt in 1959 in response to the pleas of Lavrentev to accelerate planning and construction. Design institutes representing the heating, electrical, and transportation industries, and the water, sanitation, electricity, insulation, and other trusts from the RSFSR Ministry of Construction were instructed to assist Sibakademproekt. Sibakademstroi, a huge conglomeration of parts of each of these organizations, and until recently a closed organization involved in the construction of ICBM missile silos, finally gained responsibility for the massive undertaking of the construction of Akademgorodok. Even when Sibakademstroi took over administrative control, however, it was not the independent, streamlined, and efficient organization capable of forcing suppliers and builders to toe the line that was envisaged.

Another reason for construction problems was the raw nature of the worker recruits. After Akademgorodok was declared a building site for the national Komsomol (Communist Party Youth League) in the fall of 1958, the workforce swelled rapidly with untested laborers. By the next spring the number of young enthusiasts had reached eight thousand. Several individuals with whom I spoke confirmed that prisoners were also used in construction, but it seems their numbers were not large nor were they the slave laborers of the Gulag. These were common criminals, not political captives. Sixty-one percent of the builders were younger than twenty-five, another 14 percent were between twenty-six and thirty-one. Almost twelve thousand individuals flocked to Novosibirsk between 1958 and 1962, where they received on-the-spot training. These inexperienced workers could not meet even the low standards of Soviet projects despite laboring under the slogan "We will build a city of science of high quality!" At first two-thirds of construction was rated only "satisfactory." So party activists in special propaganda brigades and agitation groups were engaged to exhort their comrades to rise to the occasion of communist construction.

Khrushchev himself took special interest in the construction's progress since he viewed Akademgorodok as a symbol of his cosmopolitan rule. He first set foot in Akademgorodok in October 1959, on his way back from meeting Mao Tse-tung in China. He dropped in to chart Sibakademstroi's progress and to issue an ultimatum: fire Nikolai Dubinin, head of the genetics institute, for having failed to give Lysenkoist biology (which rejected modern genetics) its hallowed place. Mikhail Lavrentev had no choice but to acquiesce if genetics as a field of study was to flourish surreptitiously in Siberia. The unobtrusive Dmitrii Beliaev assumed directorship of the institute, and the geneticists carried on quietly. Khrushchev also used the occasion to dress down Akademgorodok architects for including in their designs a twelve-story hotel: "Why make all this fuss about twelve floors? Why are you trying to imitate New York skyscrapers?" Khrushchev then leaned forward and snipped away the top four floors of the hotel, imitating scissors with his fingers: "That's what I think about this skyscraper."[13] The building came in at eight floors.

Khrushchev's 1959 trip gave impetus to speed up, mechanize, and economize on every aspect of Akademgorodok construction. Khrushchev himself insisted on the adoption of these methods, fascinated as he was with Taylorism—"Amerikanizm"—with its modern mass production techniques. He demanded replacing bricks with prefabricated forms in keeping with the needs of Akademgorodok and what was called the "high culture" of construction. Within two years the largest factory in the USSR for the production of reinforced concrete forms had been built, and the time-consuming and expensive bricks were abandoned. When Khrushchev returned to Akademgorodok on March 10, 1961, he was impressed with what he saw. Fourteen institutes had already opened, although several shared space. In 1960 alone, 40,000 square meters of apartments had been built. There were already ten thousand inhabitants.[14] The high culture of concrete was in evidence everywhere.

Although it may have helped to keep construction costs down, Khrushchev's meddling went a long way toward destroying the aesthetic quality of Akademgorodok. Planners and architects had hoped to create an architectural as well as an intellectual oasis in terms of numbers of architectural styles and their blending with the forest. But Akademgorodok apartment buildings have three basic styles. The older ones have ten-foot ceilings and brick construction, the newer ones have eight-foot ceilings and prefabricated concrete construction. The institutes are of two basic designs in what I call the "neo-Soviet" style, one of spacious brick buildings, the other of more cramped concrete structures. The results are bland, brown or gray, and boring. Of a total area of nearly 7,000 acres (more than 5 square miles), about 750 acres were designated for apartment buildings with living space totaling 4.5 million square feet for forty-three thousand residents in six microregions. This amounted to between 84 and 110 square feet per resident, but only 40 to 60 square feet for student dormitory rooms, and 180 square feet, large by Soviet standards, for scientists, an incentive to move from Moscow, Leningrad, Kiev, and elsewhere. Twenty percent, more than the Soviet norm, of each region was reserved for "greenery"—parks, gardens, and playgrounds. Each microregion had identical rudimentary shops ("Meat," "Milk," "Bakery," "Vegetables and Fruits," and so on), dining halls, schools, and kindergartens. Institutes were built on the remaining 850 acres. The main bookstore and department store, the "Moscow" movie theater with eight hundred seats, and the Scholars' Club (begun in 1963) were located near the center, equally accessible to all. The university was built just to the northeast of the town center.[15]

The efforts of Sibakademstroi to fashion an efficient construction industry—cement and other factories—went poorly. The Novosibirsk Factory of Large Panel Apartment Construction, which officials had intended to produce enough concrete slabs to build fifty-five hundred apartments annually, each roughly 500 square feet, met only one-fifth of that target. Bricks and cement had to be imported from Tomsk, Kemerovo, and other cities hundreds of miles away. Gosstroi, the state construction committee, criticized this lag, as a result of which "nonindustrial methods"—by hand and shovel, not jackhammer and crane, and by bricks and mortar, not huge pieces of prefabricated concrete blocks—persisted. In spite of all this, one journalist joyfully observed, "And before the workers stand new concerns. The era of the lowly brick—evidence of a low level of construction—is ending."[16]

Unfortunately the large panels were used regularly only in January 1961, after construction had been under way nearly three years. By the end of February Sibakademstroi revealed that construction "culture" remained low: many concrete forms were distorted and failed to meet clearances. Too much manual labor was still required. Finishing work was poor. Spaces between panels allowed them to freeze all the way through. Looking at photographs of construction, one sees not apartment buildings but houses of cards stacked on top of one another at right angles. It was not surprising that such buildings as these collapsed instantly during earthquakes in Armenia in 1988.

Construction of the scientific buildings was just as bad. Some researchers remained under the wing of some laboratory in Moscow or Leningrad; others moved to poorly appointed furnishings in downtown Novosibirsk while waiting for building to commence; and others doubled or tripled up, then jockeyed like mad to be given the next completed building. The buildings had specific requirements that were to be met with bricks and special forms, not standardized ones. The specific requirements clearly confused the workers, so Sibakademstroi directors were forced to erect institutes by the "high culture" industrial methods as well.[17] Construction on Lavrentev's own Institute of Hydrodynamics, begun in 1959 with the reliable brick, was essentially completed only in 1964, and when finished it had to share its facilities with five other institutes. In the spring of 1962 Akademgorodok moved toward its official November opening, which was scheduled to coincide with the forty-fifth anniversary of the Great October Socialist Revolution. Construction was perhaps only 60 percent completed, 75 percent of funds allocated in 1958 were spent, the gala opening was postponed, and the weekly banner in *Za nauku v sibiri*, which proclaimed how many days remained until completion, disappeared from the masthead. The stories of construction difficulties had moved to page 2 of the local papers, but the problems would not fade.

One source of the poor quality of construction may have been the resentment of class-conscious workers about academicians being given "cottages," that is, houses, not apartments, and certainly not barracks, located in the Golden Valley itself. A few scientists gave fuel to the charge of elitism. Iurii Kerkis, a leading geneticist, argued that "housing should be given to those needed by science." There were too many common workers, even ne'er-do-wells who had been given housing ahead of the active scholar, merely because they had friends in high places. Value to science, not allegiance to the communist ideal of the workers' state, should be the primary criterion, Kerkis argued.[18] Scientists were able to convince the builders that the cottages were not a symbol of elitism but rather a necessary perquisite to create the "material technological basis of communism." Propagandists moved in to explain the purpose of the institutes they were building. Apparently the workers were entirely caught up in the joy of their construction efforts. They waxed eloquent about their contribution to the "construction of communism." One observed, "Three years have passed between the Twentieth and Twenty-first Congresses. And what truly magnificent achievements. Tens of millions of hectares of irrigated desert! An atomic icebreaker and the Kuibyshev power station! Three sputniks and the first artificial planet! You can judge the tempo of the development of our Motherland by the example of our Akademgorodok!"[19]

Finally, the construction plans failed to take climate adequately into consideration. According to the data provided by a national engineering council on cold-resistant equipment, tens of millions of rubles were lost annually in the USSR in broken machinery, equipment, and vehicles, owing to the failure of this equipment to meet the needs of Siberia's difficult climate.[20] Sibaka-demstroi was forced to use equipment designed for use in more moderate climates. The result was tens of thousands of machine hours of idleness, constant repairs, and concrete that set poorly. Construction crews encountered the harsh Siberian winter freezes, permafrost, and the mud and mosquitoes of the spring thaws and summer rains. Some scientists initially lived and worked in hastily erected temporary barracks or wood houses in the forest. They wanted to see the construction with their own eyes. Scientists from the Institute of Hydrodynamics were already conducting expeditions and had designed a powerful water cannon with applications in mining and construction, while temperatures in the barracks during the truly Siberian winter of 1958–59 reached -50° C.[21]

In these difficult circumstances, in spite of press reports to the contrary, research was held back in most cases until the early 1960s. This was even true for research under Lavrentev's direction. Andrei Deribas, one of Lavrentev's closest associates in the early years, was a troubleshooter. He graduated from Moscow University in 1952 and earned his candidate of science degree in 1956. In his junior year he heard Lavrentev lecture on the application of explosions in the national economy. Learning about Akademgorodok, he presented himself to Lavrentev and was hired on the spot as the first staff member of the Institute of Hydrodynamics. He departed for Novosibirsk with Lavrentev. Their research focused on theoretical questions since they had no equipment for experiments. In May 1958 they moved into a small hut in the Golden Valley with Lavrentev's son, Mikhail, now director of the Institute of Mathematics, and a handful of other people. It was a hard existence, but life was rich. They even had a small laboratory for running explosion experiments in the forest. Deribas's daughter, now a research physician in the Academy of Medical Sciences, was the first baby born in Akademgorodok. Deribas became a laboratory director in the Institute of Hydrodynamics and in 1976 left to become director of a design bureau for hydroimpulse technology. Deribas recalled how he and Lavrentev had "stood at the edge of a partially dug foundation ditch and watched for a long time how two workers struck at the frozen earth with shovels. It was sad," he said. "It seemed that at that rate it might take one hundred years to build Akademgorodok."[22] Even the scientists had to pitch in. Lavrentev recalled: "We burned dead trees, sawed, split, and dragged wood and buckets of water. In one of the barracks there was a kindergarten. Aunt Varya carried on in the dining hall. She had fisherman friends. And sometimes we had fish soup of white salmon or fried white salmon. The kindergarten and dining hall stayed in existence for around two years, since they hadn't built real housing in the science city."[23] No wonder they had so little time for research.

Not only did delays, supply shortfalls, increased costs, and shoddy workmanship continue, but new problems arose. Few apartments had both warm and cold water regularly until the mid-1960s. Some apartment buildings leaked both from the roof and basement. Officially more than thirty kinds of mushrooms grew on freshly painted hallways. Safety inspectors identified fire hazards in corridors, hallways, and courtyards.[24] Officials of Sibakademstroi complained that water shortages were caused by rampant waste and overuse—washing down roads or keeping food cold with running water in the absence of refrigerators. The quality of "life under the city"—electrical wires, conduits and transformers, pump stations, plumbing and sewers, and gas lines—left entire neighborhoods with spotty service. Inadequate construction documentation often made it impossible to find the source of the disruption. The party archive holds hundreds of letters written by Akademgorodok residents to local officials complaining of faulty plumbing and electricity, and the impossibility of getting repairs done. Residents themselves contributed to these problems, for example, by opening steam pipes in their apartments in the winter. This created a nice, warm "microclimate" but denied others heat. Others in the same building opened windows in the dead of winter to escape oppressive heat.[25]

For all the discussion of using modern science in the construction of Akademgorodok, all "communications" technologies lagged. The local post office served twenty-eight different organizations from a twelve-by-ten room with seven workers, two assistants, and a telegraph operator. The telephone system dated to 1953 and served five times as many organizations and lines as it was made for.[26] It was impossible to make international calls without placing orders beforehand or calling in the wee hours of the night. To this day it can hardly handle computers and modems. Buses, too, were either packed or empty. During rush hour citizens stood freezing in orderly long lines, only to have anarchy break out in a mad scramble for the door when buses approached.[27]

Social problems of the kind one would expect in any city also haunted Akademgorodok: petty theft, delinquency, and public drunkenness; open sores of garbage; and the occasional cat tossed out of an apartment building window by a disgruntled neighbor. The lawns, parks, and gardens around apartment blocks were littered with construction debris, garbage, private cars parked with disregard for common space. Trees and flowerbeds were particularly prone to abuse.[28] These problems may have been exacerbated by the pressure to meet targets and start new projects, leaving old ones incomplete or poorly done. Some microregions had no bread, vegetable, or fruit stores. In others, residents complained of unsanitary conditions: filthy countertops, uncovered food, dirty floors. Foodstuffs, especially grains, were often stolen. In addition, since the state subsidized the cost of food, institutes often found it cheaper to feed laboratory animals with bread or flour than fodder. In September 1963 alone laboratory animals consumed 1,400 pounds of bread, 1,200 pounds of groats, 750 pounds of wheat, and 500 pounds of oats. Dining halls had huge lines, making it impossible for a researcher to spend less than two hours trying to secure a modest meal. In a word, Soviet-style infrastructure sabotaged utopian scientific visions.

Soviet reality tempered any sense of accomplishment in the area of cultural events as well. The celebration of the opening of the skating rink gave way to the realization that numerous cracks and bumps prevented skating. Others found fault with the "pretentious" and "inappropriate" names chosen for cultural establishments in Akademgorodok. Instead of a movie theater named "Moscow," why not "Era," "Mars," "Problem," "Hypothesis," "Civilization," "Andromeda," or "Taiga," a student group suggested. The beach, which had been built at a cost of more than 2 million rubles, fared only somewhat better. A cyclone in October 1959 with winds gusting to ninety miles per hour destroyed much of the shoreline and the railroad bed with ten-foot waves. Happily for the sunbathers, it was decided to move the rail bed inland and use a three-mile-long artificial beach as protection against future acts of nature. The result was the Ob Reservoir beach.[29]

Planners miraculously failed to take into account the human biology of Akademgorodok's relatively youthful residents. They provided too few day care centers or kindergartens. Citizens, like their American counterparts, were at the mercy of "market forces." Lavrentev bitterly complained to the Council of Ministers that six thousand people lived in the microregions of Akademgorodok, where the kindergarten had room for a mere 125 children. As of December 1962 every second child waited for an opening in nursery school or kindergarten, and the six operating centers had 936 children in areas suitable for 765. Builders promised that three more would open in 1963 and two in 1964, but parents were wary since two such facilities were canceled in 1962. As a rule, state-run day care facilities were cold and drafty or overheated and damp. They were havens for colds and infections. In many cases parents resorted to pooling their resources to share day care responsibilities.[30]

As elsewhere in the USSR, medical care was given low priority in Akademgorodok and failed to deliver the promise of modern science. In 1959 the "medical center" had a staff of five: a pediatrician, obstetrician-gynecologist, X-ray technician, general practitioner, and nurse. By July 1960 there were twelve doctors, ten nurses, but only fifteen hospital beds. In August a polyclinic of sorts opened, but only late in 1962 was construction completed for an emergency room, maternity ward, and dental clinic. In all, 160 hospital beds were available for a town of thirty thousand. The field of obstetrics and gynecology also fared poorly throughout the history of Akademgorodok, with declining fertility rates, increasing numbers of miscarriages, and infant mortality rates reflecting trends all too common in the USSR.[31] Infectious diseases, cancer, and heart disease, and the ecological sources of these problems, and cases of dysentery, diphtheria, hepatitis, and tuberculosis were common. The town newspaper, *Za nauku v sibiri*, often provided discussions of these diseases—their manifestations, how to avoid them, what to do if infected, and the particular dangers they posed for the young.[32]

Some of the problems were peculiar to Siberia: ticks with encephalitis and mites called "*gnusy*," indigenous to the forest, that caused painful swelling and irritation and resisted all attempts to eradicate them. Hundreds of citizens pleaded with the authorities to do something. In January 1959 a special coordinating committee made up of leading biologists, physicists, and chemists was created to deal with the nasty bloodsuckers. A local investor came up with the idea of using huge engines as aerators to spread DDT throughout the city. This killed not only the mosquitoes, gnusy, and ticks but also birds and other wild-life. Just as in the West, DDT was abandoned, although somewhat later, and wildlife was slow to return.[33]

All this notwithstanding, the Siberian city of science was far more attractive than most postwar Soviet establishments because of the preservation of the surrounding forests, the use of various architectural styles, the limited height of most structures, rarely higher than the tallest trees, and the geographical dispersal of institutes, housing, and parks. The center of the institutional zone is at the intersection of University and Lavrentev Prospects, where facing one another are a series of institutes with similar bland facades, usually of light brown or tan bricks and windows like those of the typical American high school; these are the Institutes of Cytology and Genetics, Automation and Electrometry, Mathematics, and Geology and Geophysics, with the Nuclear Physics Institute, the largest in terms of personnel, space, and size, looking out at them. Nearly four hundred acres were reserved for future building, while some eleven hundred acres of heavily forested areas were set aside in perpetu-ity. There are parks, some with illuminated cross-country ski trails to permit skiing in the dark, the long continental Siberian climate providing snow cover almost half the year. The Ob Reservoir provides a beach, boating, and fishing. Most visitors agree that it is significantly more pleasant to live in Akademgoro-dok than many other locales in the former Soviet Union. Children play freely in parks close to their apartments; most inhabitants walk to work; and the crush of Moscow and Leningrad life is absent. Yet two significant problems, central to scientific activity, remained to be solved before research could com-mence: the supply of equipment and instrumentation and the attraction of qualified personnel.

THE SOVIET SCIENTIFIC HARDWARE STORE

Equipping research institutes in Akademgorodok proved to be an even greater source of difficulty than for the typical Soviet scientific establishment. This was both because of the size of the endeavor and the short time in which it was to be completed. An overly centralized system of supply compounded the prob-lem. Fear of alleged internal and external enemies, the belief that individual initiative would result in the resurrection of capitalism in Russia, and nearly legendary faith in the ability of bureaucracy to distribute goods and services led to the creation of a top-heavy, highly centralized system for the administration and financing of R and D in the 1930s. Why political and ideological pressures should have resulted necessarily in the centralized control of all research mate-rials, from nuts and bolts to flasks, test tubes, instruments, and equipment, is unclear. But the outcome was the absence of anything remotely resembling the neighborhood hardware store and, instead, the creation within the Academy of

Sciences of "Tsentrakademsnab," a central office in Moscow to administer demand and supply on a national level, with near mythic capabilities to lose orders and delay shipments.

In Akademgorodok, centralized supply of research material and equipment was handled by the Material-Technical Supply Administration (Upravlenie material'no-tekhnicheskomu snabzheniiu, or UMTS), an organization seemingly incapable of coordinating its activities with Tsentrakademsnab. UMTS adopted the attitude that the scientists themselves were irresponsible, not to mention profligate, in their use of resources. UMTS administrators resented having to act like the local fire department, ready to put out the fires of academic short-sightedness and waste. According to UMTS, scientists stored the equipment or reagents improperly. "Their desire for equipment is unlimited," one Semenov of UMTS claimed. Only further centralization of management, rational use of resources including the sharing of equipment across laboratories and institutes, and politeness toward workers in UMTS would solve the problem.[34]

Scientists blamed UMTS for the poor state of affairs. They asserted that lazy, ignorant UMTS personnel were responsible for glacially slow response times, shortages, and the un-Marxist (epistemologically speaking) need to predict the unpredictable. UMTS paid no attention to current publications or exhibitions and had a rudimentary understanding of contemporary instrumentation, equipment, and research needs. Scientists called for the hiring of specialists from the engineering professions, familiar with domestic and foreign equipment, to work in UMTS; the creation of a special technical library; and decentralization of supply.[35]

The secretary of the party committee of the Siberian division of the Academy of Sciences, G. S. Migirenko, himself a specialist in mechanics, blamed both sides for the supply impasse. The scientists had spent three-quarters of the monies allocated, and much still remained to be built. There was a need, Migirenko claimed, both for better paperwork and rational orders. Migirenko acknowledged that UMTS, while full of well-intentioned people, moved too slowly and was a pit of disorder. At the same time N. K. Rudnev, chairman of the forerunner of the State Committee for Science and Technology, showed poor understanding of the needs of modern science. He called for standardization and faster production of various instruments, machines, and tools, and urged institutes to focus on research topics not duplicated by other institutes.[36] He seemed not to have recognized that modern science required special equipment that institutes were often unable or perhaps loath to share with other institutes.

The problem of inadequate supply drew the attention of a general meeting of the Siberian division in 1962, where it was decided to overhaul the process entirely. The secretary of the party organization of UMTS finally admitted that difficulties existed in providing gas products, dry ice, and various reagents, let alone complex equipment. He acknowledged that the current system tied up thousands of rubles of resources in storerooms, both in institutes and at UMTS, to make up for the months needed to fill an order in an "emergency"—some unanticipated but pressing research needs. Tsentrakademsnab and the Siberian division agreed finally to put UMTS on a "self-financing" (profit and bookkeeping) system so that it could rent and repair equipment for a fee. This freed assets since institutes could rely on UMTS to secure standard materials, reagents, and instruments fairly rapidly for a profit, sort of like a hardware store.[37] For complex machinery and instruments, most institutes created their own glass and mechanical workshops to build the one-of-a-kind items they needed.

This kind of technical difficulty was not new in Soviet science. In the early years of his rule Stalin prematurely announced that "Technology decides everything!" Soviet Russia would surpass the best America had to offer based on advanced industrial technology, mostly borrowed and some indigenous. Within years Stalin's slogan had been altered. "Cadres decide everything!" the party leader now acknowledged. He well knew the significance of this slogan. Ill-trained workers had been advanced into positions of responsibility in all industrial, educational, cultural, and scientific establishments solely on the basis of class origin and party membership, not such traditional criteria as merit or talent. This left industry, agriculture, and science in a shambles. After all, how could a worker, however militant a communist, master complex machinery, let alone quantum mechanics, in a matter of months? In the Soviet Union of Khrushchev's thaw, Lavrentev would not make the same mistake. But he would need to advance scientists prematurely into research and management positions.

CADRES DECIDE EVERYTHING!

Where would the hundreds of Akademgorodok biologists, chemists, physicists, and mathematicians and the thousands of laboratory technicians come from? In the post-Stalin years the total number of scientists in the USSR grew rapidly, from 162,500 in 1950 to 665,000 in 1965; the number of senior and junior specialists grew from 62,000 to 140,000 in the same period. It was clearly simpler and cheaper to turn out specialists than to build equipment and instruments. Still, it was challenging to provide staff for the new institutes because of the novelty of the idea of Akademgorodok and the hardships of life in Siberia. Yet Lavrentev had three things going for him in his effort to attract scientists to his bold endeavor. First, paradoxically, was the very novelty of the city, its promise of independence and academic freedom away from the control of Moscow's and Leningrad's scientific bureaucracies. Second was the support of a crucial mix of national political and scientific leaders. And third was Lavrentev's own boundless energy.

Although Lavrentev was brilliant in his promotion of Akademgorodok, many Academy colleagues told him outright that no accomplished scientist in his right mind would move to Siberia. While supporting the idea of Akademgorodok, the president of the Academy of Sciences, A. N. Nesmianov, and leading officials in the Ministry of Higher Education expressed doubts that Lavrentev could succeed in achieving the right mix of qualifications and specializations. Soon, however, Nesmianov came to see the Siberian division of the Academy as central to efforts to decentralize and revitalize the Soviet economy.[38]

In November 1957 Lavrentev presented his plan for Akademgorodok to members of the Academy of Sciences. In the discussion that followed, such leading scientists as Igor Kurchatov, Nikolai Semenov, head of the fusion research program Lev Artsimovich, and low-temperature specialist and future Nobel prize winner Petr Kapitsa urged their colleagues to give unconditional support to the endeavor. The western-looking Petr Kapitsa actively sought support for Akademgorodok. Kapitsa had worked with Ernest Rutherford in Cavendish Laboratory in Cambridge, England, from 1921 until 1934. Each summer he journeyed home to Russia to see family. In 1934 Stalin declined to let him return to England. Kapitsa seems to have faced Stalin down, refusing to work again until his entire laboratory was moved from Cambridge to a new institute built specially for him in the center of Moscow. Clearly never one to mince words, on the occasion of Lavrentev's speech, Kapitsa commented on the domination of Muscovites on the committees for Lenin prizes, elections for Academy membership, editorial boards, and VAK (the degree attestation committee). Kapitsa urged his colleagues to avoid personal bias and to wish the Siberians well, to treat them not "as outcasts, but beloved sons who are fulfilling a great undertaking." He concluded, "We must create the moral climate for scientists to move to Siberia."[39] With Kurchatov and others, he promoted these ideas in mass scientific-popular monthly magazines; many of the scholars with whom I spoke remember learning about Akademgorodok from these journals.

Surely some scientists would come from Siberia itself, perhaps from other branches of the Siberian division of the Academy in Tomsk, Krasnoiarsk, and Irkutsk. However, at the time a Siberian division was created in all scientific institutions of Siberia and the Far East, there were fewer than a thousand scientific workers, and nearly five hundred of these had no advanced degree and none were academicians or corresponding members of the Academy. Only 330 workers comprised the West Siberian branch of the Academy of Sciences, with 13 doctors of science (equivalent to an associate professor) and 110 candidates of science (equivalent to a doctoral or master's degree).[40] Obviously this was not a source of large numbers of Akademgorodok personnel.

Others would come from a pattern first established by the father of Soviet physics, Abram Feodorovich Ioffe. Ioffe founded the well-known Leningrad Physical Technical Institute in 1918, the leading Soviet physics institute until World War II. Beginning in the late 1920s and 1930s, several "spin-offs" from Ioffe's institute were created that used personnel and equipment drawn from his institute to complement the small numbers of local scientists whose research focused on economic problems of their particular location. Akademgorodok was built with personnel from Moscow, Leningrad, Kiev, and L'vov.

The principle was to use established laboratories in Moscow, Leningrad, and elsewhere as the kernel of the new Akademgorodok institutes. When construction crews had finished the shells of buildings in Siberia, whole "institutes" were ready to move by train.

Lavrentev's Hydrodynamics Institute was formed within the Moscow Physical Technical Institute with recent graduates and students from upper division courses. Many of them first moved to Siberia in 1958, but the move was not complete until 1965. Gersh Budker's Institute of Nuclear Physics functioned essentially intact as the laboratory of high-energy collisions within the Kurchatov Institute in Moscow until its move in 1961. The Institute of Automation and Electrometry was formed from laboratories of the L'vov Polytechnical Institute and Institute of Machine Building and Automation of the Ukrainian Academy of Sciences. Others were created from scratch in Novosibirsk, although their directors, often Moscow scholars, attracted other Muscovite colleagues to the newly established institutes.

A third path for staffing institutes in Akademgorodok was to encourage young scientists, many of whom were still students, to join Lavrentev's utopian endeavor. In December 1957 Lavrentev turned to the Central Committee of the Communist Party with a request to facilitate the move of established scholars to Siberia and to give Akademgorodok first choice of young scholars who had finished the leading universities and institutes in recent years. As many as two-thirds of these promising specialists nationwide declared an interest in moving to Akademgorodok. A resolution of the Council of Ministers of February 22, 1958, gave the Siberian division first choice on juniors, seniors, and recent graduates from all higher educational institutions nationwide and removed many of the obstacles to transferring those who already had jobs. In this way, first hundreds, then thousands of scientists, most young but including established scholars, made their way to Akademgorodok. Still, a number of scientific and ministerial organizations in Moscow and Leningrad objected to Akademgorodok's favored position. Some "categorically refused" to permit its students to leave for Siberia.[41]

Another path led to Akademgorodok, though not the chosen one. Some scientists were "settled" in Siberia through the infamous Tupolevskaia Sharaga (named after Tupolev, the aircraft designer, and *sharaga*, prison camp—those of the Stalinist Gulag reserved for engineers and scientists put to work in service of military R and D). The most prominent case in this regard, but far from the only one, was Iurii Borisovich Rumer.[42] Rumer entered Petrograd University in the mathematics department in 1917 and then transferred to the mechanical-mathematics department of Moscow State University. Upon graduation he taught mathematics in a series of technical schools and *rabfaky* (special departments for workers). In 1927 he traveled to Göttingen to study and was taken on as Max Born's assistant through 1932. During this time he worked on relativity theory, discussed his work with Einstein in Berlin, and collaborated with Weyl and Teller on the quantum theory of chemical valence. He returned to Moscow in 1932 as a docent and in 1933 became a professor.

He was awarded his doctorate of science in 1935 without a defense, becoming a senior scientist at the Physics Institute of the Academy of Sciences in the department where theoretician Igor Tamm worked and where Andrei Sakharov began his career.

Rumer was arrested during the Great Terror of the late 1930s, along with several other leading physicists: future Nobel prize winner Lev Landau, the noted theoretician V. A. Fock, and the brilliant, promising young scholar Matvei Bronshtein. The Terror hit the Leningrad physics community particularly hard, but also enveloped the recently founded Ukrainian Physical Technical Institute in Kharkov where Landau worked in 1937. In all, perhaps 10 percent of Russia's physicists perished. Landau sensed the brewing storm and escaped the hostile environs of Kharkov for Kapitsa's Institute of Physical Problems in Moscow. For Rumer, the arrival of Landau was like a holiday. The two had long been close friends, spent all their free time together, and indeed Landau lived with Rumer until his own apartment was built. On April 26, 1938, Landau and Rumer were arrested and put in the same cell.[43] This was a mistake: two days later they were separated. The arbitrariness of the purges would now work its magic: Landau was in prison for one year and then returned to the Institute of Physical Problems as leader of its theoretical department; Rumer spent ten years in prison and then suffered another exile to Siberia as a teacher.

After his arrest Rumer worked for the next ten years as an engineer in a series of aviation factories connected with Tupolevskaia Sharaga. Only at the request of the president of the Academy, Sergei Vavilov, was Rumer permitted to return to "normal life" in Siberia after the war. When released in 1948, he was given a teaching position at the Eniseisk Teachers' Institute, then became head of the Department of Technical Physics in the West Siberian branch of the Academy, director of the Institute of Radiophysics and Electronics in Novosibirsk (1957–1964), and finally joined Budker's Institute of Nuclear Physics where he remained until his death in 1981.[44]

Taking one of these four paths to Akademgorodok, the number of its scientists increased rapidly, although the primary route was that taken by the young scientists. The scientists' average age was thirty-four, five years younger than the average age of a Soviet scientist, and almost 40 percent were younger than thirty as of July 1, 1963. The presence of young scientists caused a series of problems, for example, individuals without degrees were serving as laboratory and department heads. How could these budding scientists with inadequate training and limited horizons head up important research programs? The solution was to accelerate dissertation defenses by plan: more than 500 doctors and nearly 2,100 candidates of science by 1965. But frequent changes in advisers, poor selection of topics, and formalistic attestation handicapped the process.[45] Of 153 successful defenses by 1962, all went to scientists who finished their graduate training before 1958—that is, before the move to Novosibirsk. By 1967 still only 168 doctors and 1,250 candidates had defended their dissertations. A successful defense was a cause of great pride as the front-page photo-

TABLE 1.2
Scientific Workers in Akademgorodok, 1957–1964

	1957	1958	1959	1960	1961	1962	1963	1964
Scientific Workers	331	944	1,144	1,547	1,760	2,431	2,353	2,654
Academicians	—	10	10	10	10	10	10	14
Corresponding								
Members	—	22	23	26	25	27	27	35
Doctors of Science	13	26	33	38	48	57	60	69
Candidates of								
Science	110	234	280	372	445	496	503	649
No Advanced								
Degree	208	652	798	1,111	1,232	1,841	1,753	1,887
Total	654	1,904	4,369	7,650	9,937	11,926	12,249	13,465

Source: Moletotov, "Problema kadrov," p. 345. Technical support staff are included in the "Total" in Soviet sources.

TABLE 1.3
Doctoral and Candidate Defenses in Novosibirsk Scientific Center, 1958–1964

	1958	1959	1960	1961	1962	1964
Doctoral	—	2	12	9	10	22
Candidate	22	29	43	53	86	12

Source: Moletotov, "Problema kadrov," p. 344.

graph of the young Roald Sagdeev, a specialist in plasma physics, in *Za nauku v sibiri* attests.[46] (See Tables 1.2 and 1.3.)

Another difficulty, one that grew more complex over the years, concerned ideological supervision of these young scholars. Their formative years were during the Khrushchev thaw. Many of them believed that societywide reform would be followed by the achievement of communism within their lifetimes. It would be communism with a human face, devoid of the horrors of Stalinism. The rapid promotion of these scientists meant that individuals who were reform-minded, or at the very least had a broader sense of academic freedom than their Moscow and Leningrad counterparts, had assumed positions of political and scientific responsibility in Akademgorodok. Their views and behavior often conflicted with the expectations of conservative Brezhnev-era policy makers and scientific administrators, leading to sharp disagreements over freedom of discussion—of the arts, of music—and over the very atmosphere of Akademgorodok itself.

Several factors motivated both established and younger scientists to move to Siberia: the possibility of more rapid advancement to the honored titles of "academician" or "corresponding member" of the Academy; freedom to do research away from scientists who dominated entire institutes and fields; perhaps faster achievement of doctoral degrees or candidate of science degrees; and the lure of larger apartments and marginally higher salaries. According to the bylaws of the Academy, different divisions and disciplines have a set number of "chairs" that can be filled only when vacant. Over time, new chairs were added for various reasons, for example, the general growth of the scientific enterprise. In the 1920s and 1930s the Communist Party insisted on new chairs in the social and technological sciences to be filled with Marxist scholars in order to tip the balance of control of the Academy to communists, and from then on sought to control the election process by approving slates of candidates beforehand. This tactic has not always succeeded: the Academy never removed Andrei Sakharov as a member even when he was persona non grata in government circles.

Lavrentev wrote Khrushchev personally with requests to authorize increased numbers of Siberian chairs in the Academy. In such a letter in 1961, he reminded Khrushchev how rapidly Akademgorodok had grown in terms of numbers of laboratories and institutions. Unfortunately, Lavrentev pointed out, personnel qualifications had not kept pace. Dozens of laboratory heads did not have even a candidate of science degree. Although scientists in the Siberian division comprised one-fifth of the entire Academy, they made up only 6 percent of the "qualified" scientists.[47] The Siberian division had 39 of the Academy's 111 institutes, 3 of its 10 branches, 11 of its 17 observatories, in all 57 or one-quarter of its 195 scientific institutions, but only 10 of its 157 academicians, 35 of its 392 corresponding members, and 74 of its 1,541 doctors of science, that is, only 119, or 6 percent, of 2,040 "highly qualified personnel." Opening vacancies in the Academy, Lavrentev argued, would help overcome the poverty in personnel.[48]

Owing to the physical-mathematical profile of Akademgorodok, the vast majority of the first appointments represented those fields. Most doctors of science were also in these disciplines. In the decade from 1960 to 1970 as many as twenty academicians and fifty-six corresponding members were appointed to positions at Akademgorodok. They included the leading actors of this story: the mathematicians G. I. Marchuk, A. D. Aleksandrov, and L. V. Kantorovich; the physicists G. I. Budker, S. T. Beliaev, R. Z. Sagdeev, and A. N. Skrinskii; the biologists D. K. Beliaev and G. I. Galazii; and the social scientists A. G. Aganbegian and T. I. Zaslavskaia. The creation of "cadres" in Akademgorodok had many critics. Several scientists resented the pressure to vote for Academy chairs designated specifically for the Siberian division. Others opposed what they alleged was the awarding of honorific titles and administrative responsibilities as laboratory and institute directors based on geography, not merit. In fact, because of disagreements over qualifications, three corresponding members, ten doctors, and ninety-two candidates of science, fully 14 percent of the entire staff of scientific workers, quit Akademgorodok in the early years. Although it may have been true that a few individuals could not have passed muster in Moscow or Leningrad, the record shows that as a rule highly qualified individuals were promoted.

The central issue concerned the youth movement in Akademgorodok. Many Akademgorodok officials claimed that precisely because of this youth movement—to this day scientists in the Siberian division have a lower average age than any division of the Academy—more attention had to be paid by higher party organs to ideological supervision and to such potential failings as self-promotion, inattention to others' work, and poor work habits. In January 1962 Lavrentev worried publicly about both political and scientific qualifications of young scientists. But in many cases failure to meet qualifications was the fault of someone else. In fact, from 1958 to 1962, more than two hundred young specialists were dismissed and hundreds of others demoted. Some scientists without advanced degrees had failed their graduate student charges. Acceleration of dissertation defenses left other scientists burdened with teaching responsibilities and not enough time for work.[49]

A major weakness of Soviet scientific training was the separation of research and education. Universities and institutes were isolated by wide gulfs of ministerial jurisdiction, which meant that teaching was divorced from the laboratory. Lavrentev hoped to improve on this system in two ways. First, scientists in Akademgorodok set up a physics-mathematics boarding school to provide gifted students with first-rate instruction from academicians and professors. They used "olympiads" throughout Siberia to identify talented schoolchildren. The first such physics-mathematics olympiad was held early in 1962, with chemistry following in 1964. The students were taught by leading scholars. The boarding school had a rigorous two-year curriculum based on eight hours of physics and mathematics per week with three hours of lab, two hours each of chemistry, biology, and agricultural science, and some history. The goal was to mimic a university curriculum of general physics. In 1963–64 ninety-three students completed this program. By 1965 more than six hundred students were in residence in three grades. The boarding school students were given preference for admission to the university. In 1967, for example, the university received more than 3,300 applications for 750 places. All 154 "graduates" of the boarding school were successful in their applications: 116 students were admitted to Novosibirsk University, 29 to other higher educational institutions, and 9 began work in institutes of the Siberian division.[50] To this day, most graduates stay on in Akademgorodok to study and work.

Although Soviet historians claimed that Lavrentev always endorsed, indeed first promoted the idea of the boarding school, Lavrentev seems initially to have thought there was a danger of elitism associated with it. He said at one point, "It is a mistake to create special schools for specially gifted children. For surely aptitude in physics, chemistry, and mathematics is not always expressed at an early age. Besides, the vocation of science requires not only ability but also a love of work and an inclination to the chosen speciality, that is, those qualities that are already formed at a comparatively mature age." In the egalitarian Soviet Union, admission to universities and institutes had to be guaranteed to all qualified students.[51]

Lavrentev hoped to improve on the Soviet educational system by founding a new university. The Ministry of Higher Education approved the establishment of Novosibirsk State University (Novosibirskii gosudarstvennyi universitet, or NGU) at the same time as Akademgorodok's charter was adopted. A temporary building was soon erected, and in October 1959 the university opened officially under rector I. N. Vekua. Just as scientists had to be brought in to staff Akademgorodok's institutes, so accomplished scholars had to be imported for the university. Tomsk University, the only other major university east of the Ural Mountains, was too small to be tapped, despite its history to 1880. So to make up the staff, Vekua brought a number of colleagues with him from Tbilisi, Georgia. Unlike most Soviet higher educational institutions, the plan was for university students from their third year onward to work simultaneously in the laboratories and the scientific institutes. The laboratories were no more than twenty minutes away on foot. The students' teachers were academicians, so they got science "firsthand." Virtually all leading Akademgorodok scientific personnel worked closely with students. According to an official history, 8 percent of academicians, more than half the corresponding members, 46 percent of the doctors of science, and one-fifth of the candidates of science taught at the university. The university grew rapidly. By 1961 there were 946 students, including 365 women, and within another year 1,566 students in five departments, although construction was incomplete. More than half of the first 54 graduates in 1963 remained in Akademgorodok. By 1967 of 1,107 students who had graduated from the university, 504 worked in institutes in the Siberian division.[52]

Whatever the path to Akademgorodok, whatever the construction problems, the Siberian city of science grew rapidly. In 1958 there were 1,718 scientific workers within the Siberian division of the Academy (many still residing in Moscow or Leningrad) of whom 10 were academicians, 26 corresponding members, 58 doctors of science, 473 candidates of science, and 1,151 with no advanced degree. By May 1, 1964, there were 15 institutes with 7,300 employees, including 2,200 scientists, of whom 14 were academicians, 31 were corresponding members, 49 were doctors and 551 were candidates of science. Thirty thousand settlers had arrived, hundreds of thousands of tons of dirt had been moved, and tens of thousands of cubic meters of concrete had been poured.[53]

Many argue that the quantitative indexes of growth of Akademgorodok— and Soviet science in general—masked significant problems of quality. By the 1970s a fourth of the world's scientists, half its engineers, and a third of its physicists resided in the USSR. Yet many of those listed as "scientific workers" in Soviet publications were laboratory assistants or test-tube washers, not scientists in the western sense. In addition, the rapid "manufacture" of specialists through methods of rote learning did little to ensure flights of fancy, creativity, and scientific excellence among all but the very best—as in any social or cultural setting. On the other hand, as the following chapters show, a world-class research community rapidly developed in Akademgorodok owing to the efforts of Lavrentev, the special mission of the Siberian city of science, and its youth movement.

UTOPIA IN THE GOLDEN VALLEY

Would it be possible for the scientific visionaries of Akademgorodok to over-come all the problems connected with hiring new staff, identifying and train-ing young scientists, supplying material and equipment, and building new in-stitutes, stores, and apartments? Would the impetuous Nikita Khrushchev ease the burdens or stand in the way? Could the local Communist Party apparatus be counted on to assist Akademgorodok? And would the Siberian climate co-operate? Mikhail Lavrentev had no doubt that man would soon master nature as Francis Bacon himself had suggested. Soviet scientists would outstrip the futuristic visions of Jules Verne and H. G. Wells. When Lavrentev addressed writers at a conference of authors of children's prose in Siberia in 1958, he said, "It is well known that many scientists very much love to read the works of [Verne and Wells], rereading them dozens of times, finding in them, perhaps, something useful for their thoughts, a push of some kind toward the solution of problems yet to be solved." But Lavrentev asserted that science fiction was no longer attractive because science itself had accomplished everything de-scribed more boldly. He therefore called on these Russian authors "to fantasize further, penetrate with their gazes far beyond the limits of what is already known and achieved by our science, so that our youth will be captivated by new science fiction, with bold dreams of human progress." These writings would "show how man controls nature, how he compels nature to do what he wants, and not what she wishes. We need the pathos of scientific struggle. We need new flights of scientific thought, scientific dreams in the name of the coming victories of Soviet science for the good of our Motherland, for the good of all progressive humanity."[54] Akademgorodok was Lavrentev's vision of one of the many coming victories of Soviet science.

On the surface, the ideal vision of Akademgorodok was achieved. A world-class scientific community had been created where before there had only been forest. Still, the ideal of a scientific community free of economic, political, cultural, and ideological constraints is hard to imagine in any society. The notion of such a community was even more incongruous in the postwar Soviet Union with its authoritarian political system and its Communist Party, intoler-ant of any perceived deviation from accepted behaviors in politics, culture, the arts, and of course science. The economic system was centrally planned, capable of meeting the demands of rapid industrialization but clumsy in deal-ing with the special needs of modern science and technology, including the production of exact instruments. From an ideological point of view, there may have been sufficient leeway to experiment with new subjects in biology, sociology, economics, and computer science during the period of relative, yet incomplete relaxation of controls on culture during the Khrushchev thaw. So when the Soviet leadership took a hard turn back to more extensive and overt political supervision of scientific and cultural activity during the

Brezhnev years, and insisted on accountability of scientists to state economic development programs, the Siberian New Atlantis, which to many had symbolized the ideal of academic freedom in fundamental research, had to give way to Soviet standards.

Yet the institutions and personalities of Akademgorodok combined to create a scientific environment unique in the USSR in terms of its openness and mission. Innovations in administration, organization, and education at the university and in research at the institutes permitted democratic communication between established and novice scholars, between young and old, even between academicians and schoolchildren that differed significantly from the stodgy, hierarchical scientific rule so prevalent elsewhere in the country. There were innovations in cultural organizations as well, with young scientists dominating the Komsomol, the scientists' councils, and professional organizations. They even established nightclubs where the scientists discussed recent literature, music, art, and politics in tones that surely disturbed the policy makers and administrators whose formative period had been the Stalin years. At times they simply flaunted their independence and created openness not seen again until the Gorbachev years.

In terms of personalities, from the points of view of personal histories, leadership styles, political sentiments, and research interests a wide range of individuals gathered in Akademgorodok. Some were quite conservative and reticent to consider burning political and cultural issues. Others had experienced firsthand the horrors of the Stalinist system. Still others were merely iconoclastic. But all shared a devotion to Lavrentev's plan, a love of their respective missions, and the conviction that they could not go wrong in building a new Soviet society since they were armed with the tools of science and technology. They were mistaken to think that they could carry out research divorced from the social, economic, and political problems facing the USSR. But there is no question that their new approaches to these problems were central to improving the performance of Soviet science and technology. One such approach involved Gersh Budker's efforts to build a novel particle accelerator in Akademgorodok.

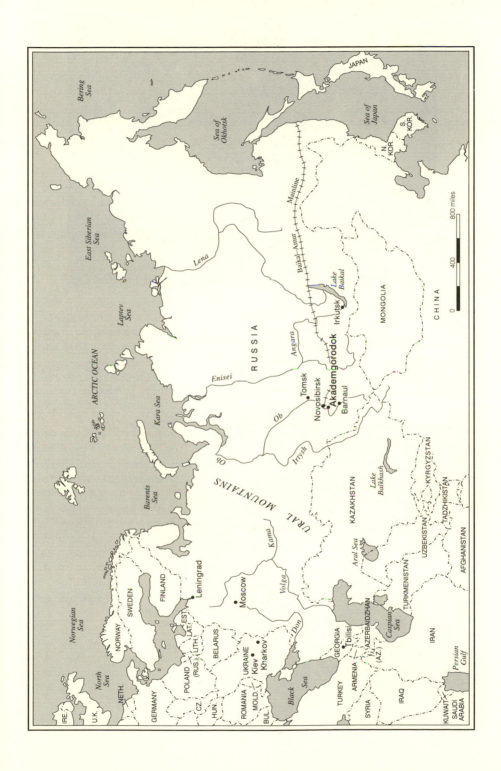

In 1957 Nikita Khrushchev, Stalin's successor, endorsed a plan to establish a city of science, Akademgorodok, in Siberia (top photo courtesy of Rashid Akhmerov). Akademgorodok would serve three purposes: lessen the concentration of scientific resources in Leningrad and Moscow; promote the development of rich Siberian natural resources; and move Soviet science and technology far from the reach of any future invasion. Construction on Akademgorodok began in 1958 in harsh Siberian weather with poorly trained workers and modest equipment (photo courtesy of Boris Mordukhovskii).

The scientists involved in Akademgorodok planned for the city to have unique aesthetics compared with other Soviet cities. But cost overruns early on forced the substitution of large concrete panels for bricks. Still, when completed, Akademgorodok differed from Moscow and Leningrad for its ideological and geographical distance from the Communist Party as well as for its attractive, college like campus with broad boulevards, parks, and gardens surrounded by forest (photos courtesy of Boris Mordukhovskii).

Khrushchev personally attended to matters of construction, visiting Akademgorodok on two occasions. On one of these trips, in 1959, he consulted with the founder
of the city, Mikhail Lavrentev (*center, with glasses*) about the presence of modern
genetics in Siberia, which still remained politically suspect among some Communist
Party figures and scientists (above photo courtesy of Rashid Akhmerov). But the
physical and mathematical profile of the city, including Lavrentev's own Institute of
Hydrodynamics and the prestigious Institute of Nuclear Physics, provided a coun-
terweight to political interference in biology (photos courtesy of Rashid Akhmerov).
The Institute of Nuclear Physics carries a slogan typical of the Soviet era and its faith
in science to achieve policy goals: "Glory to Soviet Science!"

Akademgorodok stood apart from the Soviet scientific establishment because of its democratic relations among all personnel. This relationship extended to the children chosen for special science schools who later filled Akademgorodok's institutes. Lavrentev (top left photo courtesy of Rashid Akhmerov) especially enjoyed working with young people in the Club of Young Engineers (called, ironically, "Kiut" [cute]), in the summer boarding school, and in the physics-mathematics boarding school, all used to identify promising young children in order to track them into research institutes (photos courtesy of Rashid Akhmerov).

Lavrentev, a mathematician with an interest in explosives, was a natural teacher (top photo courtesy of Rashid Akhmerov). He lived in one of the cottages reserved for the scientific elite in the so-called Golden Valley, where his son still lives. The workers charged with building Akademgorodok questioned why they were building spacious houses for the scientists when they themselves had to live in simple barracks. Lavrentev's cottage was the first "building" to be erected in Akademgorodok in 1958 (photo courtesy of Boris Mordukhovskii).

When Lavrentev pushed the notion of Akademgorodok, he counted on the support of such leading Soviet scientists as Igor Kurchatov, head of the Soviet atomic bomb project, and physicist Petr Kapitsa, the future Nobel laureate (top photo courtesy of Rashid Akhmerov). Several scientists in the Soviet Academy of Sciences opposed the establishment of Akademgorodok fearing that it would diffuse scientific resources and authority; others worried that Akademgorodok could never attract first-rate talent; and still others resented Lavrentev's ambition. But most of them, including Academy of Science President Mstislav Keldysh, who visited Akademgorodok in 1962, came around to support the endeavor (photo courtesy of Rashid Akhmerov).

Akademgorodok attracted open-minded young scientists who were deeply affected by the intellectual currents of the de-Stalinization thaw of the late 1950s. Among these scientists was the brilliant iconoclast, theoretician Gersh Budker, director of the Institute of Nuclear Physics, shown here expressing thanks for a birthday medallion with physicist Spartak Beliaev, then rector of Novosibirsk University (top photo courtesy of the Institute of Nuclear Physics). By the mid-1960s the construction of Akademgorodok was finished and more than a dozen institutes had opened (photo courtesy of Boris Mordukhovskii). Under Leonid Brezhnev, who had ousted Khrushchev in 1964, the history of the city of science took a new turn: Akademgorodok scientists were forced by the Communist Party to conform to Soviet ideological norms.

CHAPTER TWO

Colliding Beams and Open Traps

It's great to have a place that the big-wigs don't visit, and we don't permit
the little ones to join us.

—Gersh Budker

IN THE FALL of 1958 Gersh Budker, a talented forty-year-old theoretician working in the heart of the Soviet atomic energy establishment at the Institute for
Atomic Energy in Moscow, was asked by the institute's director, I. V. Kurchatov, to consider moving to Novosibirsk to head up a new institute. Budker
immediately accepted. Hard-working and brilliant, grating and misunder-
stood, he needed a new environment to further his studies in high energy and
plasma physics. In spite of the largess offered to atomic energy institutes in
Moscow, Leningrad, and Kharkiv in Ukraine, new avenues for research were
hard to find. Unless a scholar could show that short-term military benefits
might grow out of his research, his new project might have to wait. Budgets
were tight, and senior scholars dominated program planning.

For Budker, Kurchatov's proposal was an opportunity too good to pass up.
A number of leading Moscow scientists thought that his proposal to build
accelerators on the principle of colliding beams of charged particles was far-
fetched, and they wrote negative reviews. Others disliked the irrepressible
man himself. But with Lavrentev's full support, the promise of funding, and
the chance to assemble a dynamic scientific collective with minimal interference, Budker gladly moved to Siberia to test his theory of accelerators at
the newly founded Institute of Nuclear Physics. Once in Akademgorodok,
Budker and his associates astounded the world by beating Stanford University
and the Americans to the punch in bringing their accelerator on line. "We
moved from Moscow to Novosibirsk," Budker stated. "I would have liked to
see my Stanford colleagues do as well if they had had to move from California
to Alaska."

Budker was a Moses-like figure with wild, dark, long hair and a full beard.
His research program and style of leadership dominated the Institute of
Nuclear Physics. His colleagues remember how he led them in the search
through the Siberian wilderness for confirmation of quantum electrodynamics.
Budker's associates are still exploring aspects of his theories of plasma physics
and proposals for alternative approaches for controlled thermonuclear synthesis (fusion). Budker's success in creating a scientific community that operated
according to democratic principles indicates just how open life in Akademgor-

odok could be. For the physicists, "community" went beyond scientific matters. It meant freedom to discuss openly the ideas of physics, art, literature, and politics. It may have been the prestige of nuclear physics and the potential of military spin-offs from his research that led the authorities to give Budker such a free hand to pursue scientific ends. Yet the physicists were also among the leaders of the Akademgorodok social clubs.

Budker's forceful personality continues to permeate the institute. A marble carving of him with flowing hair and beard looks down on the huge round table in the meeting room of the institute's academic council. Physicists at other research centers share this reverence. Carlo Rubbia, winner of the 1983 Nobel prize in physics for the discovery of the W and Z bosons, has only one photograph in his office, that of Gersh Budker of the Siberian Institute of Nuclear Physics. Working at the European Center for Nuclear Research (CERN), Rubbia and his colleagues pushed furiously to beat Budker to the Nobel prize, for Budker had pioneered the techniques for CERN's new accelerator.[1]

Since the birth of Akademgorodok, the Institute of Nuclear Physics has been the leading institute in the Siberian city of science, both in quantitative terms (its personnel, budget, and research program) and qualitative ones (its power and influence). The huge institute covers several hundred acres, has four particle accelerators, a production facility with a massive foundry, cranes, machine tools, and dozens of laboratories. The main building is five floors with two wings. It takes ten minutes to walk at a brisk pace through the various passageways and hallways—sometimes underground—from one end to the other. The cafeteria feeds two thousand daily and is opened to workers from other institutes at 2:00 each afternoon. There are some 3,000 employees, including 4 academicians, 4 corresponding members, 35 doctors, and 130 candidates of science. In many respects, the history of Budker's Institute of Nuclear Physics is like the history of any major physics institute in postwar America. Superpower competition, military value, and public acceptance of big physics led to the rapid growth of the physics establishment. Like in the United States, particle accelerators and nuclear reactors assumed great cultural, ideological, and physical significance.

Although ultimately a success story, the history of the Institute of Nuclear Physics is also one of profound disappointment. At first Mikhail Lavrentev and the Siberian division of the Academy of Sciences wholeheartedly supported Budker's efforts in elementary particle physics. In the end, however, his demands for ever larger accelerators and their corresponding facilities, offices, and staff conflicted with the philosophy of Akademgorodok. Lavrentev wanted moderately sized institutes united by a common purpose and interdisciplinary approach. He began to resist Budker's incessant efforts to expand the institute. Other Siberian scientists were less sporting. They resented Budker's successes, his authority in the scientific establishment, and the wealth of his institute. Together with the science budgets allocated in Moscow growing tighter over the years, this frustrated Budker's plans to stay in step with big physics in the West. The tumultuous year of 1968, with the crackdown on Akademgorodok's political freedom and the invasion of Czechoslovakia, also impeded the success of the Institute of Nuclear Physics. The third international conference on plasma physics, which was held in Akademgorodok in 1968 and should have been a celebration of Soviet leadership in fusion, instead became a symbol of what might have been: an open, vibrant scientific community in the center of Siberia.

THE ATOM IN SOVIET CULTURE

Gersh Budker was a pure product of the forces of the Soviet atomic age: the fertile soil of Soviet "gigantomania," the mentality of cold war rivalry in geopolitics and science, and the cult of postwar science. The cult of science was based on the notions that science and technology were a panacea for social and economic problems. Building on the success of the atomic bomb project (the first Soviet detonation was in 1949) and nuclear power (the first Soviet reactor to produce "commercial" electricity, 5,000 kilowatts, came on line in 1954), physics benefited from the cult of science and from societywide fascination with large-scale technologies. Many nuclear physicists came to see themselves as "engineers" whose goal was to tame the speed, energy, and mass of the subatomic world for Soviet economic and social programs. They took advantage of state support for their visions of atomic-powered rocketships, automobiles, and locomotives, and peaceful nuclear explosions (to build dams, divert rivers, and mine valuable metals) to expand their research programs. They built experimental and industrial prototype slow (thermal) and fast (breeder) reactors, various fusion devices, and increasingly large particle accelerators. Soviet society willingly embraced the cultural artifacts physicists built as proof that the USSR could live up to ideological pronouncements of "reaching and surpassing the West" in all fields.

As in the United States, the cold war encouraged the expansion of the atomic energy establishment. Competing with the United States for global military, economic, and ideological preeminence, no expense was barred. From modest beginnings in the 1930s in Leningrad in the Physical Technical and Radium institutes, and in the Ukrainian Physical Technical Institute in Kharkov, atomic research spread to Laboratory Number 2 (later the Kurchatov Institute of Atomic Energy, or KIAE) and Laboratory Number 3 (later the Institute of Theoretical and Experimental Physics) in Moscow, the Joint Center for Nuclear Research in Dubna, the Physics Engineering Institute in Obninsk, to institutes in Troitsk, Melekess (later Dmitrovgrad), to the Institute of High-Energy Physics in Serpukhov south of Moscow, Gatchina outside Leningrad, Arzamas, Cheliabinsk, Krasnoiarsk, Gatchina, Troitsk, and Akademgorodok. This mirrored the development of the nuclear establishment in the United States at Los Alamos, New Mexico; Oak Ridge, Tennessee; Hanford, Washington; and later Savannah, Georgia; Fernald, Ohio; Rocky Flats, Colorado; and elsewhere.

Under the direction of the "Atomic Tsar," Igor Vasilevich Kurchatov, the nuclear physics establishment grew in leaps and bounds.[2] Kurchatov was known affectionately as "the beard" because of his distinctive cut. He wielded tremendous power in establishing institutes and in overseeing the distribution of scarce resources, at special request even allocating apartment space or medicine to needy scientists and their dependents. Soviet hagiography often obscures the truth about Soviet heroes. Kurchatov is painted as an omniscient figure and a just and capable organizer. In this case one can only agree with these characterizations. Kurchatov seldom allowed his personal feelings to intervene in scientific decisions. He judiciously allocated resources to competing programs. He shepherded the expansion of the Soviet nuclear enterprise with a clear vision of the future. In May 1951 he secured a government resolution to support controlled fusion research. In 1952 he approved the commencement of research in his institute on fuel rods for industrial nuclear reactors. In 1956, at the twentieth party congress, he called for government support for research on isotopes, nuclear-powered airplanes, locomotives, and automobiles. He cut through bureaucracy at Minsredmash (the Ministry of Middle Machine Building, the nuclear military establishment) and at the State Committee for the Utilization of Atomic Energy (Gosudarstvennyi kommitet po ispol'zovaniiu atomnoi energii, or GKUAE, Minsredmash's civilian side) in order to pry funds loose, expedite publications, and secure foreign travel. He supervised the creation of atomic centers in other union republics, personally attending to plans for equipment and personnel for nuclear physics institutes in Kazakhstan, Georgia, and Armenia.[3] Kurchatov also personally protected Budker from the hooks of the secret police and selected him to be director of the new institute in Siberia. How Budker, an irreverent, iconoclastic individual, ended up in Akademgorodok, avoiding the secret police, is one of the many startling stories of Soviet history.

Budker was born in a Ukrainian village in 1918. His mother was a simple peasant, his father a farm laborer at a mill, who was shot by the partisans. In 1935, after ninth grade, he moved to Moscow to attend Moscow University. During his entrance interview he was asked what the reasons were for the current "food problems." Rather than acting ignorant about government policies that led to a famine in Ukraine in 1932–33, millions of deaths, and the slaughter of half the Soviet livestock, Budker answered without blinking, "Collectivization." The authorities delayed his admission for a year, but at least Budker was not arrested. He completed his studies without further incident. He took his last university examination on June 23, 1941, two days after the Nazis invaded the USSR. Budker, like many other able-bodied men, volunteered for the army, and his physics career was put on hold. On August 6, 1945, when the United States dropped the atomic bomb on Hiroshima, Lieutenant Budker was serving in the Soviet Far East. A year later he was demobilized and returned to Moscow where he was conscripted to Kurchatov and the Soviet bomb project.

Budker revealed mature and varied interests from the start. He worked in the theoretical department of Laboratory Number 2 under the energetic theoretician A. B. Migdal. He completed a cycle of studies on the first reactor in the USSR in 1947, including the theory of the optimal uranium-graphite lattice and on the kinetics and control of reactors (1949). After he learned about the construction of the Big Volga (now Dubna) proton accelerator in record time, Budker turned his attention to resonance processes in particle dynamics, that is, how to use the tracks left in bubble chambers to demonstrate the presence of various subatomic particles. He developed a method for calculating shimming (correction) of the magnetic field and suggested original ways to extract the beam from the accelerator, work that earned him a state prize in 1949. He then turned to the study of the stabilized relativistic electron beam and new electron accelerators. With several future Akademgorodok associates he advanced an idea in 1950 for an accelerator of very high energy which served as the foundation for electron cooling for proton-antiproton accelerators. Migdal later recalled, "The most striking quality that put Budker in first place in his field worldwide was his inexhaustible engineering fantasies, and not only his ability to fantasize but to realize those fantasies."[4] Budker did not look for step-by-step improvements but rather sought new approaches altogether.

Budker was also a central figure in Soviet programs on controlled thermonuclear synthesis (fusion), although his contributions were arbitrarily interrupted by the xenophobic actions of the Soviet police state. At the end of 1950 Kurchatov invited Budker to work with other scientists on the creation of a magnetic thermonuclear reactor. After initial successes with a toroidal configuration for magnetic confinement of plasmas, a number of problems arose. After a year of experiments with torroids of different types and sizes, Lev Artsimovich suggested moving back a few steps, to experiment with direct electrode discharges in order to understand what kind of plasma might result in containment devices. Physicists suggested alternatives to torroids. Budker proposed to solve the problem with an original suggestion—plasma traps with magnetic mirrors, so-called open thermonuclear systems (to be discussed later).

The joy of working in the exciting fields of fusion and elementary particle research, and of heated competition with the United States for prestige, was tempered by the fact that the entire nuclear research effort was placed under the direct control of the secret police chief, Lavrenty Beria, who reported daily directly to Stalin. Scientists were accountable to the demands of Soviet economic programs, propaganda, and military technology. They were often simply co-opted for work. To refuse a transfer to the military industrial complex was unheard of. Yet Budker's participation in fusion research was suddenly interrupted when Beria dropped him from the program in 1952—even though plasma traps with magnetic mirrors had been his suggestion. The anti-Semitic "Doctors' Plot" had been hatched. Stalin and his henchmen had accused leading Kremlin doctors of attempting to poison the Kremlim leadership. Another terror like that of the 1930s was in the offing. Jews everywhere lost their jobs.

One time Budker came close to being executed at the hands of Beria. It came about because of an argument between Budker and Kurchatov. The two often argued, usually over trivial matters, but they quickly made up. But on this occasion, no doubt over an impetuous comment, Kurchatov had had enough and yelled, "Do as I say or you're out." Because Budker had already been denied access to top-secret work, being fired would have meant arrest, perhaps execution. Kurchatov thought better of his anger and backed off. Beria, too, let the matter drop. He told Kurchatov, "Budker's an especially dangerous criminal. But let him live awhile. We will still succeed in cutting off his head." When Beria was shot in June 1953, after Stalin's death, several personnel files that he personally controlled were found. One was a dossier on Budker.[5]

Kurchatov was able to arrange for Budker to continue work during this time with the help of Spartak Timofeevich Beliaev. Beliaev, later rector of Novosibirsk University, first met Budker when he was in his second year of studies at Moscow Physical Technical Institute, the newly founded training center for Russia's nuclear specialists. Budker advised Beliaev on his senior thesis research at the Kurchatov Institute. Beliaev's work involved efforts to explain the great losses in the intensity of the beam during its first orbits on the institute's cyclotron. After his senior thesis defense, Beliaev became a full-time employee of the institute. Under Lev Artsimovich and Mikhail Leontovich the institute was moving ahead full steam on fusion. At Kurchatov's request, Leontovich and Migdal asked Beliaev to work with Budker to provide him with scientific companionship during his exile from fusion. Periodically, the results of Budker and Beliaev's work were "passed on" to other physicists for proper scientific criticism. But Budker felt intellectually impoverished by his isolation. He was forced to write abstracts of articles to earn extra money during his "divorce" from thermonuclear synthesis. In one way Budker's exclusion from fusion was fortuitous for he was forced to turn to another area of research: new accelerator methods. A special laboratory with a staff of eight was created to work on these problems.

This experience had a telling impact on Budker's leadership style and on the openness of the Institute of Nuclear Physics. Budker resented the secrecy surrounding work on fusion. His first paper on open traps, written in 1954, was published only four years later in the fourth volume of *Physics of Plasma and Problems of Controlled Thermonuclear Reaction* (Fizika plazmy i problemy upravliaemykh termoiadernykh reaktsii), which was prepared for the second in a series of Geneva conferences between western and Soviet physicists in the 1950s to share results and build confidence about potential cooperation. Budker knew all too well how damaging secrecy was to Soviet science. It slowed the diffusion of ideas not only to the United States but to his Soviet colleagues. On one occasion Kirill Sinelnikov, director of the Ukrainian Physical Technical Institute, even though a close friend of Kurchatov's and head of the Ukrainian nuclear and high energy physics programs, asked Ia. B. Fainberg, a leading figure in the Kharkov fusion program, what he thought of the work of a heretofore unfamiliar physicist named Budker. Fainberg was duly impressed by the style and quality of Budker's work in plasma physics.[6] How was it possible that Kurchatov had not told Fainberg and Sinelnikov of Budker's work earlier?

Even had secrecy requirements not been so strict Budker still may have felt trapped by the stultifying environment of Moscow physics. His outspokenness rubbed too many people the wrong way. Even his friends had trouble tolerating his outbursts. Brilliant and incorrigible, after a sleepless night filled with flights of fancy, he would charge into the office with fresh ideas each morning. He would blurt out what was on his mind, often interrupting his colleagues' work. Once when he observed Migdal tackling a difficult problem, Budker called out, "Try the reciprocal."

Migdal responded, "Stop bothering me."

"Try the reciprocal!" Budker insisted.

"Get out of my office!" Migdal yelled, threw Budker out, and shut the door. Through the key hole Migdal again heard, "Try the reciprocal!"[7]

Budker constantly skirmished with the authorities. After marrying his third wife, he was called in by the Novosibirsk regional party secretary and was admonished for setting a bad example for Soviet youth. "How many wives have you had?" the party secretary inquired. Budker, who had had a number of lovers as well, responded, "Do you mean officially or unofficially?"

But it was his politics that really got Budker in trouble. He refused to sign official Academy of Science condemnations of Andrei Sakharov and probably would have been removed as director of his institute had his premature death from a heart attack in 1977 not rendered the matter moot. Budker was famous for long discussions with physicists, writers, and dramatists at the Ob Reservoir. Evening discussions of politics and literature under the moonlight became a ritual. Here Budker talked about Landau, about his "scientific uncle," A. F. Ioffe (Kurchatov's teacher). When, in 1952, Ioffe was removed as director of the institute he had founded in 1918, he spent entire days sitting in the public garden (*skver* in Russian). Ioffe's scientists and students would come running to sit around and talk with him, and this became known as "public physics" (*skvernaia fizika*), a tradition Budker intended to keep. The authorities said that Budker "stuck in your craw. This academic wouldn't join the party, he really stuck out and he was always asking for something."[8] Perhaps because of his abrasive personality, Budker failed to generate real support for his idea of a stabilized electron beam with the exception of Beliaev and other close associates.

Through the activities of the Geneva conferences Budker gained a reputation among western physicists for his fusion work. But he thought he could go no further in Moscow either with fusion or the accelerators which increasingly dominated his thinking. Budker also believed that physicists ought to focus on peaceful, not military applications. So when Kurchatov nominated Budker as the director of a new institute in Siberia and gave him a night to think it over, Budker took five minutes to accept. He was determined to take his sharp wit and unlimited creative potential to promising beginnings at the Institute of Nuclear Physics. Siberia, with its untouched beauty, far from the authorities in the capital and from physicists who impeded his research, hearkened.

A Home for Wayward Accelerators in Siberia

Of course physics was central in Akademgorodok. Lavrentev believed in the nearly limitless power of physicists and mathematicians to solve Russia's problems on its path to communism. The Institute of Nuclear Physics (Institut iadernoi fiziki, or IIaF) was created according to the typical Akademgorodok pattern. Budker's already existing laboratory for new methods of acceleration would be moved wholesale to Siberia. Initially there were discussions about whether to build a reactor in Akademgorodok as well but Lavrentev vetoed the idea: Any reactor would have to be "outside the city."[9]

Two organizational principles dominated Budker's vision of the institute: democracy in scientific disputes and the absence of bureaucracy in administration. He hated the secrecy and bureaucracy of Minsredmash and vowed to avoid its characteristics. "Scientific workers should not give power to the administrative apparatus," he said, "for no matter how good and well-intended that apparatus is, it can guarantee order and supply, but not a creative atmosphere." In all the days Budker was director, there was but one secretary. Anyone could drop in without an appointment at any time. He treated workers and scientists equally. Jokingly he proposed to fire all Soviet bureaucrats and send them to the Black Sea resorts—with full pay and privileges; both science and the economy would function more efficiently as a result.

The architecture also promoted democracy. Budker wanted an institute where scientists would constantly bump into each other, inspiring daily interaction. Toward this end he had the architects design open corridors and hallways. He was involved even in the mundane details of construction; a steam pipe next to the institute, instead of being round, has a rhomboid shape because of Budker's intervention. He also met regularly with the workers of Sibakademstroi during the construction to explain to them the significance of the institute's research in nuclear physics.[10]

The democratic rule at IIaF was signified by the "round table," a daily noon gathering of all members of the academic council and their guests from 1963 onward. The huge wooden table represented openness, community, equality, and youth. (Initially the physicists' average age was not quite thirty.) The round table was a place "where scientific and organizational ideas were crystallized, where current and future questions of life, including, it would seem, the trivia of minor administrative problems, were discussed," said Aleksandr Skrinskii, himself a thirty-four-year-old academician and current director of

IIaF. Every question, the most complex or delicate, was hashed out, with heated discussions sometimes dragging on for days. But the decision was always binding. "The 'round table' has ruled!" Budker would never force the decision. Other ideas were killed on the spot after a few hours of discussion. Budker would then respond, "It's too bad, but it was a beautiful idea." On successful days, which were quite rare, Budker's comment was, "Look! The old man can still do something!"[11] As the institute grew so did the academic council. Starting in 1971 the full academic council met once a week in rotation with other thematic councils on high-energy physics, accelerators, synchrotron radiation (high energy beams of electrons and positrons that are powerful sources of ultraviolet and X-ray radiation), plasma physics, and applications. The round table engendered the morality and freedom of discussion required of scientific decision making in a closed political system.

The equality that characterized the round table spread naturally to other spheres of institute activity. Some say there were no "nationality problems" in IIaF. Yet others, more familiar with Budker's views, confirm that anti-Semitism may have played a role in his personnel placement. Whatever the case, anti-Semitism was a major force in Soviet science. The Tsarist regime made entrance into universities difficult for Jews. Whole disciplines were essentially proscribed by spoken and unspoken rules. So Jews gravitated to mathematics and physics as two relatively young disciplines with fewer obstacles for entry. This situation continued during the early Soviet period until the Stalin years when entrance requirements for Jews were introduced to keep their numbers from growing. Over the next fifty years Jews were discriminated against, particularly in mathematics. Rarely were Jews voted membership in the Academy of Sciences in mathematics. They faced great obstacles to admission to universities and in many cases were prevented from defending theses.[12] Jews were also active in the dissident movement in Russia. Those who chose to emigrate lost their jobs, were harassed, and were denied exit visas. They became third-class citizens, so-called *refuzniks*. Of course anti-Semitism existed in Akademgorodok, and it has become more pronounced recently since nationalistic right-wing groups already strong in Siberia have become stronger still since the breakup of the USSR.

In any event, Budker needed qualified personnel for his revolutionary research. Many of the physicists who moved to Akademgorodok were, like Budker himself, independent-minded scientists who had left the Russian front during the war for the secret institutes of the Soviet nuclear establishment designated only by post office box numbers. Others were talented university students who were identified by their mentors for potential contributions to nuclear physics. These were the "Kurchatov conscripts." Budker pushed his ideas for new accelerator principles among these promising scientists all the way to Akademgorodok. Veniamin Aleksandrovich Sidorov was one such conscript. Sidorov (b. 1930) graduated from Moscow University in 1953, worked in the Kurchatov Institute until 1962, then transferred to the Institute of Nuclear Physics as Budker's deputy director and a laboratory head. A recent university graduate, Sidorov was blessed to gain admission to the Kurchatov Institute seminars. At one seminar he heard Budker discuss his idea for a "device" to accelerate electrons and ions based on relativistic effects. For Sidorov, Budker's talk was brilliant, although the concept seemed impossible to realize. In 1962, following Sidorov's defense of his candidate of science thesis, Aleksei Liapunov, a "pure" mathematician, remarked, "If physicists call this a candidate dissertation, what do they call a doctoral dissertation?" Budker responded, "The very same thing but the doctor must be a little older and the dissertation a bit worse."[13]

Aleksei Aleksandrovich Naumov (b. 1916) met Budker under slightly different circumstances. In May 1945 Naumov was ordered to Laboratory No. 2. He had recently graduated from the Moscow Communications Engineering Institute and worked for two years in the Red Army Scientific Research Laboratory on impulse and high-frequency radio technology. His experience with electronics made him a perfect foil in the development of accelerator technology for Budker's theoretical flights of fancy. In 1953 Naumov and his colleagues received a state prize for a cycle of work on accelerators, and that year Naumov joined Budker's efforts. In 1956, at the first conference on accelerators with foreign specialists participating at the Lebedev Physics Institute in Moscow, Naumov and Budker presented the results of work on stabilized closed electron beams in a non-iron synchrotron, for example, copper with a short-lived but more powerful magnetic field than was possible with iron. The presentation provoked great interest.[14]

In 1957 Naumov became IIaF's deputy director of science while IIaF was still a laboratory in the Kurchatov Institute. In the "*kharakteristiki*"—personal evaluations placed in ubiquitous Soviet personnel files—Budker wrote of Naumov at the time: "Comrade Naumov treats his work with great love and care. Very demanding of himself and comrades in work, he cultivates in co-workers responsibility for the task to be fulfilled and utilizes fully his deserved authority." Naumov's engineering intuition, sublime orientation in the world of electronic paradoxes, his talent in disseminating human and material resources all helped to formulate the proper scientific strategy to pursue Budker's irrepressible experimental ideas: stabilized electron streams, "non-iron" accelerators, and colliding beams.[15]

With personnel in place, it remained to build the accelerator. Aleksandr Skrinskii recalled the great challenges involved in building Budker's brainchild, the VEP-1 accelerator (*vstrechnye elektronnye puchki*, colliding electron beams) and in moving to Akademgorodok. Skrinskii, a third-year physics student at Moscow University, was taken on in Budker's laboratory for his senior thesis research on the phenomenon of a virtual cathode in strong electron beams in a longitudinal magnetic field. Skrinskii did not meet Budker personally until the end of 1957 when he became a member of the group that would work on colliding beam (electron-electron) accelerators. From then on they worked together closely for twenty years. Budker had only praise for Skrinskii. Before Skrinskii was voted a full member in the Academy of Sciences, Budker said of him, "[His] only fault . . . is youth. But as you all well know, this failing will disappear with age."[16]

Budker recognized that it would be impossible to compete with western laboratories, which were better equipped and had access to modern factories for standard high-quality equipment. In Russia, even in factories of the military industrial complex, shoddy manufacturing and delay were the rule. Equipment had to be manufactured specifically for each institute. At IIaF, the race with Stanford physicists was also waylaid by construction problems attributable to careless workers and drinking on the job. In the first nine months of 1963 alone there were thirty injuries, fourteen because of drunkenness.[17] So the physicists worked on into the night, and on weekends, so that eventually the presidium of the Siberian division actually had to forbid work on Sundays for all Akademgorodok personnel.

Budker believed that in the face of all these problems research had to be done using novel approaches—colliding beam accelerators and open-trap fusion reactors. He stressed independent thinking. "Choose the path less trodden," he was fond of saying. He rejected the Stalinist slogan, "Reach and Surpass America!" for his own, "Surpass, Not Having Reached!" This required IIaF physicists to build much of their own equipment. (Of course they had to first build production facilities in which to build the equipment.) New accelerator principles were needed also because of the drastic growth in the cost and size of standard models, billions of rubles and thousands of meters. Nevertheless, Budker faced pressure to build a synchrophasotron and other "standard" machines in Siberia.

Competition with the United States for scientific leadership interfered with efforts to build VEP-1. Scientists at Stanford and Princeton under Burton Richter and Gerald O'Neil, respectively, were at work on colliding beam technology. Budker's group experienced several initial successes. His B-2 accelerator with spiral storage ring and betatron acceleration served as a good injector, so it was decided, at the end of 1958, to use the B-2 at full energy after reconverting it to a one-ring accumulator to study the length of life of the beam in it, and to present the hoped-for stunning results at the next Geneva international conference. Budker, Skrinskii, and their colleagues had to focus all their efforts on this project, leaving the two-ring VEP-1 to be hurriedly designed and manufactured at the Novosibirsk Turbogenerator Factory without their complete attention. It was then to be shipped to Moscow for experiment, and later shipped back to Akademgorodok when the construction of IIaF's building was complete.

The physicists practically lived in Budker's laboratory. But the intensive effort failed to produce a single result before the conference. In fact, Skrinskii says, nothing went well. In essence, they were further from any results than they were at the start. When the VEP-1 arrived from Novosibirsk, the physicists focused all their efforts on it, again with the added pressure of a rapidly approaching conference deadline. This required that the entire VEP-1 complex, not even having begun "to breathe," let alone "surpass," had to be transferred in the second half of 1962 to the recently finished shell of IIaF's building 3 in Akademgorodok. Miraculously they got it operating by 1963 and that same year presented their first data at an international accelerator conference in Dubna, with their Stanford and Princeton colleagues in attendance. This resulted, finally, in Skrinskii's first publication, even though he had worked on the machine for nearly six years. Other promising young physicists had left IIaF, frustrated at having to tailor their research interests to projects engendered by superpower competition. Others refused to transfer from Moscow to the empty institute in Siberia. Skrinskii considers his having stayed with the VEP-1 so long to be "one of his greatest moral achievements." When the VEP-1 finally produced significant results in 1966, Skrinskii and his colleagues, thirty-year olds, young and green, all of them, understood they had overcome great odds. They had fought the doubts of fellow physicists, dealt with geopolitical pressures, and moved from Moscow to Novosibirsk while the Americans stayed put in sunny California.[18]

COLLIDING BEAMS

Budker's revolutionary idea to use colliding beams of charged particles in accelerators was realized with some difficulty. The move from Moscow, the pressure from authorities to undertake other projects in competition with the West, and the backwardness of Soviet industry created obstacles uncommon among western physicists. But in the twenty-five years from the founding of IIaF until the mid-1980s, Budker's physicists achieved a number of firsts—confirmation of quantum electrodynamics, more exact measurement of the resonance of various subatomic particles, and so on. According to modern physical theory the four basic forces of nature are electromagnetism, gravity, the strong force that holds elementary particles together, and the weak force responsible for radioactivity. Quantum electrodynamics (QED) explains the interaction between the weak force and the electromagnetic force.

The first accelerators were based on the principle of the interaction of a beam of charged subatomic particles with stationary targets. Here the reaction energy available for the creation of new particles is only a small fraction of the total energy of the particles. The problem was that according to Einstein's $E = mc^2$, the heavier the particle one tried to produce, the higher the energy one had to reach. The energies produced in fixed target machines were limited. Colliding beams offered the promise of exponentially higher energies. It was easier to accelerate lighter particles because they could be "cooled" by synchrotron radiation.

The discussion of colliding beam methods in high-energy physics centered on the hope of improving the quality of reactions between colliding particles sufficiently to gain important physical information (high "luminosity"). In colliding beam methods the reaction energy is the sum of the energy of two beams of particles that were accelerated in storage rings in opposite directions and then made to collide with each other. This necessitated building cyclic storage rings where beams of accelerated particles collide after injection for several billion revolutions. The colliding beam method of producing electron-electron and electron-positron collisions provided a significant gain in the interaction energy of particles, removed numerous quantum limits on possible reaction channels, and made feasible the study of strongly interacting particles with electron-positron colliding beams.

The IIaF's first machine, the VEP-1, an electron-electron storage ring with two adjacent ring channels each about 1 m in diameter in the shape of a figure eight, and with an energy of 2 · 150 MeV, was designed and tested in Moscow and shipped to Akademgorodok in 1962 for assembly. At around the same time Siberian and Stanford physics teams produced beams (in 1963), the first electron-electron scatterings (at small angles, in 1964), and the first experiments on the verification of QED (in 1965). It is not unusual for discoveries in modern science to occur relatively simultaneously. Akademgorodok physicists raced to be first against Gerald O'Neil at Stanford's High Energy Physics Laboratory (HEPL), a team of Italian physicists under the direction of Bruno Touschek, French physicists at Orsay, and physicists at several other national physics centers. O'Neil arrived at HEPL in 1957 on leave from Princeton with a proposal similar to Budker's for colliding beams. Wolfgang Panofsky, the HEPL director, liked the idea and convinced the U.S. Navy's Office of Naval Research, a major source of funding for university-conducted basic science, to underwrite the construction of such a collider. Like the one at IIaF, the HEPL collider worked as planned only after great difficulties. In 1965 the HEPL physicists published proofs of QED showing that electrons behaved like point charges at least to the level of 6 · 10^{-14} cm.[19]

The VEP-1 was placed "upright," giving it several advantages, but those advantages were not considered in the initial planning stages. Rather, the limited space given to Budker's laboratory in the Kurchatov Institute by GKUAE (the state atomic energy committee) required this orientation. Every fifteen seconds a special synchrotron fired electrons with an energy of 40 MeV in the lower and then the upper path. They circulated to 150 MeV as a dense, unbroken beam of particles. The VEP-1's energy was equivalent to a conventional accelerator and a stationary target at 100 GeV. A vacuum in the chamber of the storage ring provided an electron lifetime of thirty minutes in one beam—a century in the microworld—which was considered quite an achievement at the time. Scattered particles were detected by a set of optical spark chambers. The experiments on VEP-1, the main purpose of which was to verify QED at small distances and to suggest what "atoms" really were, showed that QED is valid to 5 · 10^{-14} cm.[20]

The next stage in the development of colliding beam accelerators involved electron-positron (e-e+) interactions. These interactions promised to be extremely advantageous for studying elementary particles with short lifetimes, or "resonances." The ideas for these accelerators also date to discussions in the 1950s, around the same time that physicists at Frascati under Bruno Touschek, then at Orsay, Cambridge, and Stanford, began to consider these interactions. This idea faced even more scrutiny than that of the electron-electron colliding beams. Sasha Skrinskii recalled that if the majority of specialists found electron-electron colliding beams far-fetched, "then the discussions about the electron-positron experiments were perceived by most as proof of the absolute foolhardiness of [Budker] and the rest of us." Kurchatov solicited three outside reviews on the project, all of which were negative, but he supported it anyway. And in 1967, two years after the VEP-1 experiments, the Siberians were the first to run experiments on electron-positron annihilation. Vladimir Veksler, whose Dubna accelerator at one time had been the most powerful in the world, and who had written one of the negative reviews, flew to Akademgorodok to congratulate the physicists publicly for their success.[21]

Touschek's group succeeded in 1961 in storing separate bunches of electrons and positrons in one ring, in a prototype called "AdA," for Annello di Accumulazione ("storage ring"), at $2 \cdot 250$ MeV. Budker's VEPP-2 (for *vstrechnye elektronnye pozitronnye puchki*, colliding electron-positron beams), with an energy of $2 \cdot (0.35 - 0.67$ GeV), had a maximum luminosity of $3 \cdot 10^{28}$ cm^{-2}s^{-1}. On June 18, 1964, electrons were sent through the VEPP-2 and "lived" 15 seconds. The VEPP-2 was intended to reach 2,000 BeV, while a linear accelerator of such energy would perhaps be dozens of kilometers in length. To analyze the data produced in collisions IIaF physicists had to work with backward computer systems—second-generation Polish, only later third-generation Soviet. The control room consisted of dozens of calculators, blinking lights, and vacuum tube panels, with indicators and oscilligraphs. The VEPP-2 used a sodium iodide detector. The researchers would have preferred xenon, but technology and cost prohibited this approach.[22]

The mid-1960s proved to be banner years for the institute. In the 1964 elections to the Academy, A. A. Naumov, S. T. Beliaev, and R. Z. Sagdeev became corresponding members, and Budker became a full academician. In 1965 the physicists ran a series of successful experiments on their VEPP-2 accelerator, including the first experiments on elastic scattering of electrons at wide angles at energies of up to 7.5 GeV. Theoretical and experimental research on the instability of large currents in storage rings and accelerators that permit interaction of streams of particles commenced. The physicists began construction of the VEPP-3. The institute developed and produced high-current accelerators for industrial purposes with an energy of 1.5 to 3 MeV and power of beam to 25 Kw. The physicists had produced forty-five articles, delivered thirty-five papers at international meetings, and put out sixty-two preprints. Four doctoral and eight candidate of science dissertations were defended.[23] In 1967 scientists in Novosibirsk first achieved sufficient luminosity to study pion production during electron-positron annihilation.

The year 1968 proved to be just as exciting for institute physicists. Colliding beam research stood at the cutting edge. The development of e-e+ and proton-antiproton accelerators looked promising. Powerful industrial accelerators generated income for the institute. The Siberians justly claimed world leader-

ship in the field of colliding beam accelerators. Their experiments were repeated at Orsay, while the Italian machine ADONE had yet to operate. The
re-building of the electron synchrotron in Cambridge, Massachusetts, into an
electron-positron machine also lagged, while the CERN proton-proton accelerator was a device of the early 1970s. The construction of the VEPP-3 neared
completion, with the construction of the magnetic system of the storage ring
having begun. Finally, S. T. Beliaev and R. Z. Sagdeev were elected full members of the Academy, and V. A. Sidorov, A. N. Skrinskii, and R. I. Soloukhin
were elected corresponding members.[24]

VEPP-2 experiments were conducted until 1970 and produced interesting
results. The main characteristics of φ-resonanceand the probability of its decay
into charged and neutral kaons or three pions were investigated on this machine. At the end of 1970 the VEPP-2 was transformed into an electron and
positron booster for the VEPP-2M. From 1974 to 1987, experiments on the
VEPP-2M were the major source of information on the processes occurring in
electron-positron annihilation.[25]

In the 1970s high energy research focused on verification of QED and the
development of new accelerators. A cycle of works was completed on vector
mesons and the study of their interference at narrow resonances on the range
of energies from 760 to 1,340 MeV at intervals of 0.5 MeV. Physicists determined the mass of various mesons. They studied electron-positron interactions
that reached 2 GeV, several of the first studies undertaken on strong forces
in pure conditions during lepton collisions, and Bremsstrahlung. (Bremsstrahlung is "breaking radiation," electromagnetic radiation produced by a sudden
slowing or deflection of charged particles passing through matter in the vicin-
ity of strong electric fields of atomic nuclei.) Accelerator facilities were en-
hanced by the addition of a synchrotron B-3M, a booster for the VEPP-2, and
a storage ring for the VEPP-2M at energies of up to $2 \cdot 700$ MeV. Work continued on a large magnetic ring (45 m) for electron cooling on the VEPP-2M, -3,
-4, and VAPP. The VEPP-2M and the VEPP-3 were used by various institutes
in Akademgorodok and the Kurchatov Institute for experiments on synchrotron radiation and on the X-ray structure of various biological and chemical
substances. World leadership was being challenged, however, by Stanford,
DESY, and CERN.[26]

The VEPP-2M storage ring was built at the institute in 1971–73 for carrying
out experiments in a lower energy range—$2 \cdot (0.25 - 0.70$ GeV)—with luminosity an order of magnitude higher than the total luminosity of all colliders in
that range. The first experiments commenced in 1974. Parameters of resonances were made more precise, various channels of their decay were carefully
studied, data about pion and kaon interaction in the resonance and non-
resonance end were accumulated, and new data about the multiple production
of hadrons at various energies were obtained. In 1976 work on the VEPP-2M
proceeded in a higher energy interval and the cross-section of the creation of
four charged pions was measured.[27] From 1975 to 1976 theoretical and experimental research was carried out on radiation polarization that involved the

study of e-e+ interactions with colliding beams of polarized particles. This led to accurate mass measurements of φ-mesons and charged kaons.

A second path of colliding beam research focused on high energy regions. In 1966 the physicists decided to build a new storage ring with an energy of $2 \cdot 3.5$ GeV. This machine, the VEPP-3, an electron-positron booster at an energy of $2 \cdot (1 - 3$ GeV), a strong focusing ring with a circumference of 74.5 meters, was built at IIaF during 1968–72. Work advanced slowly as it was begun at the same time that the proton-antiproton VAPP was being created. Since the VEPP-3 turned out to have low luminosity, it was used instead as an intermediate storage ring for the VEPP-4.[28]

The VEPP-4 accelerator has a circumference of 360 m. It was first designed as a proton-proton collider at 35 GeV. Several factors contributed to the decision to abandon the VEPP-3, use it as a booster for the VEPP-4, and give up the creation of a proton-proton or proton-antiproton accelerator entirely. First, owing to the technological backwardness of Soviet industry, IIaF physicists had to design and build much of the equipment they needed on site, although local industry now provides some vacuum equipment and certain standard metal parts. Budker was fond of saying that "a physicist must also be an engineer." But Budker and his associates surely grew weary of the constant struggle to meet the challenges of the Soviet supply system. Second, money was tight. Other Soviet high energy physics centers also wanted to upgrade their facilities.

Third, American and European institutes had succeeded in getting their plans in order more quickly. Why build a machine that would duplicate someone else's effort? IIaF physicists well knew Budker's admonition not to compete with the West on its own terms but to take advantage of IIaF's unique facilities and approaches. Budker also said that a good physicist enjoyed taking risks. He praised the accelerator at Dubna as a fine machine but criticized it as nothing new or risky. When SPEAR (Stanford Positron-Electron Asymmetric Rings) commenced operation in 1973, the VEPP-3 was far from completion so it was decided to use it as a booster for the VEPP-4. In 1976 the VEPP-4 at $2 \cdot 7$ GeV was started up. In the opinion of IIaF physicists, its operation compared favorably with SPEAR, DORIS, and PETRA ($2 \cdot 19$ GeV) and at Cornell CESR ($2 \cdot 8$ Gev).[29]

In the tenth five-year plan (1976–80) researchers intended to extend colliding beam work in all directions. Most important was the completion of the VEPP-4, with experiments on the VEPP-3 and VEPP-4 on μ, π, and v mesons, and research on symmetry nonconservation. In 1980 a new electron-positron storage ring for the VEPP-4 was used to measure precisely the mass at a narrow resonance of the ψ and ψ' particles through resonance depolarization of the polarized beam. VEPP-4 physicists focused on the development of a method to produce colliding electron-positron beams at energies higher than 100 GeV. The most important results of 1981 in elementary particle physics included precise measurements of the π-meson on the VEPP-2M using a cryogenic magnetic detector; the introduction of a high-frequency system using a gyrocon on the VEPP-4 to produce particles at 5.5 GeV; and the introduction of a super-conducting "snake" on the VEPP-3 for the study of synchrotron radiation. In the mid-1980s researchers conducted a variety of experiments on the VEPP-4 and VEPP-2M accelerators: to measure v mesons precisely; conduct experiments on various particle-antiparticle annihilation processes where the range of energy of cross-sections was found to exceed the predicted model of vector dominance; and continue research on violation of CP invariance.[30]

The physicists also modified accelerators with interesting results. On the VEPP-2M, with the help of the superconducting "Snake" magnet, they increased luminosity twofold. Roy Schwitters, who worked at SPEAR, wrote, "One theme that runs throughout [Budker's] productive career is the application of classical and 'well known' concepts in physics to entirely new problems and phenomena. One example of this is beam polarization in high energy e-e+ storage rings." The first experimental indications of radiation polarization were obtained in the late 1960s and in 1970 at Orsay and Novosibirsk, with unambiguous evidence for this reported by both groups in 1971. One general scheme "for providing longitudinally polarized beams was christened the Siberian Snake in honor of its development at Novosibirsk through the work of Derbenev and Kondratenko and others."[31]

ELECTRON COOLING AND ACCELERATION
OF HEAVY PARTICLES

Another direction of research at IIaF involved beams of heavy particles such as ions, protons, and antiprotons. It involved two significant innovations of Budker and his colleagues but was eventually waylaid by technical and financial constraints unique to the Siberian experience. In the late 1960s the record energy for protons was 10 GeV. Budker wanted to build a machine at $2 \cdot 10$ GeV. Although he did not succeed in this endeavor, he made two interesting contributions to high energy physics. One was the charge-exchange method of injection of protons; the other was electron cooling.[32] IIaF physicists developed the charge-exchange method for the injection of protons into accelerators and storage rings. The method permits a manyfold increase in the density of particles in the phase volume. The essence of the method is that negative hydrogen ions, accelerated in an injector, are introduced (along the tangent to the orbit) into the vacuum chamber of the storage ring, where they lose their charge (both electrons) on a thin gas jet or film. As a result, protons stabilize in an equilibrium orbit in the accelerator without betatron oscillations. Such a method facilitates the storage of the maximum proton current.

This charge-exchange method offered a number of promising but untried applications. At the end of the 1960s Budker proposed using the technology for the production of high brightness neutral particle beams to shoot down nuclear warheads. Although this "Star Wars" activity was not actively pursued in IIaF, many laboratories throughout the world set out to develop it. The method, first developed for production of high luminosity in proton-proton and proton-antiproton colliders, ultimately was adopted in a proton synchrotron at Argonne National Laboratory. After the development in IIaF in 1971 of high-intensity negative ion sources with cesium catalysis, the charge-exchange method was adopted in all proton synchrotrons.[33]

Budker's second contribution was "electron cooling." In 1966 Skrinskii and Budker suggested the possibility of attaining high luminosity proton-antiproton collisions by electron cooling with the proposed VAPNAP proton-antiproton accelerator at $2 \cdot 32$ GeV. Researchers also planned to design a method for lepton interactions at 200 to 1,000 GeV. Budker's efforts to design a proton-antiproton accelerator played a major role in the efforts of Rubbia, Van der Meer, and Kline at CERN to discover the W and Z bosons.[34] "Electron cooling" forms and stores dense narrow beams of heavy charged particles. The method introduces an effective friction into the beams. For electrons and positrons such a friction is produced by synchrotron radiation of particles traveling along a curvilinear trajectory. In the case of heavy particles, protons, antiprotons, and ions, for which there is essentially no radiation, other friction mechanisms are required. In this case an electron beam with average velocity the same as the heavy particles is put into one of the straight sections of a storage ring of heavy particles. If the electron beam is sufficiently monochromatic and the particle trajectories are parallel, the electron beam will absorb the excess energy of the "hot gas" of heavy particles, like a cooler. The experiments on electron cooling were carried out on the NAP-M installation. Electron cooling was adopted throughout the world in the 1970s and 1980s.

The center of mass energy available from the charge-exchange method and electron cooling that used so-called hadron-hadron collisions increased the levels of energies from 30 to 50 GeV to 500 to 2,000 GeV. Unfortunately for the Siberian physicists, growing costs and the opposition of the Siberian division led them to abandon their efforts to build proton-proton and proton-antiproton machines (the VAPNAP and VAPP). They opted instead to participate in experiments at Serpukhov, Russia; CERN, near Geneva, Switzerland; and the Fermi National Accelerator Laboratory in Batavia, Illinois. Design and minor construction on the VAPNAP facility was set aside when the Fermilab proton synchrotron went into operation in the early 1970s at energies close to VAPNAP potential. The resulting collaboration was based on a proposal by a IIaF physicist, Dikanskii, to study electron-cooling and stacking of proton beams at 200 MeV. U.S. and Soviet physicists at the Fermi Laboratory also jointly conducted research on the fundamental constituents of matter, using the method of electron cooling of proton-antiproton beams. The proton-antiproton program at IIaF was then refocused on electron-positron experiments on the VEPP-4 at energies to $2 \cdot 5.5$ GeV.[35]

When describing proton-antiproton experiments, the need to reach higher energies, and the increased cost and size of accelerators, Budker drew attention to the peculiar problems that the closed Soviet system and the small science orientation of Akademgorodok put in the way of IIaF to stay even with CERN:

It has been proposed to build the paths in France, while the accelerator will be in Switzerland, and the particles will fly from Switzerland into France without visa or other papers. Naturally, we do not want to get into such things, nor on such a grand scale, and we cannot do it because the entire Siberian division, the chemists in particular, would murder us, because it would take five times more [money] for such an apparatus than is allotted to the entire Siberian division.[36]

At least Budker and his colleagues did not have to worry about visas for their particles.

There were other reasons for concern. Financial pressures began to intrude on all areas of IIaF research. The competition with other Soviet physics centers for scarce resources intensified. At a meeting of the Novosibirsk party organization, a leading physicist complained about IIaF's growing difficulty in attracting fresh young blood while "the building of huge accelerators in the western part of the USSR (Serpukhov, Gatchina, Erevan, Kharkiv) has attracted practically all great Soviet physicists." The efforts of IIaF physicists to create a leading center of Soviet high energy physics where other Soviet physicists could come to work, experiment, and train students were undermined by the new facilities. Worse still, there were constant delays in scavenging from older machines and rebuilding them to improve parameters, with the consequence that "a series of important physical experiments was delayed and our institute lost its leading position, achieved a year ago, having fallen behind French physicists."[37] These difficulties were rooted in the growing number of disagreements between Budker and Lavrentev over the profile of IIaF.

As the institute grew, as it experienced other scientific successes, as its reputation grew, and as its physicists planned larger accelerators and fusion devices to stay on the cutting edge of international physics, tension developed between Budker and Lavrentev, big science and little science, fundamental and applied research, and between the institute and the Siberian division. It was clear to Budker that research in high energy physics required more advanced and larger colliding beam accelerators. He hoped that the Siberian division would provide funding. When funds began to dry up, he saw no option but to use the institute's facilities to produce equipment for sale to industry: small high-power accelerators and synchrotron radiation devices with industrial applications. This approach, in turn, offended pure-blooded communists who smelled the scent of capitalism.

BIG SCIENCE AND INDUSTRIAL ACCELERATORS

Lavrentev envisaged a city of science where contact among scientists from the various sciences would be facilitated by moderately sized scientific collectives. Lavrentev recognized that high energy physics and fusion research required extensive facilities and a large staff of qualified personnel. He did not initially fathom how the physicists' logical drive for higher energies might put pressure on him to abandon this organizational principle. But Lavrentev supported

Budker fully in all matters until the late 1960s, when the relationship between them soured over big science. Some accuse Gurii Marchuk, then head of the Computer Center and soon to become chairman of the Siberian division, of instigating the problem by planting doubts in Lavrentev's mind over Budker's devotion to Lavrentev's vision. Marchuk apparently disliked Budker's iconoclasm, perhaps envied his stature, and certainly disliked his bravado. Whatever the case, in Budker's last years, as his health deteriorated, many of his initiatives for the IIaF research program were blocked or delayed. It seemed impossible to mediate increasingly poor relations between Lavrentev and Budker.[38]

Not even the institute's special financial circumstances helped. Such research centers as the Institute of Nuclear Physics had the moniker of an Academy of Sciences institute but operated with significant financial support of GKUAE. GKUAE was the Soviet equivalent of America's Atomic Energy Commission. It provided both the largess and blanket of secrecy needed to keep the program's successes and failures out of the public domain. Such non-academy nuclear institutes that were located in the atomic cities of Tomsk-7, Arzamas-16 where Sakharov helped design the hydrogen bomb, Krasnoiarsk-26, and elsewhere, were subordinated directly to Minsredmash. Budker detested Minsredmash and GKUAE equally for their militarism and secrecy. He intended to accept only money from GKUAE, not its secrecy. But since other institutes also had designs on GKUAE's money, Budker could not get the money he needed to build the multikilometer circumference accelerator he had in mind.

Budgetary problems dated to the construction of production facilities in the institute. The facilities built equipment and components for accelerators. The institute's directors produced a study that called for the expansion of production facilities even before the institute had opened. In the study the physicists noted that the production facilities were the key to the future of larger, more powerful colliding beam accelerators. The institute had the space, qualified workers, experience, and the tradition of the Kurchatov Institute behind it. To make further advances, the directors contended, an additional 200 to 250 workers were needed for a second shift, along with ten to twenty more lathes, milling, screw cutting, and other machine tools, and hard currency for foreign equipment purchases. Academy president M. V. Keldysh, and the chairman of the State Committee for the Coordination of Scientific Research, N. K. Rudnev, endorsed this study.[39] The Siberian division reluctantly approved the increases in personnel, but the centralized Soviet supply system waylaid the delivery of equipment and completion of construction for some years.

The government encouraged institutes like IIaF to expand their production activities through policies adopted in the early 1960s that were intended to accelerate the introduction of scientific advances into the production process. Officials hoped that bringing science and production under one roof would foster innovation. Budker readily embraced this opportunity to expand his production facilities. It was better to build one-of-a-kind equipment on site than to trust Soviet branch industry to produce it for him. Early on, Budker met with A. A. Nezhevenko, who was the director of the turbine factory Sibelektrotiazhmash (Siberian Electrical Heavy Machinery), to discuss the purchase of equipment for IIaF. He was impressed by Nezhevenko's no-nonsense attitude and his managerial skills, and invited him to become deputy director of the institute for production.

On the example of the production facilities at IIaF and several other institutes throughout the USSR, a new reform was introduced with great fanfare: the scientific-production association (nauchno-proizvodstvennoe ob"edinenie, or NPO). NPOs existed primarily to solve a narrow technological problem in a specific branch or sub-branch of industry, but they were seen by central planners as key to the modernization of the Soviet economy. NPOs consisted of a lead scientific institute and laboratories and factories from branch industry. The institutes were usually the main force behind the creation of an NPO. Some 150 NPOs were created by 1978. Unfortunately the State Planning Commission, Gosplan, treated NPOs as separate organizations, not the sum of their parts. NPOs often became poorly coordinated mechanical conglomerations. Parochial ministerial interests, concern over the level of the financial independence of NPOs, and the absence of capital limited their effectiveness except in major scientific centers like Leningrad where they were backed by the party and the scientific apparatus.[40] Fortunately for IIaF, the experiment was a great success. The institute produced equipment for its experiments and industrial accelerators for sale. In an analysis of the potential of NPOs, Akademgorodok economist Abel Aganbegian considered IIaF production an indication of what was possible within the narrow band of reforms permitted during the Brezhnev era.

Like the FAKEL ("Torch") enterprise created within the bowels of the Computer Center (see chapter 4), the production facilities of the Institute of Nuclear Physics were a forerunner of the independent enterprises of the post-Gorbachev era. In the face of great resistance among economic planners who feared a resurgence of capitalism, FAKEL and the IIaF production foundry managed to establish "businesses" that ran in the black. FAKEL was disbanded by central Communist Party officials in part because of its great financial success. IIaF also faced opposition. The huge foundry at IIaF should have symbolized for Lavrentev precisely the "innovation beltway" of production firms drawing strength from science that he hoped to see arise around Akademgorodok. Instead, it only served to remind Lavrentev and many other institute directors of Budker's power and independence. In spite of their opposition, the foundry was permitted to operate since its profits were plowed back directly into research activities, and accelerators were far more tangible to economic planners than FAKEL's computer software.

Another policy directed toward improving the economic responsiveness of scientific research was based on an administrative order for institutes to increase the amount of funding they earned from contracts with various organizations throughout the nation. In April 1961 the Council of Ministers required institutes to raise the amount of income generated by contracts to 3 percent of their budgetary requirements. In 1962 institutes of the Siberian division garnered only 0.5 percent of their budgets through contracts, but by 1966 they had reached 7.5 percent. This progress encouraged the State Committee for Science and Technology (GKNT) and the Ministry of Finance to establish a level of 14.5 percent for 1968. M. Belousov, the head of the planning and financial administration of the Siberian division, described this order as "totally unfounded." It would require institutes to turn to "useful" contracts and move away from fundamental research. Belousov pointed out the pitfalls associated with production activity within a fundamental science institute. He argued that it was inappropriate to push the level of production upward yearly according to some arbitrary percentage. Similarly, it was difficult for the institutes to fix the term for completion of research leading directly to application. Contracts should grow naturally out of research, he argued, and should not be intended solely to produce income.

Against this backdrop the Institute of Nuclear Physics found the wherewithal to expand production facilities and produce money-making synchrotron radiation sources (storage rings of relativistic electrons), detection equipment, and industrial accelerators. The institute became the main center of synchrotron radiation in the USSR when the VEPP-2M and VEPP-3 commenced operation. Substantial income was also generated from research and production on synchrotron radiation for a variety of scientific production associations, ministries, and institutes with applications in X-ray metallography, holography, structural analysis, spectroscopy, and microscopy, and in isomers and geology.[41]

More important from a financial point of view were industrial accelerators. In 1966 the State Committee for Science and Technology approved a proposal by IIaF physicists to set up a small-scale serial production facility for high power accelerators that ran on 220 current and when turned off were no more dangerous than a typewriter. The institute used the profits earned in the sale of the accelerators for basic and applied research. Rather than hindering fundamental research, an institute financial officer reported, "The income earned permits [us] to commence construction of an unprecedented accelerator for colliding proton-antiproton beams."[42]

The parameters of IIaF industrial accelerators cover the energy range from 0.3 to 2.5 MeV, and power ranges from hundreds of watts to hundreds of kilowatts. Three kinds exist: the ELIT based on high-voltage pulse transformers, the ELV operated on commercial frequency, and the ILU-type accelerators based on RF (high-frequency) resonators. The ELIT are primarily for experimental studies, whereas the others have found broad application in industry, agriculture, and medicine. For example, the ELV-1 and ELV-2 accelerators were produced for the cable industry with applications in polyethylene cable. Irradiated polyethylene insulation has increased resistance to thermal breakdown. It can withstand heat to 250°C., whereas untreated it is safe only to 80°C. Other industrial applications included an ELV-6 for the "Iskozh" factory in Kirov for the production of artificial "leather" for shoes, furniture, and cars; an ILU-8 for "Ufimkabel'" in Ufa for the modification of insulation of microelectronics.[43] Other applications included irradiation of grain to destroy insects. Introduced experimentally at the Novosibirsk Grain Elevator of the Siberian branch of the All-Union Grain Research Institute, this type of accelerator was installed at the Odessa port elevator and in five other installations on the Volga and Don Rivers. The accelerators are used for the preparation of animal feed from radiolysis of inedible plant material by the NPO of the Main Microbe Industry Administration, "Gidrolizprom." Intense electron beams could be used for cutting and welding, cutting and smashing hard rocks, smelting, for plasma and laser engineering, pulse radiolysis, and in agriculture for seed disinfestation, sterilization, pre-sowing radiation, and radiation induced mutations for breeding.[44] Over an eight-year period ending in 1984, the economic effect of these accelerators was calculated to be more than 140 million rubles, with 250 million rubles forecast for the eleventh five-year plan.

The "bread accelerators" earned the institute enough money to expand laboratories, build two apartment buildings, and provide more frequent staff bonuses. "You are capitalists," Budker often told his American colleagues, "but I generate income, not you." One U.S. physicist observed, "Our country is the most democratic in the world. Why does the most democratic scientific laboratory in the world turn out to be Budker's in Siberia?" But his successes, his democracy, and finally his profit-making endeavors provoked the envy and enmity of several of his neighbors, that is, directors of the other institutes. They tried to "prohibit" or at the very least sharply curtail the initiative of the nuclear physicists. They used the tried-and-true method of calling in verification commissions, in this case to ensure that "profiteering" did not violate socialist norms. Budker was always vindicated, but he took the investigations personally and was often rude to the authorities on the commissions.

As the high-energy physicists produced one success after another, Budker sought out even more funding to expand his institute's research program and facilities. In January 1968, in a presentation before an intergovernmental committee on nuclear physics, Budker argued for additional resources. He described the research results of his colleagues in glowing terms, outlining the basic physical principles of colliding beam accelerators and the method of electron cooling. He drew comparisons with other research groups at SLAC, Orsay, Cambridge, and CERN to demonstrate the leading position of the Soviet Union. But the future could be secured, he explained, only by building more powerful electron-positron complexes, and proton-antiproton colliders to create π-mesons and muon pairs. Unfortunately the Siberian division provided meager support for his plans. Budker promised to match the new proton-antiproton accelerators to be built at CERN at a substantially lower cost. Savings would be achieved in the usual Soviet fashion: by forcing the pace of construction, restricting all comforts, and finding every labor surplus. Budker requested 30 million rubles which included funds for a cryogenics laboratory for work on superconducting magnets for the proton-antiproton storage ring to boost its power to 100 GeV.[45] But this could not be done without additional funding.

A commission on nuclear physics whole-heartedly endorsed Budker's program for the institute's expansion. Bruno Pontecorvo reported the committee's deliberations. Pontecorvo, who was affiliated with the Institute of Nuclear Problems and then the Joint Institute of Nuclear Research at Dubna with V. I. Veksler, I. M. Frank, and others, had worked with Fermi in Rome in the 1930s, and then in France, the United States, Canada, and the British nuclear facility at Harwell, before defecting to the USSR in 1950. His works ran the gamut from nuclear and high-energy physics to the physics of weak interactions, neutrinos, and astrophysics. Pontecorvo called Budker's school "first rate." He observed the great potential in colliding beam accelerators. He cited IIaF's international reputation. He complimented IIaF's design and production facilities while endorsing supplementary funding. The commission therefore resolved, first, to support IIaF's efforts to develop over the next two years a proton-antiproton collider at $2 \cdot 25$ GeV, a revamped electron-positron collider at $2 \cdot 3.5$ GeV, and a cryogenic laboratory for the development of a superconducting storage ring at higher than 100 GeV. Minsredmash and the Academy of Sciences were asked to provided an additional 25 million rubles and to supply the top-of-the-line Soviet-made BESM-6 third-generation computer by 1970. In order to attract qualified personnel, both short-term and long-term, Soviet and foreign, the commission proposed constructing guest facilities and a pension. Unfortunately Minsredmash did not allocate additional funds, and the Siberian division refused to provide further resources. So IIaF was obligated to expand the manufacture and sale of industrial accelerators. Once again, in March 1975, the national scientific establishment endorsed Budker's effort to expand research and production facilities.[46] But once again the endorsement fell on deaf ears.

These repeated requests to expand research facilities and build larger accelerators created tension between big and little science and drove a wedge between the institute and the Siberian division. The latter refused to provide additional funds for capital investment or new research. The institute's directors complained to Academician Markov, at the time academic secretary of the Academy of Sciences division of nuclear physics. That the institute had to focus on production activities to meet expenses, the directors argued, "practically signifies the cancellation of scientific activity of the institute in fundamental research for the next five-year plan" and for the future. The decision to deny funds was "contrary to the commission's decision." The institute therefore repeated its request for 20 million rubles for the VEPP-4, a new electron-positron accelerator, 25 million for the VAPNAP, a proton-antiproton facility, 6 million rubles for the synchrotron center, and 16 million for fusion, a total of 67 million rubles. That there were bitter feelings toward the Siberian division, there can be no doubt. The complaint to Markov concluded by highlighting the Siberian division's "artificial delay" in providing resources for a new expanded production facility about ten miles away from Akademgorodok in Pravye Chemy. The institute was denied even the right to relieve the pressure on its overburdened production foundry to generate additional income.[47]

It was natural that the Institute of Nuclear Physics competed both with other physics research centers throughout the USSR and with other institutes at Akademgorodok. To be sure, each Akademgorodok institute was forced to compete for funding with Moscow and Leningrad centers in its discipline. However, the huge cost of modern physics equipment—fission and fusion reactors and accelerators—made competition for resources between biology and chemistry institutes mild in comparison. Whenever physicists wished to upgrade their facilities, they had to make their case for millions of rubles in competition with other physicists who hoped to move their institutes in new directions, to be first in the USSR, if not the world.

As their understanding of the physics of the microworld broadened, physicists required more complex, large-scale, and expensive apparatuses to push to the next level of energy—and knowledge. Each school or institute adopted its own approach. Each wanted a more powerful and versatile accelerator or reactor. With the support of the prestigious Physics Institute of the Academy of Sciences in Moscow, Vladimir Veksler sought a new 10 BeV synchrophasotron—a proton accelerator—for the Joint Center of Nuclear Research in Dubna. Ultimately, as the machines physicists built reached miles in diameter at CERN in Europe, Batavia in Illinois, and Serpukhov, there was no way governments could fund them all. The Soviet Union had an additional problem: Its economy had begun to show real strains as a result of the arms race. The costly effort to achieve parity with the United States restricted funds for the consumer sector, health care, education, and science. The Soviet economy had naturally begun to slow down from the rates of growth achieved during the reconstruction of the immediate postwar years.

Denied additional funding by the realities of budget appropriations and the relative poverty of Soviet science, the physicists turned to their production facilities for additional revenue. The shift to applied research and production was rewarded by the presence of a large number of willing customers. But there was a downside: Fundamental research had to take second seat, and never again would IIaF physicists rival their western counterparts as they had when the institute was first founded. Moreover, the production facilities did not earn enough to allow for the necessary expansion.

FUSION AND THE PROMISE OF VIRTUALLY UNLIMITED ENERGY

As he had for high energy physics, so Budker provided the paradigm for fusion research at the Institute of Nuclear Physics based on an idea he first suggested in 1952 or 1953. The fusion of lighter particles, ions or atoms, say deuterium, into heavier particles, for example lithium, would release tremendous quantities of energy. If the energy of the hydrogen bomb could be tamed in a fusion reactor it might power a series of small cities. The problem was that fusion could occur only at a temperature like that in the center of the sun, around 100 million degrees. Physicists in fusion research, also called controlled thermonuclear synthesis, have tried to figure out how to contain a plasma of the lighter particles in a reactor, fuse them into heavier ones, and transform the energy produced into electricity. Their research involves the study of the physics of plasmas and various experimental fusion devices—tokamaks and stellarators, which use magnetic fields to hold the plasma and keep it away from the walls of the reactor—and studies on shock waves, high frequencies, and lasers, among other things, to ignite the plasma.[48]

Budker's idea was to create plasma traps with magnetic mirrors, so-called open thermonuclear systems. His concept first saw light in an article published in 1958 just before the second international conference on the peaceful uses of nuclear energy, which was attended by Soviet and U.S. specialists. As the conference approached, Kurchatov heard rumors that the American side would come prepared "to show their leadership in fusion," having been beaten badly at the big science propaganda game by Sputnik the previous year. So Kurchatov told Budker to proceed with a project to realize his idea. The Budker apparatus, OGRA (*odin gramm neitronov v sutki*, one gram of neutrons per day), was built within a year and was followed by the OGRA-2, -3, -3B, and -4. On the basis of these achievements, the Soviets considered themselves victorious once again in the big science race with the United States.

When Budker and his associates moved to Siberia, they pursued research on open traps in keeping with his general rule to work in areas unexplored by other scientific collectives. Using open traps, his plasma specialists would focus on areas of plasma physics less studied: "super cold" plasmas (less than 1 million degrees), super hot plasmas (at temperatures greater than 5 billion degrees, also known as relativistic plasmas because their particles move with speeds approaching that of light), and super dense plasmas. The IIaF plasma program, under the leadership of Roald Sagdeev, Gennadi Dimov, and Dmitrii Riutov, was strong in theory. In practice, however, as for all fusion programs in the world, the plasmas subjected to study in experimental apparatuses experienced great instabilities, failing to follow the predictions of the theorists.[49] But this did not mean physicists gave up their study.

As in the case of high energy physics research, fusion drew strength from superpower competition for prestige and potential military applications. Fusion research grew out of defense programs connected with the development of hydrogen bombs. It seemed that the powerful stream of neutrons which accompanied a thermonuclear reaction could be used for the production of a large quantity of fissionable material, like plutonium, for nuclear weapons. After several years it became clear that this could be accomplished more cheaply through fission, and the main focus of controlled thermonuclear synthesis turned to its "peaceful" applications. Many scientists believed that the mistrust between the superpowers generated by fears of military technological breakthroughs could be partly overcome by opening physicists' research programs to joint scrutiny, first accomplished publicly at the 1950s Geneva meetings. For the Soviet side the meetings had another purpose. Having pioneered the tokamak fusion reactor concept, Soviet physicists wished to retain their

leading position and be recognized for their work. Some of the older generation grew to resent the frosty and skeptical reception they encountered from
their western counterparts.

The Americans, for their part, could not believe that Soviet physicists,
Budker among them, were ahead of them in fusion research, but they took
comfort in learning that technological breakthroughs giving one side a military
advantage were unlikely. And by the second international conference in 1958,
U.S. and Soviet theoreticians and experimentalists developed working relations, compared results, and learned more details about the work of Sakharov,
Tamm, Artsimovich, Budker, and others in various areas of fusion. The scientific relationship between the United States and the former Soviet Union in
fusion research persists to this day, but of course it was always subject to the
political winds blowing in Washington and Moscow. High-level exchanges
would give way to small, periodic ones, especially after the Soviet invasion of
Afghanistan. Clearly the race "to be first" dominated U.S.-Soviet relations in
science and technology until the collapse of the USSR.

In addition to superpower competition and potential military applications,
two other factors influenced the long-term support that fusion commanded in
the Soviet Union. First, the uneven geographical dispersal of energy resources,
industry, and the labor force generated interest in such renewable resources as
fusion. The USSR had almost half the world's fossil fuel reserves and 12 percent
of its hydroelectric potential. But a tremendous energy imbalance existed.
Seventy percent of Soviet energy consumption (and population) was in the
European part of the country, whereas 90 percent of the basic fuel resources
were in Siberia and Soviet Central Asia. Two-fifths of all rail transportation
involved coal. The cost of transporting the energy thousands of miles from the
east to the west, either in its primary form or as electricity, was quite high. The
expense of building industry, infrastructure, and housing in Siberia was perhaps higher still. To counteract these problems, Soviet planners and policy
makers pursued an aggressive program of rapid commercialization of nuclear
power stations located near the most populated centers of the European USSR,
with the goal of producing 20 percent of Soviet electricity by 1990, perhaps 50
percent by the end of the century. Fusion was the hero of the scenario. Thermal
fission reactors would carry the USSR to the 1990s. A network of liquid metal
fast-breeder reactors, which doubled their plutonium fuel every eight to ten
years, would come next. Finally, by 2020 or so, fusion reactors would come on
line to produce virtually unlimited energy. The Soviet nuclear industry never
came close to hitting any target.

Another reason for the continued high-level support for fusion research in
the USSR was the special position of the Kurchatov Institute in Soviet history
and the political favor that Kurchatov's successor, Anatolii Aleksandrov, commanded. As the home of the Soviet atomic bomb and the birthplace of the
military nuclear complex, the institute historically garnered great resources. Its
scientists were represented in the upper echelons of the Academy of Sciences
and other organizations important to science policy. Moreover, as president of

the Academy and member of the Communist Party Central Committee, Aleksandrov was in a position to ensure that conditions remained favorable for fundamental and applied research at KIAE.

In fact, the Kurchatov Institute of Atomic Energy had the leading Soviet program for plasma physics research, and the programs of most other institutes were born within its walls. Work at KIAE on plasma heating and confinement in torroidal magnetic fields (the so-called tokamaks) dates to 1955 when Igor Golovin put the first tokamak, the TMP, into operation. Soviet physicists built a large number of tokamaks of different parameters over the next thirty-five years ending with the T-15. (What happened with the T-15, a superconducting tokamak, is symbolic of the current funding crisis in physics and in science in general in Russia today. Four years late in construction, it is now waylaid by long downtimes, shoddy equipment and construction, and helium and nitrogen shortages, and sits idly.)[50]

The Leningrad Physical Technical Institute (LFTI) commenced high-temperature plasma research under Boris Pavlovich Konstantinov in 1958 at I. V. Kurchatov's initiative. LFTI also worked on tokamaks and was involved in diagnostics of hot plasmas, research on plasma diffusion in a magnetic field, and the study of the interaction of high-frequency waves with plasma. At LFTI, too, fusion research lags. Furthermore, LFTI tokamaks are more than twenty years old, as is the research program, and there is little prospect of gaining new results from them.[51]

After initial successes in the Soviet Union in torroidal magnetic confinement of a plasma, physicists realized that plasmas were far more complex than they initially thought. In attempting to achieve a break-even point in thermonuclear reactions, physicists encountered all sorts of diagnostic, experimental, and physical obstacles. But they continued to embrace fusion's promise of relative environmental safety with respect to other energy sources and its potential for producing unlimited energy. Therefore they pursued several alternatives to tokamaks: stellarators; composite apparatus using electromagnetic confinement; reverse field pinches; and mirrors, or open confinement systems with magnetic mirrors. The two major actors in fusion research, LFTI and KIAE, staked their programs to tokamaks, leaving other institutes to pursue research on the alternatives. The most important "alternative" centers of fusion research in the USSR were the Physics Institute of the Academy of Sciences in Moscow, important in stellarator research, laser controlled fusion, and Z-pinch apparatuses; the Ukrainian Physical Technical Institute in Kharkiv, which also focuses on closed magnetic traps like stellarators; and IIaF.[52]

IIaF physicists focused on various open-trap, multicell, multimirror machines where the plasma is created by the injection of fast molecular ions into a chamber. (The change mass injection of molecular ions later gave way to the charge-exchange injection method for proton storage rings.) Then the plasma is heated by various means: lasers, shock waves, or relativistic (high-speed) electron beams. Early research led to the development of several unique approaches: tandem mirror (ambipolar) confinement advanced by Gennadi

Dimov with the goal of building a reactor; gas dynamic traps under the direction of Dmitrii Riutov, where one of the major goals was the creation of a good neutron source; and long open traps known in Russian as DOL, which were multicell devices created by Eduard Krugliakov for fundamental research into the nature of plasmas. The first two seemed to be promising enough, from the standpoint of military spin-offs, and were awarded funding and a new building from the construction wing of the nuclear arms ministry, Minsredmashstroi.

Since IIaF experimentalists first focused on high energy physics, most of the early achievements at IIaF in plasma research came from the theoreticians. The achievements were played up frequently in the Soviet press. The average Soviet worker learned in his daily paper, *Trud*, in 1968, that the amount of energy in the hydrogen in one liter of water was equivalent to 300 liters of gas, and from deuterium (hydrogen with a neutron) in one liter of heavy water—one million liters of gas. He also read that physicists were getting closer to tapping that energy through controlled thermonuclear synthesis under the leadership of Roald Sagdeev, a young plasma specialist at the Siberian Institute of Nuclear Physics. Sagdeev, one of the great early success stories in Akademgorodok, made great strides in understanding the capricious plasmas. Sagdeev was born in Moscow in 1932. He graduated from Moscow University in 1956 and for the next five years worked in the Kurchatov Institute of Atomic Energy before transferring to IIaF. At the age of twenty-nine he already headed the laboratory of plasma physics. At thirty-two he became corresponding member of the Academy of Sciences. Sagdeev then moved to the Institute of High Temperatures for three years before becoming director of the Institute of Space Research in 1973. He resigned this post in 1989 to take a position at the University of Maryland in the physics department. His research concerns controlled thermonuclear synthesis, magnetohydrodynamics, and space science, including the theory of stability of plasma, the physics of nonlinear oscillations, and the turbulence of plasma.

Unlike high energy physics where experiment and theory closely coincided, in plasma physics not one apparatus operated according to theory because of the instabilities of plasmas. So IIaF researchers like Sagdeev strived to turn the rough approximations of theory that had been outlined into something more rigorous. The research at IIaF took several directions: the physics of turbulence of plasma, the physics of dense, relativistic, and super cold plasmas, and theoretical plasma physics. At first (1961–65) the IIaF effort focused on the realization of old ideas (traps with magnetic mirrors that used shock waves for heating). By the late 1960s small apparatuses were being built to develop an understanding of fundamental physical principles.

As the research turned more toward the experimental devices, Sagdeev fretted that IIaF had prematurely abandoned theory for experiment. In its best days early on, the theoretical laboratory (Laboratory 6) had ten staff members and occupied a leading position in the country. In other institutes with more developed fusion programs, theoreticians bided their time waiting for experimental results. At IIaF a number of theoreticians unfortunately had moved into other fields, having grown impatient for something to feed their interests. A small group worked on theoretical astrophysics and lasers, others moved to high energy physics, and some transferred to other institutes.

Sagdeev, too, grew impatient with this state of affairs. Moreover, he wanted to be in the center of attention in Moscow. When Budker suggested creating a cryogenic laboratory to work further on superconducting magnets, both for accelerators and plasma apparatuses, Sagdeev warned that the prospects were unclear for a several reasons. In as much as the USSR lagged behind the United States in technology, it was doubtful that results would be won easily or that the USSR would get there first. Sagdeev suggested building devices that did not require superconducting magnets. He was also troubled by the lack of contact between younger physicists and the older generation. The strength of IIaF and Akademgorodok in general had been an almost American informality between professors and students. With few exceptions, Sagdeev complained, "Now the young have fewer possibilities to communicate either with Budker or with any physicists who stand on the next level." More had to be done to reinvigorate seminars, unofficial meetings, and inter-institutional contacts.[53] Ultimately Sagdeev tired of inadequate facilities for fusion research and declining manpower in his laboratory. He especially resented Budker's opposition to the stellarator he hoped to see built to test his theories; Budker demanded more novelty and this kind of machine was already being built at Princeton, Kharkov, Oak Ridge, and Moscow. When a position at the Institute of High Temperatures in Moscow opened up in 1971, he jumped at the opportunity to move. Sagdeev was surely an accomplished if egotistical man, and people were sad to see him go.

Budker chose just this moment to return to plasma physics. Sagdeev's impending departure required that he provide leadership. True, Budker continued to be occupied with the design of a facility for proton-antiproton collisions using his technique of electron cooling. But when IIaF hosted the Third International Conference on problems of physics of plasma and controlled thermonuclear synthesis in August 1968 Budker recognized the need for better leadership at IIaF in plasma physics; he had had little to add to the proceedings. Budker had considered focusing on tokamaks but then decided, as usual, to have IIaF go its own way—to apparatuses with mirror traps. Even after suffering a heart attack in 1971, while bed-ridden he kept up his interest and invited the young plasma physicists to his home for discussions. Physicists in other laboratories noted his growing interest and naturally gravitated toward the rejuvenated effort.[54]

At IIaF, until 1968, "fusion" meant the study of high temperature plasmas. Physicists encountered properties of plasmas that they did not expect. They elaborated and verified the theory of confinement of charged particles in adiabatic magnetic traps. They examined plasma behavior in closed traps. They discovered microinstabilities so that their plasmas had lower than critical density. They worked on the neoclassical theory of transition processes for tokamaks and stellarators. They carried out pioneering work on burst compression of a magnetic field to obtain a super dense plasma.[55]

In the pivotal year of 1968 plasma physicists were determined to construct installations, perhaps a stellarator, to verify theoretical predictions. They succeeded in producing a plasma of 100 million degrees and a speed of several thousand kilometers per second. The Komsomol's daily newspaper trumpeted the news that "space is an ideal vacuum. Producing such a plasma on the earth is a challenge being met by young physicists at IIaF." Who were these young scientists? Iurii Nesterikhin had just returned from the United States and Geneva where he had met with plasma specialists. Roald Sagdeev had recently defended his doctoral dissertation and had been elected to the Academy. His student, Alberg Galeev, had just graduated from Novosibirsk University, where his senior thesis was presented as a published article. The State Attestation Commission considered Galeev's thesis the level of a candidate dissertation. To prove that a plasma can move in the form of a shock wave, Sagdeev and the others had created an apparatus to accelerate the plasma to interplanetary speeds. This was the UN-3 (for "udarnyi nagrev," or shock wave heating), a small device consisting of a long glass tube with several magnetic rings between which so-called magnetic traps arise. Deuterium is injected into the glass tube, compressed by shock waves, and then heated by an artificial solar wind to millions of degrees to simulate a solar flare.[56] For plasma heating in direct installations, powerful electron beams were used with currents of hundreds of kiloamperes and an electron energy of about 1 MeV. From 1970 to 1972 the first experiments in the world were carried out on plasma heating with relativistic electron beams. A quantitative theory was also worked out for the interaction of such beams with a plasma.

In the 1970s research moved in two directions—the confinement and heating of a dense plasma in direct pulse installations, and the development of quasi-stationary open traps. In 1971 the physicists decided to shift from a direct magnetic field to a corrugated one, under which the installation was transformed into a set of "mirrors" connected through their ends. The specialists developed a detailed quantitative theory of multimirror confinement. This showed the possibility of reducing the reactor length by an order of magnitude or more. This was confirmed in experiments on multimirror confinement of cesium plasma (1972–73). The experimental installation for this purpose, the GOL-1, a gaseous open mirror, a multimirror system with pulse, not steady-state relativistic electrons with a corrugated magnetic field, was built in about six months with the help of the Efremov Institute in Leningrad and was used by plasma physicist Evgenii Shunko with the support of theoreticians Dmitrii Riutov and Vladimir Mirnov. Riutov has since further developed the idea of gas dynamic traps.

Gennadi Dimov then proposed a modification of an open trap in which thermonuclear plasma is confined in the transverse direction by the magnetic field of a direct solenoid and an ambipolar electric field in the longitudinal direction. At a workshop held in Novosibirsk in 1975 and attended by scientists from Lawrence Livermore Laboratory, Dimov revealed that he had developed the idea of ambipolar traps to overcome problems of confinement loss in other mirror machines. For verification of the operational principles of this trap, the experimental AMBAL-1 with ambipolar mirrors (yin-yang coils with two windings) was developed. An advantage of the system is the possibility of producing a reactor of relatively small power.[57] The tandem mirror fusion reactor concept was proposed independently by Fowler and Logan at Livermore Laboratory in California.

Unfortunately the construction of the AMBAL-1 suffered from cost overruns and technical failures. A leak and a bad magnet prevented AMBAL-1 from ever operating as planned. Dimov was distraught over the failure. Since the building for the device was incomplete and his laboratory space was tight, he used the one good yin-yang magnet and a good ion source to get the AMBAL-U ("U" for "south") working in 1980. In the meantime Japanese and U.S. physicists (the latter at Livermore with TMX and at MIT with a device called "Tara") had made great strides. So Dimov designed the AMBAL-M ("M" for "modernized") in 1983. There continued to be problems with financing and parts because of nationwide budget cuts in fusion. Only in 1991 did he receive the equipment necessary to rebuild the AMBAL entirely.

When Sagdeev left the institute in 1971, leadership in theoretical plasma research fell to Dmitrii Dmitrievich Riutov. Riutov, too, moved from the Kurchatov Institute to IIaF, although somewhat later in 1968. A charming, erudite, and attractive man, Riutov brought wit and determination to the problem of controlled thermonuclear synthesis. He was born in Moscow in 1940 and graduated from Moscow Physical Technical Institute in 1962. He worked at KIAE for six years and was introduced to Akademgorodok by Margarita Kemoklidze Riutova, a theoretician specializing in solar physics who is now his wife. Attendance at the third international conference on plasma physics in Novosibirsk in 1968 convinced him to leave the Kurchatov Institute for Siberia. Riutov's works focus on the theory of plasma turbulence, the physics of nonlinear waves, the physics of powerful electronic and ion streams, and the theory of the processes of transfer in thermonuclear devices. With Budker and others he proposed a series of novel ideas for plasma containment.

Under the leadership of Riutov and others, IIaF physicists focused on experimental verification of a method of containment of a dense high temperature plasma using a multimirror trap. In 1971, at the fourth international conference on plasma physics in Madison, Wisconsin, they reported on their first experiments on plasma heating by a powerful beam of relativistic electrons. They described their theoretical research on a new method of containment based on a trap with a corrugated magnetic field. Their theoretical calculations were completed on multimirror confinement of a cesium plasma in a strong magnetic field. Systematic experiments on the GOL-1 showed the promise of this approach.[58]

Who knows how far the Siberian physicists' research would have gone had national fusion programs not suffered severe budget cuts in the early 1980s. Saved from even more drastic cuts at special plenary sessions of the Communist Party Central Committee, even the Kurchatov Institute program was slashed. Temporary salvation came from an unlikely source, Mikhail Gorbachev. At the Geneva Summit in 1985, to demonstrate "perestroika" in foreign policy, Gorbachev proposed that the United States and the USSR undertake joint fusion research, in part through ITER (the international thermonuclear experimental reactor project). The project was a pet of Evgenii Velikhov's. Acting as Gorbachev's unofficial science adviser, Velikhov, by then the third director of the Kurchatov Institute, convinced Gorbachev of the efficacy of the ITER program and of its importance as a confidence-building measure in arms negotiations with the United States.[59] Yet the Institute of Nuclear Physics could not take full advantage of Gorbachev's fusion gambit since ITER was a tokamak project that excluded de facto IIaF physicists and their magnetic "mirrors."

The Chernobyl disaster also served as a temporary boon to fusion research during this period of the budget crisis. Since the Three-Mile Island accident, a number of Soviet nuclear engineers had raised concerns about the safety and siting of Soviet reactors. They suggested improving the containment vessels. In a country inadequately concerned with environmental matters they showed striking awareness of both environmental and safety problems including waste management and disposal, legal framework, low-level power plant radioactive emissions, and public health. Fusion specialists like Dmitrii Riutov argued that their reactors were inherently safer than fission reactors. There were fewer products of fission. "Meltdown" was impossible. The amount of tritium used in a fusion reactor was relatively small so that a safety zone of no more than 1 km would be sufficient to ensure the safety of citizens in the case of an accident. Soviet plasma researchers, the Siberian physicists included, took advantage of the concerns raised by Chernobyl to get government attention—and modest increases in financial support—for their research.

Yet the strains of budget shortfalls and equipment bottlenecks continued to tell upon IIaF's research program. With the breakup of the Soviet Union, the situation has grown only worse. Institutes go months at a time unable to pay salaries, and programs have been eliminated because of the economic decay and political uncertainty now enveloping Russia and its scientists. Riutov spent much of his time lobbying policy makers, professional acquaintances, and friends in Moscow for funding for IIaF, and not enough time on his research. He defended IIaF alternatives to tokamaks. He argued that the ITER project was very expensive and no more certain than open traps, pinches with a rotating field, stellarators, nontraditional tokamaks, and inertial systems.[60]

To understand fully the current status of the Siberian Institute of Nuclear Physics we must return briefly to the year 1968. In that year two crucial events in the life of the institute and Akademgorodok occurred. One was the third international conference on plasma physics, attended by more than four hundred participants from twenty-four countries. The conference was a turning point in fusion programs throughout the world. Until that time, since the hydrogen bomb, research had moved largely from inertia. At the August conference, Lev Artsimovich, a physicist from KIAE who had directed the Soviet fusion program since Kurchatov's death, disclosed the first big successes on the tokamak T-3 apparatus with confinement eight to ten times better than previously achieved. His comments caused a sensation, especially among the English delegation who at first thought it was impossible to achieve such parameters at temperatures much lower than in a bomb. Budker's concluding remarks to the conference participants showed how caught up he was in the excitement. He called for moving beyond smaller experimental devices to the immediate construction of a prototype demonstration reactor. Speaking without notes, he talked too rapidly for the translators who admonished him to slow down. Budker responded, "This is the only way I can speak." Within months, around the world and in the USSR, huge amounts of money were put into fusion, and a new look at tokamak programs emerged.

For IIaF there was great symbolism in being the host of the conference. IIaF researchers proudly served as guides and translators. They showed their guests everything that Akademgorodok had to offer: picnics at the Ob Reservoir, long evening discussions at the social clubs, strolls through the forests around the institute. At long last the physicists felt they had earned the respect of their foreign counterparts and that Akademgorodok's mission had been confirmed. The VEPP accelerators had secured IIaF's international reputation. Here was recognition that its plasma specialists were world class, too. Most important was the symbolism of the conference from the point of view of international scientific relations.[61]

Up to this point, in spite of the international reputation the physicists commanded, they were frustrated by the absence of normal foreign contacts. Foreign travel was closely regulated by the Communist Party and the KGB. In this most international of fields, rarely more than a dozen IIaF physicists managed to go abroad in any one year, and scores more filled out the appropriate papers and got all the signatures, only to be denied a visa arbitrarily at the last moment. Foreigners visited IIaF much more frequently, but these trips were mostly short term and usually for nonscientific reasons. The researchers were successful in publishing abroad, but no more than one-quarter of their articles appeared in any visible foreign journal annually. Granted, cooperative long-term projects were discussed or established with CERN, SLAC, and the National Accelerator Laboratory at Batavia. But only a handful of IIaF physicists actually managed to set foot in those facilities.[62] In a word, IIaF physicists were starved for international contacts and thrilled by the recognition they gained by hosting the 1968 international conference.

But from a political point of view things were bleak. Problems had been building all spring and summer, and the euphoria of international success quickly gave way to despair when Soviet forces invaded Czechoslovakia. The crackdown on political and economic liberalism in Czechoslovakia had begun in the spring (the so-called Prague spring) and its repercussions were felt in Akademgorodok.

Akademgorodok Spring (with Apologies to Czechoslovakia)

Could the cult of big science permit a political crackdown? What might tarnish the success of Akademgorodok physicists? Would academic freedom always persist in Akagdemgorodok? So long as the physicists were not openly political, the Communist Party had tolerated their social clubs. Surely questions of funding and the institute profile would be expected to have an impact on the life of any scientific research center. Yet the party apparatus had paid attention to ideological indoctrination in Akademgorodok institutes from its founding and grew increasingly concerned about the seemingly growing political disrespect Siberian scientists paid to party directives. This was especially true since the party apparatus had diligently strived to put its operatives in all laboratories and foundries in the institute.

By Akademgorodok standards, the Institute of Nuclear Physics was a huge organization. Accelerator construction required a large physical plant and hundreds of ordinary workers. This created significant problems of ideological control. The workers were drawn primarily from the Novosibirsk region. Few had finished high school. Three-fourths of them had only a middle school education or less. When the institute was first under construction, its party organization served primarily to force the pace of work and expedite supplies. Then it turned its attention to communist indoctrination. The scientists who were imported from Moscow were expected to play a role in this process. The institute ran five methodological seminars and six political education circles in which the physicists took part. Institute scientists also lectured at the university and at local technical and higher educational schools. By June 1962 fully 40 percent of the workers had been drawn into the Komsomol or the Communist Party. At a time when plans were rarely fulfilled, when workers wasted time, materials, and equipment, when paint was already peeling, and when repairs on unfinished buildings were constant, why did the party apparatus waste so much manpower and energy on "ensuring the proper worldview of workers"? Still, the institute and its party committee grew rapidly, and by December 1963, of a staff of eleven hundred, two hundred were communists and nearly three hundred were members of the Komsomol.[63]

Neither the physicists nor the workers seemed to appreciate the indoctrination effort. The political education seminars were large and unwieldy. Scientists failed to participate in discussions or did so "mechanically." Party officials complained that some scientists had the "unfounded opinion that educational work is a function of social organizations," not of the institute and its party committee. The institute's primary party organization therefore resolved to pay more attention to the selection and training of scientific workers as propagandists. To raise productivity, "socialist competitions" would be held for such lofty awards as the "Communist Laborer" prize.[64] Not surprisingly, these measures failed to excite the scientists. They had physics to do.

Then a letter that appeared in the American press on March 23, 1968 called for the Soviet courts to reexamine the case of the dissident Aleksandr Ginzburg and others. This provoked a crisis of ideological damage control. The letter was signed by forty-six members of the Siberian division of the Academy of Sciences, including four doctors, ten candidates of science, and five Communist Party members. The forty-six became known as the *podpisanty* (signatories). That so many of them were leading scientists and communists brought the KGB and party apparatus down on Akademgorodok. As described in greater detail in chapter 7, the letter was sent by certified mail to several officials but was leaked by the KGB to provide a reason for cracking down on Akademgorodok's political and social life. Physicists played a prominent role in the protest as they did in other Akademgorodok cultural activities. Physicists Andreev, Zakharov, Zaslavskii, Fridman, and Viacheslavov had signed the letter, as had Khriplovich, Komin, Shunko, and Tselnik, although their names were not listed in the western press. On March 27 the Voice of America read the text of the letter, and then it was reprinted in a number of foreign papers. The IIaF Communist Party organization secretary proclaimed apoplectically that the foreign press had been used "to drive a wedge between the intelligentsia and workers, to undermine the trust Soviet power had toward scholars."[65] Many of the podpisanty were dismissed, others censured, and the social clubs of Akademgorodok were closed. Academic freedom had ended.

Luckily another letter of protest, one with more than 250 signatures, was not sent. It was torn up, preventing further fallout. Budker had counseled the physicists not to send such a "stupid" thing. He thought both letters were a provocation that could lead to the institute's closing. At IIaF the response to the scandal was less severe than in other Akademgorodok institutes. Perhaps it was the physicists' prestige in the USSR and their world-class reputation that the authorities did not wish to tarnish before the upcoming August plasma conference. In any case, at an April 1, 1968, joint meeting of the institute's party organization and academic council, at Budker's suggestion, it was agreed to censure Andreev, Zakharov, Zaslavskii, Fridman, and the others, hold a series of meetings to inform institute personnel of what had transpired and the nature of the punishment, and hope that the matter would be dropped.[66]

But this did not still the storm. On April 10, at a closed party meeting, tensions were heightened. Many communists, particularly those with a working-class background, were troubled by what they saw as a mere slap on the wrist. They believed the signatories should be fired, not censured. They saw the actions of the podpisanty not only as political immaturity and blindness but as treason. Sibakademstroi representatives demanded exile for the offenders. They were "not indifferent to the question of who lives in the city they had built."

Other communists admitted, in fact, that letter writing was not prohibited by Soviet law. But when anti-Soviet slogans fell into the hands of enemies and was used for propaganda, then in their view it became a criminal act. The letter was also evidence of the institute's insufficient ideological vigilance of its party organization—perhaps of party organizations throughout the region. Akademgorodok scientists always took great liberties. They persisted in inviting artists and even bards to entertain them. Some of these representatives of culture were acceptable. Others, like folk singer Aleksandr Galich, had lashed out against Soviet power. One confirmed communist declared that since the podpisanty defended scum, they were scum. The letter was evidence of a conspiracy that had to be rooted out.

Theoretical physicist G. I. Dimov, secretary of the party organization, argued that citizens had the right to write to any government official on any issue. He admitted that the letter was a blow to the government, but he called for reeducation and censure, not dismissal. Fortunately a majority of party members agreed.[67]

Budker had hoped to head off tension between the scientists and the local party organization by encouraging leading physicists to join it. Some 10 percent of them already belonged and a few more joined. Budker hoped they could tip the scales against growing pressure to control science and not just politics. He remembered when his university admission had been blocked for a year for his naive criticism of Stalinist policies. He had gotten off easily while millions had perished in the purges. He saw this crackdown as the party's attempt to return to the mindset, if not the coercion, of the 1930s. Yet he himself caused further trouble by refusing to sign an official Academy of Science condemnation of Andrei Sakharov for his dissident activities. Over the next ten years Brezhnevism, with its close monitoring of any activity deemed political, its anti-Semitism, its xenophobia toward the West, and its emphasis on applied research swept over Akademgorodok, dousing the city's revolutionary spirit.

SIBERIAN PHYSICS IN THE 1990s

In November 15, 1989, I took my first tour of the VEPP-3 and VEPP-4 facilities. Fire had destroyed the VEPP-4 five years earlier, and it was only then coming back on line. The physicists still ached about their loss. They described the smoke-damaged walls, the charbroiled wiring, the water that reached a depth of four feet. I then attended a celebration given for IIaF physicists upon their being awarded a state prize for their experimental verification of quantum electrodynamics. Of course the collective was pleased about the state prize but disappointed that it took the government "so long to notice us!" What hopes did the award bode for the future?

"If you look to the past," one exclaimed, "you can't lose your optimism."

Another rejoined, "You can lose only that which you have."

Gone is the excitement that carried the physicists through their first decade. Budgetary, programmatic, and political pressures conspired to deflate the enthusiasm of the early years. The normal aging of the institute's personnel led to a conservatism in research orientation that probably was to be expected. Yet nothing could have prepared them for the revolutionary political and economic changes that rocked the foundation of the scientific establishment under Gorbachev's perestroika. Traditional sources of funding for research dried up. Bureaucracies responsible for science policy disappeared, to be replaced by other organizations whose functions and reach is unclear. The general economic crisis forced many talented young scientists to leave research for the business world.

Under the direction of Aleksandr Skrinskii and others, however, the Institute for Nuclear Physics enters the mid-1990s with reason for hope. It has achieved a healthy balance between theory, experiment, and applications. Fifty percent of the IIaF research program is devoted to high energy physics, with 25 percent each given to plasma physics and to applications. The applications bring in a great deal of money. Japanese, European, and U.S. firms buy industrial accelerators and synchrotron radiation devices to generate income. IIaF physicists maintain a healthy reputation in the West. Alan Bromley, Yale physicist and science adviser to the former U.S. president George Bush, visited the IIaF complex in 1991 and was impressed by the level of physics and enthusiasm. Russian scientists and officials reject "go-it-alone" ideological pronouncements about the superiority of Soviet science. They welcome participation in international science through individual, institutional, bilateral, and business agreements.

Things are less certain on the home front. IIaF's current size, with more than three thousand workers, would have worried Lavrentev. "Big science" simply costs too much in a time of economic decay and political uncertainty. In June 1991 Boris Yeltsin visited IIaF and pledged his support. But that was before the failed August coup, the subsequent outlawing of the Communist Party, and the dissolution of all former Soviet ministries and committees. The new bureaucracies that have arisen in their place, the Russian Ministry of Higher Education, Science, and Technology Policy (itself recently demoted to sub-ministry level), the Ministry of Nuclear Power, and so on, have yet to establish clear lines of authority, priorities, or responsibilities. Their budgets are inadequate to support research during hyperinflation. The Russian Academy of Sciences and the ministries are pointing fingers and feuding over funding. Yeltsin has no authority to keep basic science afloat single-handedly. But even with the loss of income associated with the cancellation of contracts for equipment for the Superconducting Supercollider in Texas, the Institute of Nuclear Physics almost alone carries the Siberian division by underwriting apartment construction, social programs, and the development of infrastructure.

What is more, society has turned against "big physics." Opposing scientists' newfound academic and political freedom are wide-ranging antiscientific attitudes, mistrust, and fear of anything radioactive, known in the Russian press as "radiophobia." From a nuclear power industry in critical health to spectacular space failures to daily press reports of newly discovered toxic waste dumps, the public has increased awareness of the potential social costs of unregulated science and technology. In fact the broad coverage of nuclear energy has led to the the the so-called Chernobyl syndrome,[68] in which the public equates the physics of Chernobyl with that of fundamental research.

To combat political and economic uncertainty, Dmitrii Riutov spends much of his time in Moscow knocking on doors and abroad doing research at Italian or U.S. facilities. Nonetheless IIaF physicists have great hope for the future. They feel Budker's presence as they push on into new areas of research. They have focused on a new linear accelerator, the VLEPP, and the so-called φ factory, a new e-e+ colliding beam machine with resonance in the range of $2 \cdot 510$ MeV. VLEPP, an e-e+ linear collider at an energy of $2 \cdot 1$ TeV and very high luminosity, is being constructed four time zones away at Protvino near Serpukhov in conjunction with the "UNK" 3,000 GeV collider. Although VLEPP will be designed and manufactured at Akademgorodok, it will be built and operated at Protvino where a branch of IIaF has been established.[69]

Perhaps the real shortfall at the Institute of Nuclear Physics is in the realm of leadership. The role of the individual and of personal relations in Soviet history cannot be underestimated. Contacts, pull, and connections were vital under Stalin and Khrushchev and continued to be important under Brezhnev, even when personalities were overwhelmed by routinized bureaucratic politics. Faceless administrators ruled the scientific enterprise and encouraged change through piecemeal administrative measures, not profound reforms. Whole fields of research became dominated by individuals and their institutes. Risky endeavors, those most promising to unlock something new, were turned aside.

Everyone knows that Aleksandr Skrinskii is an able shepherd. But they also know that Soviet physics has suffered irreplaceable losses. Most crucial was the death of Kurchatov in 1960. And then the death of Artsimovich in 1973 deprived fusion research in the USSR (and the world) of a brilliant scientist and a superb administrator who had protected the fusion program from the vagaries of Soviet politics.[70] Gersh Budker did his best to step in. He was the inspiration of all institute programs. He provided leadership well into the 1970s for continued expansion of institute programs on electron cooling for proton-antiproton experiments and e-e+ experiments. He knew that opposition from Moscow physicists and the Siberian division would be decisive in blocking new devices in Akademgorodok. A joint venture had to be arranged with the Institute of High Energy Physics in Serpukhov. Budker was in Crimea in the fall of 1976, resting on the advice of his doctors. But he ordered a colleague to stay with him and keep him posted on the negotiations—negotiations he took part in. This colleague served as the "opponent of [Budker's] fantastic ideas and as his personal secretary." He took almost twenty pages of notes that served as the basis for the IIaF project presented at a national conference for accelerators in 1976.[71] Budker's final vision is only now being realized.

CHAPTER THREE

Siberia—Land of Eternally Green Tomatoes

STALINISM led to a scientific diaspora. Many scientists, along with ordinary citizens, simply disappeared into the Gulag, never to return, and were granted only posthumous rehabilitation for their imagined crimes. Some served in slave labor camps created specifically for scientists and engineers, the so-called Tupolevskaia Sharaga after Andrei Tupolev, the great aviation engineer who fell into one of the camps portrayed so vividly in Aleksandr Solzhenitsyn's *The First Circle*. Other scientists were banished to the far ends of the Soviet empire to assume posts anonymously in collective farms, car parks, or teaching positions having little to do with their expertise. More biologists may have suffered these fates than any other scientific specialists. Geneticists, whose careers had flagged in this diaspora, welcomed the chance to gather in Akademgorodok, where there was the promise of political reform and the opportunity to participate in the rebirth of modern biological and agricultural science.

The word *Siberia* evokes images of vast expanses of forest, wide plains of brush, snow, and ice, mighty, yet frozen rivers—hardly a hospitable place for agriculture. Yet postwar Soviet economic plans called for the development not only of industry but of agriculture east of the Ural Mountains—corn, wheat, and rye would be grown, sugar beets cultivated; foxes and minks raised for their fur; pigs, sheep, and cattle crossbred, selected, mutated with chemicals and radiation to adapt them to Siberia and increase their productivity.

Similarly Akademgorodok, with its primarily physical and mathematical profile, seems at first glance inhospitable for the life sciences. Of the Siberian division's twenty-two institutes, only four represent the life sciences (the Institute of Cytology and Genetics, of Biology, of Soil Science and Agronomical Chemistry, and the Central Siberian Botanical Garden).

Nevertheless the biologists, plant breeders, cytologists, and geneticists had reasons for hope that Siberia would provide fertile ground for their endeavors. First, the short growing season, early frosts, and only average soils required the input of modern techniques to make Siberian agriculture modern, productive, and Marxist. Second, Mikhail Lavrentev insisted that mathematical approaches such as those in population studies and genetics, and physicochemical concepts such as the gene and DNA, play an integrative role in modern, molecular biology. And, third, Stalin was dead, and Soviet scientists hoped that Lysenkoism died with him. Would genetics and drosophila find a fruitful environment in Siberia? Armed with modern techniques, guided by communist ideology, freed from Lysenkoist precepts, Siberian genetics would experience a renaissance. New agricultural products, better than nature itself produced, would be the proof of its practice.

Modern genetics promised new plants and animals that could thrive even in Siberian conditions: grain with a shorter growing season, but higher productivity; sugar beets with 20 percent more sugar; pigs of greater fertility, cattle with higher butterfat milk content and more meat; minks and foxes with desirable fur color, and strange crossbreeds of Siberian cows and Jerseys; and even domesticated rare species indigenous to Siberia. Indeed, one can buy rather hearty, occasionally flavorful, but always small green tomatoes, hot-house grown, at any time of the year in Novosibirsk.

Lavrentev invited Nikolai Petrovich Dubinin to set up the new Institute of Cytology and Genetics. Since Stalin's death, Dubinin, an egotistical if accomplished scientist who had studied with founders of Soviet genetics, had struggled to free biology from the grasp of Lysenkoism. He was joined by Dmitrii Beliaev, Iurii Kerkis, and other specialists who had bided their time in obscure posts throughout the empire in Lysenko's shadow. Dubinin might have succeeded in Akademgorodok had Khrushchev not intervened on Lysenko's behalf. With too many institute researchers refusing to abide by the fiction of Lysenkoism, privately embracing genetics instead, Khrushchev sacked Dubinin and threatened to close the institute. Deprived of its own facilities for six years, the institute was forced to lead a half-legal existence in shared quarters. The achievements of Siberian genetics were left to Dubinin's colleagues: Beliaev and Kerkis, Raissa Berg, Petr Shkvarnikov, and Rudolf Salganik. These unique, head-strong personalities had somehow survived twenty years of Lysenkoism, undertaking genetics research underground. How they managed to create an oasis for genetics in the middle of Siberia in the face of the ideological battles over Lysenkoism is the subject of this chapter.

Most of the narrative deals with the early years of the Institute of Cytology and Genetics, as these were the most exciting days in the life of the institute as well as the most frustrating from the scientists' standpoint. After familiarizing the reader with the difficult road the geneticists traveled to Novosibirsk, I turn to the ongoing research program, and then only to demonstrate that decades of Lysenkoism and of access to rudimentary experimental apparatus never allowed genetics to prosper. Indeed the Brezhnevite emphasis on applied agricultural research necessary for what in the West is called "agribusiness" stifled fundamental research. The biologists were expected merely to be engineers of plants and animals in order to raise productivity "scientifically" beyond the levels achieved by failed Soviet farm policies. For all these reasons the geneticists' return from scientific diaspora was waylaid at every step.

THE NINE LIVES OF TROFIM DENISOVICH LYSENKO

The biologists in Akademgorodok struggled mightily to escape Trofim Denisovich Lysenko.[1] A mean-spirited, unaccomplished, pedestrian plant breeder of peasant background, Lysenko began a long journey to the top of the biology profession in the 1920s. He convinced party leadership that he held the key to increased agricultural production. Lysenko believed that acquired characteristics are inherited and that the environment is primary in bringing about changes in traits. Through such techniques as the "vernalization" of winter wheat or peas, that is, soaking seeds in water and keeping them cold or warm depending on his whim, Lysenko believed he could manipulate the length of the vegetative period and turn winter wheat into spring wheat. He also used various techniques on animals. Lysenko's biology was based on so-called Michurinist concepts, after I. V. Michurin, often called the Russian Luther Burbank, a horticulturist who selected plants and created hybrids and who accepted Lamarckian notions of the influence of the environment on heredity.

Lysenko's belief in the inheritability of acquired characteristics dove-tailed nicely with the needs of the regime. Just as the Bolsheviks intended to change human nature with a new social structure and political system, he promised to create new plants and animals with changes in temperature and moisture. Rather than waiting for the science of genetics to bring about changes through lengthy study and evolution, he would revolutionize agriculture with peasants in the fields. Lysenko's promise of quick results and his devotion to the regime stood in stark contrast to the alleged "ivory-tower theorizing" of drosophila counters and corn hybrid specialists. This was a case of "socialist" versus "bourgeois" science, which enabled Lysenko to attack scientists' alleged "Weissmanism," "Morganism," or "Mendelism," that is, their allegiance to concepts of modern genetics. It mattered little that Lysenko's techniques failed to produce the desired results, let alone that they shared little of the replicability, controls, and rigor of "bourgeois science." His biology enabled the state to introduce a new scientific system for agriculture in the countryside, noteworthy more for extending party control and organizing peasants into attentive laborers (which increased yields ever so slightly) than for its science. However, one Lysenkoist goal, to acclimatize exotic plants and animals such as minks and silver foxes for economic purposes, remained a central feature of post-Lysenko genetics and was put into practice at the Institute of Cytology and Genetics.

Ultimately Lysenko's teachings became dogma for the entire biology community. At a series of agricultural conferences in the late 1930s leading biologists found their modern scientific concepts out of favor. Genetics was outlawed and removed from texts, competent specialists lost their positions, and many were destroyed in the purges. In 1948, at a session of the All-Union Lenin Academy of Agricultural Sciences (Vsesoiuznaia akademiia sel'sko-khozaiastvennykh nauk imeni Lenina, or VASKhNIL), genetics was prohibited outright. Until Khrushchev's ouster late in 1964 Lysenko remained in power, although after Stalin's death geneticists made some progress in reestablishing their discipline, both in atomic physics institutes and in the Institute of Cytology and Genetics in Siberia, far from Lysenko's brown thumb.

Lysenko's staying power in the Khrushchev years was surprising on two counts: first, the fascination of a number of Soviet leaders with corn hybridization, and, second, Lysenko's failure to deliver on his promises. When his methods increased yields it was because he was able to organize peasants to work more efficiently in the hated collective farms rather than because of scientific advances. Few farms had modern tractors or combines; Russian agriculture had changed little over the centuries technologically. Better organization increased yields. Owing to his political fortunes, Lysenko always had greater resources at hand—more equipment, fertilizers, better strains or breeds to start with. That he was a simple peasant in class-conscious Russia, whereas many of the leaders of the genetics community were identified with the bourgeoisie and their science, also helped to outrun, if not obscure, his essential failures.

Soviet leaders expressed interest in modern scientific agriculture from the start. During the difficult winter of 1921–22, Lenin asked Gleb Krzhizhanovskii, head of the state electrification agency, GOELRO, to explore the "advantages of corn." The Council of Labor and Defense was ordered to propagandize corn and teach "corn culture" to the peasants.[2] Stalin was more interested in waging a war against the peasantry, especially in Ukraine, and in harvesting every grain, every kopek of investment capital from the peasant, than in introducing the wonders of modern agriculture. The countryside was subjugated to Soviet power by bands of armed party workers. The cost was eight million dead of famine and the slaughter of half of all Soviet livestock by peasants who refused to turn it over to the state. Lysenkoist biology was called on to overcome the resulting agricultural crisis.

Through all this the geneticists managed to live a precarious existence in small laboratories and collective farms throughout the USSR, and hence were prepared to renew the field of genetics when called upon. Russia had a deep tradition in genetics, population studies, and evolutionary biology, as well as in corn hybridization, on which to build, dating to turn-of-the-century interest in the best strains of American corn. The great agronomist and political liberal, Kliment Arkadevich Timiriazev, who fought with Maxim Gorky to free Russia from cultural backwardness through the application of "the positive sciences," saw in modern agricultural methods the way to free the peasant from the whims of Russian climate and land.

Nikolai Ivanovich Vavilov, a specialist in the study of botanical populations, was the main defender of modern science and Lysenko's chief rival. Vavilov studied under William Bateson, one of the founders of modern genetics, at the University of Cambridge. From 1917 to 1921 he was professor of botany at Saratov University and director of the All-Union Institute of Plant Science. He then became president of VASKhNIL. He traveled throughout the world collecting tens of thousands of specimens of wild plants and wheat for his studies on breeding. Vavilov organized a series of agricultural experimental stations: in the north near Murmansk, Vladivostok, Aktiubinsk, Uzbekistan, and Turkmeniia, in the Caucasus, Crimea, Ukraine, Belarus, and near Leningrad.[3] On the basis of his observations Vavilov advanced the theory that the greatest genetic divergence in cultivated plant species could be found near the origins of these species.[4] In Leningrad Vavilov was joined by Iu. A. Filipchenko, the leading Soviet geneticist who created several laboratories of genetics and eugenics which led to the creation of the Institute of Genetics of the Academy of Sciences under Vavilov in 1934. Even before Lysenko's fall, Nikolai Vavilov was hailed in Akademgorodok as the leader of Soviet genetics.[5]

The late 1920s and 1930s was a period of intense debate over the philosophical conflicts between modern science and the Soviet philosophy of science, dialectical materialism. Philosophers, Stalinist ideologues, and eventually scientists themselves discussed epistemology, methodology, even the role of class struggle in all fields, from physics and chemistry to biology and the social sciences. In physics, relativity theory and quantum mechanics were attacked as idealist. In biology, Lysenko and his allies denounced Vavilov and genetics as "Morganist-Mendelist-Weissmanist," which meant they accepted heinous theories of genes and chromosomes as the stable material of heredity. This led to Vavilov's arrest, imprisonment in the Gulag, and death in 1943. In one of the more bizarre twists in the history of Soviet science, his brother, Sergei, became president of the Academy of Sciences in 1945.

After Stalin's death Khrushchev wasted no time in putting forth modern science, and corn, as the savior of Soviet agriculture, particularly for the production of fodder but also for higher-quality food for the Soviet people. The question was how to hybridize corn, or even wheat, without genetics? In 1952 he invited several ministers and the head of the agricultural department of the Central Committee to an experimental plot outside Moscow, where he argued that increased corn production would lead to better fodder and thence to higher slaughter weights. In September 1953, at a special plenary session of the party, the post-Stalin leadership acknowledged that meat, milk, and agricultural production over the preceding two decades was uneven at best and declared that corn would feed the communist future.[6] Year in and year out draughts, crop failures, and falsified data about yields made the experiment with corn a failure, and the USSR returned to its reliance on wheat. Although climate played a significant role in this, so too did Khrushchev's bizzare behavior.

For all his faith in the power of the Soviet state to construct communism within his lifetime, Khrushchev was prone, as those who removed him from office claimed, to "hare-brained schemes." Whether these were the result of untempered enthusiasm, ignorance, or flights of fancy, they fell flat in a Russia so recently used to coercion and administrative measures to achieve reform. One such scheme involved Khrushchev's effort to supplant grain with corn as the staple of the Russian diet after a trip to an Iowa farm reinforced his belief in the glories of corn. By March 1956, 70 million acres was planted with corn, almost seven times the previous level. Almost immediately, there were problems with the corn in the northern regions as well as tremendous expenses in transporting the corn for animal feed.

Nor were the biological sciences immune to Khrushchev's touch. Geneticists had assumed that the Khrushchev thaw would enable them to forego Lysenkoist interference in their work, but they were disappointed. Khrushchev remained firm and would not free biology from Lysenko's grasp. This delayed the development of genetics another ten years, and condemned recombinant DNA research to backwardness to this day. Despite Khrushchev's love of corn, Academician Boris Sokolov, a specialist in selection, was instructed by the Academic Council of the Corn Institute "to cease work on the creation of corn hybrids from self-fertilizing lines."[7] Even in Akademgorodok, where, at Lavrentev's urging, Khrushchev approved the creation of the Institute of Cytology and Genetics, Lysenko's reach managed to interfere with research.

GENETICS UNDERGROUND

These challenges notwithstanding, de-Stalinization permitted the revitalization of genetics research throughout the USSR, including in Akademgorodok. Like other physicists and mathematicians, Lavrentev recognized the importance of a modern biology discipline in Russia. Mathematicians, for their part, saw the power of mathematics in serving as a language for genetics. Physicists knew that mutations of genes and chromosomes, concepts disallowed in Lysenkoist biology, resulted from the action of radiation. Leading physicists, especially those in nuclear physics, shielded geneticists from Khrushchev's and Lysenko's ire. Igor Kurchatov, head of the Institute of Atomic Energy in Moscow, set up a Department of Radiation Genetics within its walls. Kurchatov organized a talk for Nikolai Dubinin at the presidium of the Academy. Dubinin's *Questions of Radiation Genetics* (Problemy radiatsionnoi genetiki) (1961) and *Molecular Genetics* (Molekular'naia genetika) (1963, but begun in 1947 at the suggestion of the president of the Academy Sergei Vavilov, just before Lysenko rose to supreme power) were published only through the offices of V. S. Emelianov, head of the State Committee on Atomic Energy, and its publishing house, Atomizdat, at Kurchatov's request. In October 1963 the State Committee sponsored a meeting on the use of radioactive sources to generate mutations which was attended by more than two hundred specialists, including P. K. Shkvarnikov and I. V. Chernyi, both of the Institute of Cytology and Genetics, who discussed their work on grain, potatoes, and tomatoes. Dubinin also received support from M. V. Keldysh, then director of the Institute of Space Research, and S. P. Korolev, head of the space program, to conduct research on the genetics of plants, animals, and human beings for the Soviet space program. Another leading physicist, Lev Artsimovich, leader of the Soviet fusion program, returned from the 1956 Geneva conference on the peaceful atom, carrying drosophila, since no good collections could be found in Soviet laboratories at the time.[8] In a word, geneticists worked under an atomic shield.

Theoretical physicist Igor Evgenievich Tamm, Nobel prize winner in 1956, gave the first public talk in Moscow about the significance of the double helical structure of DNA, also in 1956. Tamm repeated his comments over the next years, including a presentation in April 1962 at Novosibirsk University and in smaller seminars at the Institute of Cytology and Genetics, not so much to familiarize the scientific community with modern genetics and molecular biology but to offer vocal, if indirect criticism of Lysenko.[9] Tamm's talk was a milestone for Akademgorodok geneticists, for a widely respected scholar had openly reviewed the recent history of biochemistry and molecular biology. He described how most proteins, the building blocks of the genetic code, were known, and he went on to say: "Like all objective knowledge about how the world is put together, these discoveries create new prospects for the mastery of nature, for the control of living processes."[10]

Nikolai Timofeeff-Ressovsky carried the mantle of genetics in the uncertain days of the Khrushchev thaw. He conducted an informal summer school that helped train many future Akademgorodok geneticists and cyberneticists. Like many Russian scientists who traveled to the West in the 1920s, Timofeeff-Ressovsky studied in Germany. He was among thirty Russians to receive Rockefeller Foundation International Education Board fellowships. He worked at the Institute of the Study of the Brain in Berlin. In 1929 he was invited to head the Institute of Sugar Beet Genetics in Ukraine, or perhaps to a new institute in Pushkin, near Leningrad. On the advice of N. K. Koltsov, who informed him that Soviet physicists, chemists, and particularly biologists were under assault from Stalinist ideologues, Timofeeff-Ressovsky remained in Germany. Indeed, that very year the laboratory of S. S. Chetverikov was razed and Chetverikov exiled to Sverdlovsk. Timofeeff-Ressovsky was caught behind Nazi lines during World War II and then captured by the Soviets. Avrami Pavlovich Zaveniagin, the director of Magnitka (the "Gary, Indiana," of Russia, Magnitogorsk), the builder of Norilsk, the huge Siberian metallurgical center on the Enisei River, and the deputy commissar of internal affairs under Secret Police Chief Lavrenty Beria, intended to offer Timofeeff-Ressovsky "normal working conditions" under the umbrella of the atomic bomb project. How could there be a bomb without radiation genetics? But Timofeeff-Ressovsky disappeared for nearly a year in the Stalinist Gulag, where he was a wardmate of Aleksandr Solzhenitsyn, and by the time Zaveniagin found him he was emaciated, nearly dead, but still willing to work on the problem of biological defense against radiation.[11]

After the 1948 VASKhNIL session, Timofeeff-Ressovsky was exiled to the Urals. He was essentially left alone and allowed to work with drosophila at a time when the popular weekly *Ogonĕk* ran an article entitled "Fly Lovers— Man Haters." That he was dealing with nuclear weapons, reactors, power stations, and biological protection against radiation provided him a constant research focus. Yet rumors that he collaborated with the Nazis persisted, and he found himself the object of disdain in the scientific community. The nuclear physicist Artsimovich refused to shake his hand upon meeting him. But his brilliance, his knowledge, his compassion for his work, in a word, his scientific reputation preceded him. Past and future geneticists alike heard of Timofeeff-Ressovsky's laboratory in Miassovo in the Urals, and joined a pilgrimage to undertake summer work with him there. When they arrived at the train station, they learned it was another twenty-five kilometers to Miassovo. Many set out on foot.[12]

It was outrageous to hear open talk about genetics in those days. But Timofeeff-Ressovsky had no reservations about "talking science." He considered himself as liberated as the atomic scientists. He made doctors and students alike sit naked in shallow water of the lake at Miassovo and listen to the lecturer standing on the shore. Unlike other scientists who dressed stodgily in the obligatory bland suit and tie, the atomic physicists felt liberated by their successes. They wore their "shirts outside their trousers," without ties, played ping pong at work, argued with their bosses, and sheltered geneticists.[13]

Miassovo's informal summer sessions may have served as the inspiration for the Akademgorodok boarding school for gifted Siberian children. At Miassovo the mathematician Aleksei Aleksandrovich Liapunov, who would be closely involved in the creation of the boarding schools, explored cybernetics with the summer students when cybernetics was being attacked by Stalinist philosophers as a bourgeois invention. Joined by the mathematicians Kolmogorov and Sobolev, Liapunov steadfastly defended cybernetics. The languages of cybernetics, biochemistry, biophysics, mathematical methods, and systems theory became "the Aesopian language of genetics. A 'unit of hereditary information' sounded less anti-Lysenkoist than a 'gene,'" one biologist pointed out. Liapunov next boldly joined with Aksel Berg to found a new journal, *Problems of Cybernetics*. In Akademgorodok Liapunov orchestrated the renaissance in cybernetics. He brought linguists and humanists together to teach, recognizing the importance of semantics for computer science. Like Lavrentev, he stressed the reading of fiction to encourage flights of fancy among young scientists.[14] And he participated in the organization of the boarding school, for the Miassovo experience convinced him of the importance of rigorous formal and informal training among the young.

Igor Tamm invited Timofeeff-Ressovsky to lecture on genetics at the Institute of Physical Problems, at Petr Kapitsa's famous weekly seminar. This would have been impossible at a biology institute, but physicists had great clout. Tamm's plans for talks on Watson's and Crick's double helix and Timofeeff-Ressovsky's radiation genetics and the mechanisms of mutations went ahead amid a host of rumors. Some had heard that Khrushchev forbade Timofeeff-Ressovsky to speak, but Kapitsa confirmed with Khrushchev personally by telephone that the seminar was not banned. On February 8, 1956, at 7:00 P.M., Tamm and Timofeeff-Ressovsky lectured an audience overflowing to the streets about recent advances in genetics. Loudspeakers had to be set up to accommodate the masses.[15]

Cybernetics and genetics played a mutually beneficial role in the resurrection of modern biology and computer science. The idea of the organism as a feedback mechanism so popular in cybernetics explained its central place among geneticists, physicists, and cyberneticians. A. A. Malinovskii, son of A. A. Bogdanov, known as the father of modern systems theory, and foil for

Lenin in the dense tome *Materialism and Empirio-Criticism*, devoted his later years to the defense of systems theory, cybernetics, and his father's name. Timofeeff-Ressovsky served as a bridge between genetics and physics, attracting the attention of Schrödinger, Bohr, and Delbrück to the molecules that made up chromosomes. Schrödinger's book, *What Is Life?*, translated by Malinovskii and published in Moscow in 1947, found its inspiration in Timofeeff-Ressovsky. In December 1958 Raissa Berg, a future employee of Akademgorodok, chaired a seminar in Leningrad on "Cybernetics and Genetics" with the support of the university rector A. D. Aleksandrov. Timofeeff-Ressovsky and Malinovskii participated. Aleksandrov later came to Akademgorodok as rector of Novosibirsk University.

Timofeeff-Ressovsky influenced the direction of research in Akademgorodok. When the focus of the Institute of Cytology and Genetics on animal and plant genetics, and radiation genetics, expanded toward modern molecular biology in the 1970s, it was assisted by the development of mathematical biology. Mathematical biology unexpectedly sprang up in the institute in the 1960s under the capable direction of V. A. Ratner. Ratner, a physicist by training who grew up in Khabarovsk, in the Far East on the Amur River near China, stopped in Moscow in the late 1950s on his way to Leningrad. He had heard about the newly organized Institute of Nuclear Physics. Out of curiosity, he sought out Gersh Budker at the Kurchatov Institute. On the spot, Budker invited him to join the Institute of Nuclear Physics. Ratner excitedly journeyed to Novosibirsk to await the opening of the institute. Continued delays in its organization and Ratner's increasing frustration over his lack of work led him to Dmitrii Beliaev's door. The Institute of Cytology and Genetics had many vacancies, if no future. Ratner wanted to do science, however, joined the staff, and for ten years served as the only mathematical biologist in the institute. Timofeeff-Ressovsky's summer school at Miassovo brought together a broad spectrum of scientists, young and old, for discussions of modern genetics. Here Ratner first met Liapunov and began to use mathematics to understand mechanisms of control in genetic systems. So unique was Ratner's specialization in Akademgorodok that in 1969 he had to travel as far geographically and psychologically as the closed nuclear city of Obninsk to his mentor, Timofeeff-Ressovsky, to defend his candidate dissertation.

N. N. Vorontsov, future Minister of the Environment of the Soviet Union, also fell under the spell of Timofeeff-Ressovsky and the mathematician Liapunov during the early years of the Khrushchev thaw. He graduated from Moscow State University in 1955 with a degree in zoology and comparative anatomy. He participated in the rediscovery of genetics and cybernetics in Liapunov's Moscow circle. Liapunov had close ties to the mathematicians Sobolev and Lavrentev who were at the forefront of early computing in the postwar Soviet Union. (Liapunov's daughter, the biologist E. A. Liapunova, was excluded from the Komsomol at Moscow University for being a member of a "Morganist circle" at the very moment Khrushchev was reading the secret speech. Vorontsov and Liapunova later married.)

Lavrentev invited Liapunov to Akademgorodok in 1957, but the future president of the Academy, M. V. Keldysh, initially would not let him go. In December 1961 Liapunov finally received permission to transfer to Akademgorodok, accompanied by his daughter and Vorontsov. Vorontsov had moved to the Zoological Institute of Leningrad University which had the advantages of a good museum collection and relative freedom from Lysenkoists, before journeying to Akademgorodok. A short time later he was called back to Moscow to teach biophysics, biochemistry, and biomedicine under the protection of the Soviet space program. To ensure the health of cosmonauts during flight, after all, the study of the impact of cosmic and solar radiation on human physiology and genetics was required. Vorontsov taught one of the first courses in Moscow on genetics at the Medical Institute. But the situation in genetics, critical through 1964, the absence of research and expedition possibilities, and the Nuzhdin affair made Vorontsov anxious to leave Moscow.

The Nuzhdin affair was a watershed in the resurrection of genetics. Early in 1964 the Academy met to vote to fill a vacancy, or chair, in genetics. The Lysenkoists hoped to see hack scientist N. I. Nuzhdin elected. Since the establishment of party control over the Academy in the late 1920s, only rarely had the Academy rejected a candidate for membership who came from the approved nomenklatura. In part because of secret ballot, however, cases of academic autonomy were not unheard of. For example, the Academy never expelled Andrei Sakharov even after his exile to Gorky. In this case, such leading scientists as the biologist V. A. Engelgardt and the physicists Igor Tamm and Andrei Sakharov all spoke out publicly at the election meetings against Nuzhdin. Nuzhdin was rejected. When Khrushchev heard the result of the ballot, he lost control. He ranted. He threatened to appoint a commission with Lavrentev as chairman, with the purpose of sending genetics back to the 1940s, and even to disband the Academy of Sciences.

At this juncture Khrushchev was ousted from party leadership. Lavrentev, Liapunov, and S. T. Beliaev, rector of Novosibirsk University, the latter two of whom had recently become corresponding members of the Academy, invited Vorontsov to Akademgorodok to become academic secretary of the biological sciences of the Siberian division. Vorontsov arrived in September 1964. Before moving to a short tenure in Vladivostok, his responsibilities included traveling around Siberia to reestablish genetics departments in existing institutes and universities, and to create new ones. Vorontsov preferred work in the presidium of the Siberian division to the tense atmosphere in the Institute of Cytology and Genetics under the "Napoleon" Dmitrii Beliaev. Beliaev, not wanting to have a rival, squelched Liapunov's efforts to invite Timofeeff-Ressovsky to the institute. He later maneuvered to see Timofeeff-Ressovsky's nomination for Academy membership tabled at his institute. Beliaev also tried to limit Vorontsov's input by putting his group under the jurisdiction of Raissa Berg's laboratory. It turned out, however, that Berg was on difficult terms with everyone but Vorontsov, and they got along famously. Vorontsov, Ratner, Liapunov, Berg, Aleksandrov, and others, all crossed paths in the underground genetics and cybernetics communities in Moscow, Leningrad, and Miassovo before meeting again professionally in Akademgorodok.

Raissa Berg, Ninel (Lenin spelled backward) Khristoliubova, and Zoia Nikoro supported the nomination against charges that Timofeeff-Ressovsky had betrayed the USSR when he remained in Nazi Germany. The institute's party and scientific leadership would not allow the nomination to pass. Hearing Nikoro's pleas, Petr Shkvarnikov said, "It's very bad that speeches such as that should be heard at our Science Council." Rudolf Salganik claimed that Timofeeff-Ressovsky should have returned to Russia well before 1937, before the real danger: "Fascism apparently was more to his liking than the land of burgeoning socialism." Citing the difference of opinion, Beliaev ruled that no one should be nominated.[16] Timofeeff-Ressovsky had offended too many people with his honesty. In spite of his central role in the resurrection of genetics, he remained an outcast. In his last days he worked in Obninsk, an atomic city outside Moscow set up to support the Soviet nuclear power program, in the Institute of Medical Radiology. In 1971 he lost his position there, too, and his laboratory was disbanded. He gained employment only as a consultant to the Academy of Medical Sciences, in spite of the protestations of such western scientists as Max Delbrück, Nobel prize winner, who appealed directly to Academy president Mstislav Keldysh. But in his turbulent career, Timofeeff-Ressovsky had provided instruction and inspiration, directly and indirectly, to a series of biologists who saw to the rebirth of geneticists in Moscow, Kiev, Leningrad, and Akademgorodok.

Nikolai Dubinin's involvement in the resurrection of genetics in Akademgorodok was of more immediate importance than that of Timofeeff-Ressovsky. Dubinin lacked Timofeeff-Ressovsky's modesty. He actively sought to rehabilitate genetics under Khrushchev while establishing a hallowed place for himself in the pantheon of Soviet scientists. He wrote widely in theoretical, biological, and philosophical journals. He spearheaded the effort to find an institutional home for modern genetics in the postwar Soviet Union. Dubinin (b. 1907) was the son of a farm-laborer who was killed at the front during the civil war. Though his mother was alive, and worked in the orphanage in which he was raised, Dubinin spread the story that he had been raised as an orphan and arrived in Moscow in 1923 with but a blanket and ten rubles in his pocket. He entered the university, graduated five years later, and fell into one of the leading genetics communities in the world, in the Moscow circle of Koltsov, Serebrovskii, and Chetverikov.[17]

From the very start of his university training as a student of A. S. Serebrovskii, head of the Department of Genetics at Moscow University, along with S. M. Gershenzon, P. F. Rokitskii, and B. L. Astaurov, he seemed to be without peers. When Dubinin visited Iu. A. Filipchenko, head of the Leningrad school of population genetics, he made a significant impression. Dubinin ended up working in the Institute of Experimental Biology under Koltsov. After Serebrovskii was exiled to Sverdlovsk, Dubinin was appointed to head an expanded genetics section of Koltsov's institute. It was important for this post that Dubinin had the correct social origins as a peasant during this period of *vydvizhe-nie* (career advancement) in the early 1930s, when class origin meant every-thing. The institute was the center of genetics research in Russia, and Dubinin's sector did good work on population genetics.

During the heyday of Lysenkoism, Dubinin found placement in relatively quiet circumstances in Alma Ata and then in the Institute of Forestry in the Urals, where the director, Academician Sukachev, a geneticist, enabled Du-binin to work on the genetics of birds. In 1956 Dubinin returned to Moscow and, with genetics seemingly on the road to recovery, was permitted to orga-nize a laboratory of radiation genetics in the Institute of Biophysics. Here he gathered the lost generation of geneticists who would replace Vavilov, Koltsov, and Filipchenko. The Lysenkoists were far from ruined, however. Instead they had gained a "lightning rod" for their attacks, especially since a number of foreign academies of sciences and scientific societies began to bestow Dubinin with their attention and awards.

Shortly after Stalin's death, with the support of the Academy presidium, Dubinin circulated a proposal setting forth measures necessary to rehabilitate genetics. The proposal pointedly indicated that since 1948 no progress had been made like that in the West where genetics had raised agricultural yields, made strides in cancer research, and so on. He called for the establish-ment of an institute of experimental and theoretical genetics, staffed by physi-cists, chemists, mathematicians, biologists, cytologists, and plant breeders, with fourteen laboratories, including biochemical, physiological, radiation, mathematical, and evolutionary genetics, the study of viruses, and plant and animal selection. He insisted on the right to reestablish graduate programs and grant degrees, publish a new journal called *Genetika*, as well as monographs, textbooks, popular literature, and translations, to travel abroad, and to pur-chase equipment.[18] Only in 1965, after Lysenko's ouster, were most of these steps realized.

This project was followed by the so-called letter of the three hundred to the Central Committee of the Communist Party that was signed by 298 leading scientists from all fields. The letter requested permission to reestablish genetics as a discipline. It criticized Lysenko's science, his methods of polemics, his betrayal of Marxism-Leninism. "Without the intervention of the Department of Science of the Central Committee," the signatories concluded, "it is impossi-ble to publish in biology journals."[19]

In 1958 Dubinin grew bold enough to circulate a form letter from his labora-tory of radiation genetics in which he asked respondents to inform him of short- and long-term research plans in their institutes. He would summarize the information for inclusion in a reborn Academy of Sciences program on "the physical and chemical basis of inheritance."[20] He asked for reports of research results that might be presented at meetings in the coming year, and he an-nounced two conferences: one on radiation genetics, the other to coordinate efforts. In his appeal Dubinin stressed that modern genetics had immediate applicability in all regions of the economy.[21]

At the same time, from all corners of the literary and academic community, calls for the rebirth of genetics were heard: in *Novyi mir* (New world) under the editorship of Andrei Tvardovskii, *Nash sovremennik* (Our contemporary) under the editorship of Oleg Pisarzhevskii, who also propagandized cybernetics, in *Voprosy filisofii* (Problems of philosophy), and even *Tekhnika-molodezhi* (Technology—for youth!) with its runs of a million copies. In 1957 alone, in *Technology—for Youth!*, S. I. Alikhanian, B. L. Astaurov, D. K. Beliaev, Dubinin, Liapunov, and Timofeeff-Ressovsky all actively touted the achievements of modern biology against Lysenkoism.

With the May 1957 decision to organize a genetics institute in the Siberian division, it was natural that Dubinin was appointed director. Even before the decision to build Akademgorodok, Lavrentev called Dubinin to ask him to consider becoming director of a center in Siberia with "unlimited possibilities for the development of genetics." Dubinin says he accepted without pause. Some of his colleagues insist that he initially intended to stay the course in Novosibirsk, that he was businesslike in all his dealings with the new institute. Yet he managed to convince enough biologists that his allegiance was divided between Moscow prominence and Siberian challenges, that he had no real intention of moving to Novosibirsk, and did more damage to the institute in its first days by his inattention than did Khrushchev and the Lysenkoists with their incessant efforts to close the place down. Dubinin once told Iurii Kerkis, a colleague and specialist in radiation genetics, "If I move to Akademgorodok there will be no one to represent genetics before the central government or to visiting dignitaries."

Responsibility for organizational activities hence fell on Kerkis, Shkvarnikov, and Salganik. Often Shkvarnikov or Kerkis had to journey to Moscow for Dubinin's signature on important documents or to beg him to show up in Novosibirsk to fight off detractors, especially since making a phone call was no simple matter in those days. In spite of Lavrentev's best intentions and the care he bestowed on the institute, Dubinin's prolonged absences had made it easier for the Lysenkoists and their representatives in the Ministry of Agriculture to interfere in normal operations. For many years the institute was denied its own building or experimental plot. Efforts to carry out research on a broad scale were thwarted. Relations between all scientists grew tense on this rocky soil.

Dubinin also earned the enmity of his coworkers when he astonished the first session of the Siberian division with a talk on the genetic dangers of small doses of ionizing radiation, carrying on for twenty minutes as if it were a great discovery. The talk was based on work Kerkis had done. In 1958, with Kerkis and Lebedeva, Dubinin published on article based on experiments with cultures of human tissue cells which produced quantitative data on the influence of radiation on chromosomes. The data generated funding from the State Committee on Atomic Energy for further research.[22] Unfortunately, in the views of his associates, this performance was the essence of Dubinin, attributing primary importance to everything connected with his name and losing interest in everything else.

Dubinin's wife, Tatiana Aleksandrovna Toropanova, generally regarded as a sweet, tactful, and intelligent woman, had some moderating influence on her husband. She interceded where possible to secure his cooperation, answering letters and fighting off various intrigues. When she was killed in a hunting accident, or perhaps committed suicide, Dubinin quickly remarried. The personality, ambitions, and behavior of Dubinin's fourth wife, Lidia Georgievna, meshed fully with his. She had been merely a laboratory worker but was transformed into the self-important wife of an academician and treated everyone as if she were the "second director" of the institute.

As soon as he became director of the Institute of Cytology and Genetics, Dubinin showed his allegiance to principles of modern genetics, which he asserted in an article published in 1958 in *Sibirskie ogni*, the leading Siberian literary journal and the oldest continuously published one in Soviet history. He also defended the honor of fundamental science, relying on the authority of physicists who seemed to be able to solve any problem. Atomic energy and chemicals would be used to manipulate genetic codes in all living things to solve problems in agriculture and medicine. Corn, milk, and meat production would undoubtedly increase. The laboratory of A. N. Lutkov and V. V. Sakharov had already successfully increased sugar beet production. Similar successes were anticipated with sables and minks. Dubinin called for an end to disagreements among Soviet scientists concerning genetics, for their ideological unity, and for their devotion to communist ideals.

From the start Lavrentev personally cared for the health of genetics in Akademgorodok. With the help of the president of the Academy, A. N. Nesmeianov, the head of the science department of the Central Committee, V. A. Kirillin, and local party officials, Lavrentev defended the institute when it struggled under the pressure of "review commissions" sent by Lysenkoists in Moscow to investigate charges of "obscurantism" and to root out "enemies of the people." The commissions inevitably returned verdicts of "not guilty," but Lysenko and Khrushchev would not be appeased.

Publication in the *Botanical Journal* (Botanicheskii zhurnal) of several articles criticizing Lysenkoism drew a sharp response from the party and was a pivotal event in the life of the institute. An editorial in *Pravda* on December 14, 1958, rebuked what it called "the false positions of *Botanicheskii zhurnal*." In essence, *Botanicheskii zhurnal* had had the temerity to suggest that Lysenkoism was pseudoscience and that the failures of Soviet agriculture had anything to do with it, and thus had broken the unspoken truce between the Lysenkoists and geneticists and violated the spirit of the Soviet public lie.

Members of the party bureau in the Institute of Cytology and Genetics gathered to discuss the *Pravda* article. Kerkis urged his coworkers to maintain the public lie. Leave them alone, and they will leave us alone, he urged. Like Dubinin, Kerkis used the Aesopian language of physics to defend molecular biology and genetics in the institute, claiming that "time will show who is right."[23] I. I. Kiknadze, a junior scientist joined with Kerkis in speaking for an end to "tendentious criticism" as a tactic against Lysenkoism. She pointed out that classical geneticists propagandize and popularize their achievements poorly, whereas the Lysenkoists use all available media to make their alleged achievements known. Thus, over the next few years, geneticists at the institute actively popularized their achievements, at forty, fifty, even sixty public engagements annually, and rarely criticized Lysenkoism directly.

The *Pravda* editorial and December 1958 Central Committee Plenum had also called for greater emphasis on applied biology. This concerned Kerkis who grew worried about "narrow practicalism." Basic science was vital. On the other hand, he noted that his colleagues would have greater success with their opponents if they learned to push their achievements into production with the same persistence as the Lysenkoists. From this point on, institute biologists made a point of demonstrating their devotion to "applications" and to proving how their science was verified by "practice."[24] As an indication of these achievements, the institute reported that it had satisfied all its "socialist obligations" in honor of the twenty-first party congress in a wide range of areas: animal selection, especially pigs and fur-bearing animals; work on corn hybrids and sugar beets; and preparation of a genetically engineered nuclease for the treatment of a retinal virus.[25]

Considering the precarious position of genetics in the country, it is not surprising that the institute party committee resolved to consider the *Pravda* editorial completely correct in its criticism of the *Botanicheskii zhurnal* article. The style of criticism was called "inadmissible in the clarification of argumentative questions of biological science," and even dangerous to the development of science.[26] The resolution indicates the extent to which geneticists were unwilling and unable to reject Lysenkoism outright. As difficult as it was for them to listen to Lysenko's untruths, they remained silent. De-Stalinization had not gone far enough for scientists to throw off the Lysenkoist legacy.

Still, the matter would not fade. A commission under M. A. Olshanskii, a leading VASKhNIL academician and Lysenkoist, arrived in Novosibirsk with the intention of changing its profile or even closing it down. The commission members met with leading Akademgorodok scientists in Lavrentev's office to present the results of their investigation. Suddenly, the special Kremlin phone on Lavrentev's desk rang. He answered it saying, "Yes, I couldn't agree more." And then, after a pause, "You are right, we will be able to work things out satisfactorily." He turned to the commission members, saying, "I suppose this finishes our business." The commission returned the conclusions that "the collective of the institute works not only well, but more than that it works with great effort, with great enthusiasm." Lavrentev never told anyone if the call had been staged.[27] Then in June 1959, at a plenary session of the Central Committee, Khrushchev accused Dubinin of being a falsifier of Michurinist biology whose research was divorced from agricultural practices. The institute fell under microscopic examination. A month later the institute party committee met in open session to consider the quality of Dubinin's work and discuss his tenure. The committee also set out to determine if the charges of anti-Michurinist activities in the institute were true. Iu. P. Miriuta, secretary of the party organization and a Stalinist at heart, reported that in terms of securing equipment Dubinin had done his job well. Staffing remained a problem, however, since many senior scientists wished to await the post-Lysenko flowering of genetics in Moscow, leaving mostly novices to carry on in the institute. Further, facilities had yet to be secured, let alone apartments for the workers. Yet despite these challenges, which were clearly not Dubinin's fault, the party committee concluded that promising research in areas of importance to the national economy on plant selection and animal breeding had commenced, and therefore endorsed his continued direction.

Not content to allow outside groups to determine their future, institute biologists wrote the Central Committee and Novosibirsk Obkom (regional party committee) directly. In one such letter, most likely written in November 1959, director Dmitrii Beliaev and Rudolf Salganik, then secretary of the party organization and head of the laboratory of nucleic acids, stressed the consonance of institute programs with the goals of "Communist construction": the important role of genetic methods in increasing harvests and productivity of domesticated animals—from grain and beets to meat, fur, and eggs. They called for the Central Committee to adopt a series of measures at its December 1959 plenary session intended to expand the application of genetic methods for the solution of agricultural problems. Research on polyploidy, hybrid vigor, and other forms of hybridization, work on corn, radiation selection, and so on, was "essential," promised a "great increase" in productivity, and "important results."[28] The utilization of these "progressive" achievements in genetics, they informed party officials, was the only way to create modern agriculture.[29]

Having tasted the freedom of the thaw, having worked with western geneticists before Lysenko, and having attracted enthusiastic students from Moscow, Leningrad, and Kiev, the older generation of geneticists was unwilling to sit idly by when Khrushchev tried to close the institute three times in its first four years. Young and old alike came together. Many entered the Communist Party to defend genetics from within.[30] The institute's party committee expanded rapidly. Having had only a dozen or so members at the end of 1958, by July 1959 twenty-five members gathered to discuss the institute's research program in light of Dubinin's recent ouster. By 1964 there were forty-five Communists in the institute's party organization, and by mid-1965 there were sixty. Only one member of the institute, B. A. Lipskii, consistently defended Lysenko before the institute's scientific collective.

Still, Dubinin felt the need to defend himself personally against Khrushchev's charges. In comments before the institute's party committee, he tried to persuade the members that pursuing the biochemical basis of genetic information promised results as significant as those that nuclear physics had for nuclear energy. He acknowledged the party's correct line in calling for a closer tie between theory and practice. "However," he said, "this does not mean we

should forget the significance of theory, which illuminates the path of practice."[31] Indeed the institute's research program was vital in several areas of
practical importance—hybrid corn, triploids sugar beets, animal hybrids with
higher milk fat or meat production, Beliaev's work on minks and other fur-
bearing animals, and so on.[32]

In an effort to head off his removal as director of the Institute of Cytology
and Genetics, and to defend genetics in general, Dubinin wrote Khrushchev
directly sometime in 1959. Dubinin praised Khrushchev for his role in creating
the Siberian division and of course the Institute of Cytology and Genetics, and
promised to deliver great achievements in agriculture and medicine on the
basis of the institute's research on cell structure, DNA, radiation genetics, and
mutagenesis. He begged his indulgence, and a little more time to solve pressing
scientific problems. The first of these was the need to expand the institute's
research facilities. He then described the great difficulties in printing research
results since the journal *Cytology and Genetics*, which had long been approved
by the Siberian division, had yet to be published. (Only in 1965 would a journal—*Genetika* appear—and it was published in Moscow.) Dubinin concluded
the letter by noting the institute's "Marxist-Leninist spirit" which was directed
toward solving the greatest problems of science and fulfilling the seven-year
plan. And then he proposed that Khrushchev meet with him personally.[33]
Khrushchev was unmoved.

Worse still, the regional party committee had grown increasingly envious of
Akademgorodok's material position and the personal ties between Lavrentev
and Khrushchev. The secretary of the Novosibirsk Obkom once confiscated a
huge hundred-car freight train bound for Akademgorodok. Lavrentev was able
to win it back after simply telling him, "We need those things." Regional party
officials tried to undermine Lavrentev's authority—and perhaps the city of science itself—by writing Khrushchev about the "Weissmanist-Morganists" in
the institute. Lavrentev implored Khrushchev to keep the genetics institute
open, but Khrushchev would have no part of it. Their relationship temporarily
soured. "Don't even bring up the notion of family," Khrushchev told him. On
his way back from China in 1959, Khrushchev stopped in Akademgorodok
and ordered Lavrentev to close the institute or remove Dubinin.[34]

Ultimately Khrushchev became inflamed by his hatred of Dubinin and genetics. Lysenko's allies in the Ministry of Agriculture put Khrushchev up to
publicly denouncing Dubinin, and Lavrentev had no choice but to fire him. To
save the institute, he asked Dubinin to resign for "personal reasons" in October
1959. Beliaev took over, for which Dubinin never forgave him. He believed that
he alone could direct the institute and that after his departure it should be shut
down. He actually began to criticize both the institute and Beliaev.[35]

Even after Dubinin's ouster, commissions put up by Lysenkoists passed
through Akademgorodok to examine Siberian genetics in microscopic detail.
The presidium of the Academy of Sciences sent an investigative commission to
Akademgorodok in September 1961 to familiarize itself "with the state of sci-

entific research in the institute and prospects for its development." The commission found the status of the institute beyond reproach in terms of its research program and cadres, but pointedly indicated serious problems that slowed the work. There was a need "to force the pace of construction." The absence of a vivarium and greenhouse inhibited work in biochemical virology, cancer research, and radiation genetics. The unusually small southern experimental plot precluded an accurate evaluation of results. Nothing was in order regarding materials and equipment. In particular, radioactive isotopes for experimental mutagenesis were in short supply.

Yet the commission determined that the future offered limitless possibilities in all areas of research, from work in the laboratories of nucleic acids (R. I. Salganik), cytology (I. I. Kiknadze), cell structure (N. B. Khristoliubova), and radiation genetics (Iu. Ia. Kerkis), to the laboratories of polyploidy (A. N. Lutkov), hybrid vigor (Iu. P. Miriuta), genetics of selection (V. B. Enken), and experimental mutagenesis (P. K. Shkvarnikov). All aspects of the physical, chemical, and structural bases of inheritance were covered. In addition, the institute had established all the appropriate scientific contacts, in some cases including financial remuneration, with a number of state and collective farms, as well as with numerous other major scientific institutes. The commission concluded that the institute's work proceeded "on a contemporary scientific level on the basis of dialectical materialism" and that it could solve all problems placed before it "with the newest achievements of physics, chemistry, and mathematics."[36]

After Lysenko was dismissed from the Academy of Sciences, Dubinin was elected academician, became chairman of a new Academy council on problems of genetics and selection, and director of the Institute of General Genetics, which previously had been Lysenko's home. Dubinin set out to create his own legend, which earned him the enmity of all but his closest associates. His autobiography, *Perpetual Motion* (called by some *Perpetual Self-Promotion*), is filled with myths contributing to this personal hagiography and totally ignores the contributions of others. Dubinin had earned the respect of Soviet scientists with his constant defense of genetics. But after he returned to Moscow from Akademgorodok, after the rehabilitation of genetics, and especially after the appearance of his autobiography, he lost all the goodwill and respect he once commanded and was now viewed as merely self-serving and self-aggrandizing. His behavior led the presidium of the Academy and its biology division to censure him and remove him as head of the Academy's Genetics Council. The request of many of his staffers to transfer to Astaurov's Institute of Plant Selection was granted.[37]

At least Dubinin left a joke behind when he returned to Moscow: "What's the difference between Dubinin and Lysenko?" they ask in Akademgorodok.

"That's easy," the answer goes. "According to Dubinin, the proof of genetics is when the son looks like the father, while according to Lysenko, the proof of vernalization is when the son looks like the neighbor."

"NAPOLEON" TO THE RESCUE

"We are building animals," Beliaev once declared. His appointment as director of the Institute of Cytology and Genetics after Dubinin's ouster rescued it from certain collapse. Beliaev was only a candidate of science with a strong experimental leaning, a capable researcher but not an outstanding scholar, a veteran of World War II but not a party man. He was an innocuous choice from the younger generation, when the appointment of an old-school geneticist would have meant more problems for nascent Siberian genetics. As it was, the institute's precarious position—the absence of rudimentary equipment or even its own building until 1963—made Beliaev's job a thankless task. But he was a capable leader, a Napoleon who tolerated little administrative dissent, and a rousing public speaker who was accused of "hypnotizing" his audience. He knew how to pose questions and provide penetrating analysis from a broad knowledge of fields seemingly far from his own.

He was, moreover, convinced of the power of his science to improve on nature's gifts: "Utilizing methods of experimental mutagenesis, the plant and animal breeder does not wait for the kindness of nature. He actively directs the heredity of organisms with the help of external influences." He also operated from a reductionist perspective. For him all biological concepts could be explained in physical or molecular terms. He believed that biology had become an exact science like physics, chemistry, and mathematics, once the unit of measurement of biological phenomena, the gene, had been established. "Genes," he wrote, "are our foundation, and we are now talking about the building which is erected on that foundation, about the productive process, about its technology. Indeed, the technology is so simple as to be elementary."[38]

Dmitrii Konstantinovich Beliaev was born on July 17, 1917, in Kostroma County, east of Moscow. His father was a village cleric, his mother, the village "favorite" who offered advice and provided medicinal cures. The family maintained a library, a rarity in the Russian village, which reflected its respect for knowledge and culture. Three talented biologists grew up in this environment: Dmitrii himself; a brother Pavel, an agronomist; and a brother Nikolai, a geneticist who perished in the purges. Beliaev grew up around animals, and his love for them continued throughout his career.[39]

At ten he was sent to Moscow to live with his brother Nikolai while attending school. Nikolai worked in the laboratory of S. S. Chetverikov, founder of population thinking in Russia, in Koltsov's Institute of Experimental Biology. The esteemed figures of Soviet genetics passed through the doors of this institute: B. L. Astaurov, P. F. Rokitskii, S. M. Gershenzon, D. D. Romashov, and N. V. Timofeeff-Ressovsky. Contact with the Chetverikov seminar known as SOORY (Sovmestnoe oranie s zhestko ogranichennym chislom uchastnikov i neogranichennymi ser'ezneishimi diskussiiami, or Joint Oration with a Sharply

Limited Number of Participants and an Unlimited Number of Serious Discussions, also known as "Droz Soor," the Drosophilist Screeching Society), summer expeditions, and the availability of literature attracted the younger Beliaev to a career in biology. Unfortunately his contact with his brother was curtailed when, on Koltsov's advice, Nikolai left to work in a Tashkent laboratory in 1929.[40]

Beliaev worked for a few years as a lathe operator in the Moszherez wagon-repair factory. His plans to study biology in a university were thwarted by entrance requirements that favored individuals from other social strata—workers and peasants, not village intelligentsia. Even though he faced virtually no competition for entering a university's biology department, he was forced to matriculate at the Ivanov Agricultural Institute from which he graduated in 1938. Here he was fortunate to study with B. N. Vasin, a student of A. S. Serebrovskii. He found employment in the Central Scientific Research Laboratory of Fur-bearing Animal Husbandry of the Ministry of Foreign Trade. The first tentative steps of the industrialization of fur-bearing animal husbandry occurred in the late 1920s when twenty farms were organized. Beliaev joined one of them in Tobolsk as a junior scientist. Here he completed his first scientific works on the influence on mutability and inheritance of the selection of such polygenic traits as the intensity of silver in the fur of silver and black foxes. For the rest of his life he focused on the biology of selection and domestication of animals. World War II interrupted Beliaev's career. In 1941 he was called up to serve at the Kalinin front. Wounded twice, he demonstrated qualities of leadership, courage under fire, and initiative and rapidly moved through the ranks, becoming a major by the end of the war.[41]

At the request of the minister of foreign trade, A. I. Mikoyan, Beliaev was demobilized in order to return to his research on fur-bearing animal husbandry in an effort to rebuild this export industry for needed hard currency. Beliaev's work initially fell under the direction of the main administration of animal husbandry (Glavzverovod). The close association between applied and fundamental research in this area of cytogenetics, which was hard to duplicate in the Academy of Sciences or VASKhNIL because of Lysenko, enabled researchers to achieve rapidly a manyfold increase in the number of animals and the production of fur. At this time Beliaev also taught a course at the Moscow Fur-bearing Institute in a department chaired by P. F. Rokitskii. This work led to his candidate dissertation on "The Mutability and Inheritance of the Silveriness of Fur of Silver-black Foxes," which he defended in July 1946.[42]

He was then appointed head of a department of breeding at the All-Union Scientific Research Laboratory of Fur-bearing Animals and Reindeer Husbandry of the Ministry of Foreign Trade. At the beginning of 1948 Beliaev published an important article in a major journal on the genetics of minks. Even though the article was sharply criticized at the Lysenkoist VASKhNIL conference, it served as the basis of recommendations for the selection of minks. After all, true science was required to generate hard currency in order to rebuild the economy that had been destroyed by the Nazi invasion. However, at a special session of his laboratory council, attended by workers from the Department of Animal Husbandry of the Moscow Fur-bearing Institute, representatives of Glavzverovod, and a number of biologists, Beliaev was found to be under the influence of "Mendelian-Morganists." He lost his positions as department head and teacher, his salary was cut in half, and his assistants were removed. He was, however, allowed to keep his laboratory.

This experience at once isolated Beliaev and drew him into close contact with other biologists who had suffered at Lysenko's hands—B. L. Astaurov, V. V. Sakharov, P. F. Rokitskii, E. T. Vasina, and others—and he quickly found a place among them as one of the "lost generation," a young geneticist among the veterans of pre-Lysenkoist days. "This was a tragic period in biology," Beliaev recalled at a philosophical seminar of the Siberian division in March 1965, "for, as Astaurov correctly remarked, all of biology was enveloped by an earthquake, and only the epicenter was located in the area of genetics."[43] Coming after the purge of his brother in 1937, this experience could only increase Beliaev's antipathy both for Stalin and Lysenko, and he would never join the Communist Party. Despite that decision, his relationship with local party authorities was unique. Although refusing to join the party, he never hesitated to turn to the Obkom to serve the institute's interests or to defend the Soviet social order.

Throughout the 1950s he continued work in various animal-husbandry state farms in the Baltic, Altai, Far East, and Moscow, in spite of having been censured. Stalin's death triggered the first grudging steps toward the resurrection of genetics. In 1954 Beliaev presented a report to the agricultural department of the Central Committee on prospects for the development of the fur-bearing animal industry. In this empirical study, he suggested several ways to intensify the industry. He showed how minks could be bred so as to produce desirable recessive genes, in this case fur color. In papers he presented in 1956 and 1958 he offered further evidence of the place of genetics in animal husbandry over the protestations of M. A. Olshanskii, the Lysenkoist president of VASKhNIL, and O. B. Lepeshinskaia, the Stalinist pseudobiologist who claimed to have created life. In this period several genetics laboratories opened under the umbrella of nuclear physics.[44]

Beliaev's efforts in his field never wavered. In 1958 he prepared a cycle of lectures on the genetics of foxes, which were published as a classified brochure. He served one year as department head in another organization established for yet another furry creature—the jet-age sounding NIIPZiK, or the Scientific Research Institute of Fur-bearing Animal Husbandry and Rabbit Husbandry—before he was called to Akademgorodok. He accepted Dubinin's invitation to organize an animal genetics laboratory (later called the evolutionary genetics laboratory) and served as deputy director for science at the Institute of Cytology and Genetics.

THE INSTITUTE OF CYTOLOGY AND GENETICS:
PERSONALITIES AND GROWING PAINS

The Institute of Cytology and Genetics grew in fits and starts amid Lysenkoist attacks, difficulties with its directorship, and miserable working conditions. Because the geneticists would not even have their own building for another four years, they had to start from scratch in all fields. With or without Dubinin, the scientists faced open hostility from Khrushchev and the Lysenkoists, making it nearly impossible to move the institute ahead on the right foot. An interim director, Iurii Iakovlevich Kerkis, although an improvement over Dubinin, managed to alienate his coworkers with his heavy-handed managerial style. Beliaev, director from 1961 until his death in 1985, capably steered the foundering institute through political, financial, and scientific uncertainties. Yet he, too, nearly failed to save the biological sciences in Akademgorodok in the Khrushchev years.

No sooner had Khrushchev been deposed than Beliaev joined with other leading biologists to publish popular scientific articles touting the glories of modern genetics. His first major article appeared in *Pravda*, in which he acquainted readers with DNA, indicated the importance of genetics for increasing agricultural yields, and castigated VASKhNIL for arbitrarily dismissing the science of genetics and causing it to lag in the Soviet Union. He demanded that a genetics journal be published and that Soviet genetics be reintegrated with western science.[45]

Beliaev's capable direction had a negative, domineering side. Hoping that the party would leave his institute alone, he expected uncritical acceptance of Soviet dogma from his staff, demanding they simply do their work and keep quiet. Many of the geneticists, themselves survivors who had been forced into exile, were unwilling to sit idly by in the face of Beliaev's oppressive administrative style.

For example, in January 1964 at a meeting of the institute's Academic Council, Zoia Sofronievna Nikoro, head of the laboratory of cattle selection, condemned party agricultural policies. A member of the older generation, Nikoro was born in Petersburg in 1904 and spent her childhood in Bessarabia (later a part of Romania, and then Moldavia, conquered by Soviet armies during World War II). She returned to the USSR after the revolution to attend the Leningrad Agricultural Institute. She then worked with Chetverikov and Romashov in genetics, taught widely, did research at various experimental stations, and finally became chairwoman of the biology department at Gorky University, where she was one of the first Soviet candidates of biological science. She kept a low profile during the Lysenko years but gladly transferred to Akademgorodok in 1958, where she commenced research on animal breeding, headed up two laboratories, and attracted a large number of graduate students. While waiting for her transfer papers to come through, Nikoro spent the summer of 1958 in Kharkov at the Iurev Institute of Plant Breeding where V. K. Shumnyi, current director of the institute, had just commenced research on hybridization of corn and other vegetables,[46] and he decided to move to Siberia as well.

Nikoro maintained that party policies had endorsed unscientific, Lysenkoist fodder, feed, and breeding techniques and were therefore at the root of production lags, underweight animals, weakened stock, and poor feeding practices. In response to a motion to create communist labor brigades in the institute to improve the agricultural situation, Nikoro angrily accused the scientists of contributing to this situation. She said they were "unworthy" of "communist labor brigades" since they had kept silent during the party's endorsement of ignorant agricultural practices that had brought about "hunger in the land." Beliaev tried to silence her, and Iu. P. Miriuta, who still believed that many of those dismissed in 1948 were in fact guilty of state crimes, called her "a wrecker and immoral." Three days later, at Beliaev's instigation, the institute's party bureau censured Nikoro for demagoguery and for wrongly accusing the collective of "compromising relations toward liquidation of fallow land, plowing up meadows, and other mistakes in soil science."[47] In this dogmatic environment it would be decades before the damage done by seventy years of faulty agricultural policies, thirty of them under Lysenko, could be undone.

Another example of Beliaev's heavy-handed administration of the institute centered on his stormy relationship with Raissa Lvovna Berg, although in this case some found Berg equally at blame. Berg, enfant terrible of the genetics community and world-renowned specialist in population studies, was invited to Novosibirsk in 1963 by Kerkis and Beliaev to head the laboratory of population genetics. Once there she quickly alienated Beliaev by refusing to sit in on his seminar on Marxism-Leninism with its discussion of Engels's *Dialectics of Nature*. She organized an alternative one steeped in the language of cybernetics: "Control Mechanisms on Various Levels of the Organization of Life." It was well attended, whereas Beliaev's seminar died a quick death.[48] Sharp-tongued, Berg on occasion managed to offend even her friends.

When the party ultimately cracked down on Akademgorodok, Berg and Beliaev played prominent roles in the drama. Berg, a dissident, offended the entire party apparatus by signing the letter of the podpisanty, for which Beliaev fired her after a stormy special session of the institute's Academic Council. (She is now a research professor at the University of Missouri in St. Louis.) As is discussed at length in chapter 7, the so-called podpisanty affair marked the beginning of the end of Akademgorodok's glory days.

A final illustration of Beliaev's Napoleonic side was his unwillingness to see rival centers of biological research established near Akademgorodok, and even to delay their founding. In this he had Lavrentev's assistance. V. I. Zhadin's Institute of Hydrobiology was pushed off into the outreaches of Siberia; the Institute of the Biology of Lakes was founded, but in Magadan and only in 1969; the Institute of Biological Problems of the North lagged in development; and only with difficulty was the Institute of Soil Science and Agronomical Chemistry founded in Novosibirsk in 1968. The environmental sciences were sent as far as Vladivostok, Irkutsk, and Krasnoiarsk where the Sukachev Institute of Forestry and Wood Products was relocated from Moscow in 1958. Finally, the Institute of Biophysics in Krasnoiarsk, established to study the ocean's economic potential and test long-term isolation in preparation for trips to Mars, was founded only in 1981.

Although Beliaev's personality found detractors, his scientific leadership was just what the institute needed. He took over where Dubinin and Lavrentev left off, inviting friends and associates to join the staff. Over the next few months the institute welcomed talented specialists like I. I. Kerkis, zoologist and specialist in sheep-breeding; V. V. Khostova, geneticist; Z. S. Nikoro, a student of Serebrovskii; P. K. Shkvarnikov, a specialist in mutagenesis; and R. I. Salganik, a molecular biologist. Beyond these talents a great gap remained to be filled. According to a contemporary, so few students had any genetics background that teaching genetics, statistics, cytology, and molecular biology resembled the anti-illiteracy campaigns of the 1920s.[49]

Two scholars who played a role in training these aspiring geneticists were Petr Shkvarnikov and Iurii Kerkis, men of quite different temperaments who knew each other from the 1930s and shared party membership and deputy directorship of the Institute of Cytology and Genetics. Shkvarnikov was born in 1906 in a peasant village of Kiev Province. After graduating from the Plant Breeding Institute in 1927, he entered the Ukrainian Genetics Selection Institute in Odessa, beginning a life-long career in experimental mutagenesis. From 1930 to 1936 he worked in the cytogenetics laboratory in the Timiriazev Biology Institute. He then transferred to the Institute of Genetics under Vavilov, where he met Kerkis and where he served as deputy director for science in 1939–40. After the war Shkvarnikov worked in the Institute of Cytology, Histology, and Embryology of the Academy of Sciences. After the Lysenko conference in 1948 he was forced to move to the Simferopol Crimean Filial of the botany division of the Ukrainian Academy.[50] Shkvarnikov applied X-rays, chemicals, and slow and fast neutrons to produce mutations for new strains of wheat and beets. He is the co-inventor of the spring wheat Novosibirsk-67 which was to be introduced in 1967 to celebrate the fiftieth anniversary of the Revolution but was planted widely only in 1972.

In 1957 Dubinin called Shkvarnikov and after only a brief conversation appointed him deputy director of organization, pleased to have a devoted Communist in such a prominent position. Shkvarnikov arrived in Novosibirsk early in 1958. He worked out of three rooms at 20 Soviet Street with a few colleagues. This group remained in Novosibirsk for eighteen months, before moving to a series of temporary quarters in Akademgorodok. His major focus at the time, Shkvarnikov recalled, was merely "bringing the institute up to strength." He cited weaknesses in the chain of command as a major failing. At a party meeting in 1962 he criticized his colleagues' inattention. For example, the laboratory of polyploidism, in his view, ranked first in importance, yet its head, A. N. Lutkov, remained in Moscow, leaving no one in charge. (In fairness to Lutkov, it should be noted that he was merely waiting for the institute to equip his laboratory, as well as provide him with a greenhouse, before he made the trek back to Akademgorodok. Ultimately he contributed to the institute's work on triploid hybrids of sugar beets. Unfortunately, although the triploid sugar beets had 15–18 percent more sugar, they were infertile, and their planting always lagged behind plans.[51]

Much of the institute's work in introducing its new techniques and crops lagged owing to continual problems with its experimental stations. Shkvarnikov's laboratory also suffered from the absence of an experimental plot for new crops and a lack of clear plans. Of course Khrushchev's failure to give the geneticists carte blanche in their work added to the institute's unsatisfactory state. It did not help that K. P. Anufriev, in league with the director of an experimental station, was found to have given away nearly three tons of the institute's hay to his wife's uncle. But the geneticists would not be denied. The institute party committee called for scientists "to strengthen ties with breeding and artificial insemination stations in the Novosibirsk region" and "to mobilize the party organization and the entire collective for the organization of an experimental state farm of the Siberian division." They demanded better instrumentation to measure protein, starch, sugar, and fat in the new agricultural products. Slowly but steadily the institute progressed in its research on hybridization, selection, and polyploidy: Shkvarnikov on radiation selection of barley, grain, rye, and potatoes; Miriuta on corn; V. B. Enken and Shkvarnikov on disease- and frost-resistant wheat. But the experimental farms would come into their own only in the late 1970s.[52] Shkvarnikov, having grown weary of his organizational responsibilities, gladly departed for his homeland in Kiev in 1966. Once there, genetics having been reestablished, he headed up the Department of Experimental Mutagenesis of the Biology Institute of the Ukrainian Academy of Sciences, where he founded and edited the first post-Lysenko Ukrainian genetics journal, *Tsitologiia i genetika.*

Like Shkvarnikov, Iurii Kerkis (1917–1977) steered his way through the minefield of Soviet genetics, moving from the center of activity in Moscow in the 1930s to an isolated sheep-breeding farm in Tadzhikistan, before finally settling in Akademgorodok in the late 1950s. He was a student of Iu. A. Filipchenko and a close associate of Nikolai Vavilov in the Institute of Genetics. He was influenced by the work of Theodosius Dobzhansky, the Russian-American founder of evolutionary genetics who left Leningrad for the United States in 1927 on a Rockefeller International Education Board fellowship. Kerkis studied with noted geneticist Herman Muller, who had fled Nazi Germany to work in the institute from 1933 to 1937. In 1937, just twenty years old, Kerkis was keenly aware of the danger geneticists faced from the Lysenkoists. In that year hundreds of outstanding scientists perished in the purges, including geneticists G. A. Levitskii, G. D. Karpechenko, and G. A. Nadson who may have produced mutations in flies with X-rays somewhat earlier than Muller did.

Kerkis, a headstrong, impetuous, and talented young man, chose at this time to disprove a major Lysenkoist theory. He demonstrated that experiments Lysenko conducted to show that grafting one plant onto another resulted in the inheritance of acquired characteristics were done without controls at best and fudged at worst. In his experiments Kerkis worked with two kinds of tomatoes. In all he completed 377 grafts, including grafts from one kind of tomato to the other as well as to the same kind of tomato. He produced 1,152 fruits that grew from grafts and 1,774 that grew from ungrafted plants. His analysis of the offspring showed that any divergence was morphogenetic with no inheritance occurring. Kerkis offered photographs as proof. In 1948 he was shocked to see reproductions of his photographs in Glushchenko's *Vegetative Hybridization* (Moscow, 1948). The caption under the photos, a gift to Glushchenko from Kerkis, read: "Plants of the fifth generation from the graft of the Humbert tomato onto the blue eggplant delicacy."[53]

The Lysenkoists may not have noticed the article in which Kerkis published his results. But they did not ignore his next refutation of their science. In 1938 Lysenko was determined to disprove Mendelian laws. A scientist in the Genetics Selection Institute in Odessa, N. I. Ermolaeva, was given responsibility to repeat Mendel's bean experiments. Counting such traits of pea seeds as yellow versus green, smooth versus wrinkled, and so on, Mendel determined that the ratio between dominant and recessive traits in the second generation was about 3:1. Mendel's laws were shown to apply universally to many characteristics in most organisms. Ermolaeva confirmed the basic law of division, meiosis, and the manifestation in the second generation of both dominant and recessive traits, which are hidden in the first generation. But then, following the mistaken views of Lysenko, she focused on the concrete ratios in progeny of two groups of plants, the carriers of antagonistic traits. She tried to show that the ratio varied "during sowing and in crossbreeding in different periods," so that the ratio of 3:1 did not obtain. Ermolaeva contended that inheritance of acquired characteristics had occurred since the ratio varied, disproving the firm mathematical regularity of fixed genomic heredity.

Kerkis used experiments on drosophila to verify the laws of meiosis and proved that the occasional divergence from the Mendelian ratio 3:1 is explained by a number of factors, including the vitality of the class of individuals and the stability or mutability of normal genes. In an article written with the budding cybernetician Liapunov, he offered the mathematical rational for this conclusion. Using the correct variational-statistical analysis, the famous mathematician A. N. Kolmogorov, who in 1938 had personally defended Kerkis against charges of "Weissmanism," showed that Ermolaeva's work in fact fully supported Mendelian laws and the 3:1 ratio, and saw to it—with great difficulty—that the Kerkis-Liapunov article was published.[54]

As was the rule at this time, Lysenko still had his way. Kerkis was forced to give up his experiments and was fired from the Institute of Genetics on the trumped-up charge of having left work fifty minutes early. He had in fact left to meet with Liapunov to discuss their work. Kerkis then transferred to the Zoological Institute in Leningrad, was evacuated to Tadzhikistan when Leningrad was blockaded during World War II, and remained there until 1957 as director of animal husbandry at a state farm specializing in Hissarian sheep. Living on a state farm in the Tadzhik countryside under the watchful eye of the party was indeed a challenge for Kerkis being both an intellectual and a Jew. The Tadzhiks resented "great Russian nationalism" and took out their frustrations on Kerkis. He managed to stay one step ahead of them until May 1957, when, shortly after receiving an award for his devoted service to Tadzhik animal husbandry, he was fired. Friends in Moscow succeeded in gaining him invitations to become director of the Murmansk Biological Station or senior researcher in the laboratory of radiobiology at the All-Union Institute of Animal Husbandry in Podolsk. But Kerkis accepted the third invitation to come his way—to head the laboratory of radiation genetics in the Institute of Cytology and Genetics in Akademgorodok. For two months the Tadzhik authorities held up signing off on Kerkis's personnel papers. But on August 15, 1957, only two months after Akademgorodok was formally proclaimed, Kerkis was released. He flew to Novosibirsk at the end of the month. "I was the third employee of the Siberian division to live permanently in Novosibirsk," Kerkis recalled. "I flew in with Lavrentev, Khristianovich, and the construction director."[55] Kerkis, a specialist in entomology, population genetics, and radiation genetics, worked at the institute until his death in 1977. In addition to training dozens of students at the university, Kerkis worked with Beliaev in preparing one of the first middle-school textbooks produced in the USSR after the fall of Lysenko. On June 14, 1965, the biology and soil science department at Leningrad University awarded Kerkis his doctorate at long last for an illustrious career stretching nearly four decades.

The Institute without Walls Opens Its Doors

Amid personality conflicts and party intrigues, the Institute of Cytology and Genetics struggled to open. Three major problems persisted: a shortage of personnel, insufficient facilities, and inadequate equipment. It was a challenge to locate young talent. The physicists, who carried the most weight and authority, as well as the mathematicians, seemed to have no trouble attracting students. To their mind biology was a diversion. Besides, who would want to study genetics in the uncertain days of the Khrushchev era? The first students to arrive at the institute were the "foster children" of Moscow and Leningrad Universities, young biologists without even an elementary knowledge of classical genetics who would become today's laboratory and institute directors. They studied intensively with the Old Guard to achieve the level of a master's degree. In 1961 the geneticists established a biology department with a genetics division at Novosibirsk University. Beliaev directed the department for twenty-five years. New courses had to be created as there had been no new textbooks since 1948. Most foreign literature was inaccessible. There was no university laboratory. It was even hard to find drosophila. Fortunately the institute was able to acquire the personal library of Academician A. S. Serebrovskii. By the summer of 1964 Berg and Kerkis were each lecturing forty to fifty students. There was concern, however, that the quality of teaching was uneven and that some laboratory students showed little interest in science while others worked well but had only meager laboratory facilities. Only in 1967 were the first advanced seminars in evolutionary biology offered by V. V. Khvostova. Shortly thereafter, a modern genetics curriculum was introduced in the physics-mathematics boarding school where high school students received the fundamentals of Darwinism, cytology, genetics, and so on.[56]

In an effort to attract workers and defend their discipline, the scientists extended their teaching activities beyond the university. In the Novosibirsk area they offered short-term courses to the Siberian populace: Kerkis lectured on medical matters, Nikoro on animal husbandry, and Shumnyi on selection stations. Kerkis spearheaded the effort to reinvigorate the biological sciences through lectures, bulletin boards, radio, and the local papers. In 1961–62 institute scientists delivered sixty-four public lectures at local collective and state farms and at factories, and delivered four radio and three television addresses. Many of the lectures were given under the auspices of the Znanie (Knowledge) Society, an all-union organization devoted to the dissemination of politically correct popular science information.[57] This recalled the tradition of the first years of the revolution when leading scientists were encouraged by the Commissariat of the Enlightenment to offer public lectures to the illiterate masses in the ways of modern science and culture. The biologists' lectures supplemented a series of newly released popular science films about genetics, for example, *In the Depths of the Living* (V glubinakh zhivogo), a 1966 Leningrad release on molecular biology.

Having attracted students, there was still the problem of having a qualified staff of scientists. By September 1961, of the entire institute staff of 280, there was but 1 doctor of science, 31 candidates of science, of whom 21 were senior scientists and 10 junior scientists, 41 junior researchers without an advanced degree, and 83 laboratory workers—nearly a third of all employees—with only a secondary education.(But the staff did include as many as 37 Communist Party members.) Although 39 candidate degrees and 8 doctoral degrees were to be defended by 1965, the record through the mid-1960s remained abysmal. And none of the degrees could be awarded in genetics until after Lysenko's fall in 1965.[58]

Once VAK (the Higher Attestation Commission) again acknowledged, on the party's orders, that genetics was a science, scholars were allowed to receive candidate and doctor of science degrees in genetics for the first time since the 1948 VASKhNIL conference. Like other biology institutes in the USSR, the Academic Council of the Institute of Cytology and Genetics immediately voted to award doctor of science degrees to the institute's leading scientists (Beliaev, Shkvarnikov, Lutkov, and Kerkis) "without defense," that is, on the basis of illustrious careers interrupted by Lysenkoism.[59] (Once before in Soviet history, in 1935, scientists had been permitted en masse to receive higher degrees without defense, ending another seventeen-year period during which they were deprived of academic titles which had been considered a legacy of bourgeois culture.

After its shaky start, the institute grew steadily. Having begun with only about twenty people spread out in five rooms, by the summer of 1964 the institute had grown to 342 individuals working in seventeen laboratories, and its staff included 1 doctor and 34 candidates of science, 86 junior scientists, and 12 graduate students. By 1968 it had achieved worldwide renown and its staff had grown to 465, of whom 11 were doctors of science and 48 were candidates of science.[60]

Despite the steady increase of qualified scientists, inadequate government support for machinery, equipment, and facilities hindered efforts to begin research. On three occasions the institute was promised its own facilities only to see one of the other institutes placed ahead of it at the last moment. Until 1964 the Institute of Cytology and Genetics did not even have its own building and was forced to work out of a few rooms at 20 Soviet Street, and then occupy space at the institutes of geophysics, organic chemistry, thermal physics, and inorganic chemistry.

Finally Beliaev went on the offensive: one weekend staffers took over an empty facility that had been promised to the Computer Center. Having occupied the building in a bloodless coup, the scientists now had the thankless task of motivating Soviet laborers to complete the finishing work—wiring, painting, shelving, and racks—and to build a bunker for manure. Office space remained tight and ventilation was poor, causing many workers to get sick. Scientific instruments were at a premium—not only electron microscopes or phytotrons but even cameras. The vivarium and greenhouse remained unfinished. Centrifuges were broken. Repairs were a nightmare. Within five months storeroom employees broke 221 pliers, wire cutters, and other instruments, 298 drills, and 710 hacksaw blades. Fifty percent of the budget was being spent on equipment. Finances were precarious. Shortages of instruments, chemicals, radiation sources, and experimental facilities interfered with research on mutations and polyploidy. "Under these conditions we cannot work," reported the institute's party committee.[61]

More important, it was impossible overnight to rescue genetics from the backwardness brought about by two decades of Lysenkoism. Leading institute scientists criticized the work of many laboratories for "phenomenological results" that did little to clarify mechanisms, for lags in publication, for poor use of the institute's potential, and for failing to ensure that their few real advances with corn, sugar beets, or wheat were introduced in the field.[62]

The absence of agronomists, zoologists, and engineers, as well as lack of materials and money, hindered the opening of the institute's experimental plots: the Novosibirsk Experimental Field and the Ust Kamenogorsk Strong Point, several hundred miles south of Novosibirsk on the Irtysh River, not far from the nuclear polygon at Semipalitinsk. In May 1960 party members discussed efforts to help regional state farms organize experimental agriculture. Worse still, N. N. Getmanov, head of the experimental plot, pleased no one. Of six hundred acres of land, he succeeded in cultivating fewer than a hundred.

Some hinted that he had illegally sold the plot's allotted peat. Efforts to test genetically engineered products in the field lagged, while the attempt to increase agricultural productivity by applying fertilizer made up of ideological pronouncements went nowhere.[63]

It seemed that an escape from all these challenges might be at hand in the new agricultural policies adopted under Leonid Ilich Brezhnev. Although better known for pushing the military-industrial complex to achieve parity with the U.S. nuclear arsenal, Brezhnev's kinder and gentler side was seen in his food program. Always the sore spot of the Soviet economy, agriculture saw redoubled attention in the 1960s and 1970s, with agricultural investment growing faster under Leonid Ilich than any other sector. All agricultural organizations were instructed to take advantage of modern science and technology in order to control capricious nature and put it at the service of the Soviet state and society. The result would be larger harvests and less dependence on western agricultural imports. For ministries of water resource management and their research institutes, this meant efforts to build dams, irrigation streams, and canals to ship water wherever it was needed and thus turn arid land into lush gardens (see chapter 5). For the Ministry of Agriculture and VASKhNIL, this meant increased use of modern farm equipment and fertilizers to improve yields. For the USSR Academy of Sciences, this meant applied science in the service of agri-industry. And for the Institute of Cytology and Genetics, it meant that successful research would be judged largely by its contribution to agriculture. Fundamental research, which had suffered for decades under Lysenko, now lost favor to applied science.

The Industrialization of Agriculture

The first order of business for the research program of the Institute of Cytology and Genetics was to bring both methodology and experimentation into the twentieth century, a process that was readily accomplished. The institute quickly became the third most productive genetics institute in all of the USSR in terms of the number of refereed articles that appeared in the major Soviet journals. The second task was to mediate between the researchers in Siberia and their Lysenkoist exiles in Leningrad, Kiev, Crimea, and Tadzhikistan and to train the next generation of geneticists and cytologists, a process that had largely run its course by the early 1970s. The final goal was to ensure that the institute's research program meshed sufficiently with national policies so that fundamental research moved forward hand in hand with applied research, that is, the production of new plant hybrids and animal breeds needed to build socialist agriculture in Siberia. But just as agriculture had been a sore point for Khrushchev, so too was it for Brezhnev: several times he had to buy grain from the West with coveted hard currency. A substantial increase in agricultural investment failed to contribute to an equally substantial increase in the use of modern machinery and fertilizers, and the result was a 50 percent decline in agricultural productivity as measured per worker or per acre. Harvests shrank even in the most fertile area.

One reason for the decline in production was a poorly developed infrastructure—dirt roads, lack of refrigeration, and unreliable farm machinery—so that much of the harvest rotted in the fields or en route to the market. Moreover, collective farm workers had few incentives. Their pay was low and village stores offered little in the way of goods. In a word, there was an enormous disparity between quality of life in the countryside as compared to that in the city (as documented by Akademgorodok sociologist Tatiana Zaslavskaia; see chapter 6). Private plots were significantly more productive. There was little biologists could do without revolutionizing the entire agricultural organization, including allowing a modest degree of privatization. Yet it was their job to rectify the situation.[64]

By 1962, in spite of the institute's half-legal status, its scientists had made some strides in molecular biology, cytology and cytogenetics, immunology, theoretical mathematical biology, and radiation and physiological genetics. An annual report to the Academy of Sciences showed results in two main areas: the physical, chemical, and structural bases of life phenomena and inheritance, and control of inheritance in plants, animals, and microorganisms. In the first area, geneticists studied the mechanisms of the biological effects of ionizing radiation and mutagenesis with denatured DNA, the effects of nuclease on viruses, and the cytogenetic effects of small doses of radiation on organisms and skin. In the second area, researchers focused on the laws of mutation for a better understanding of hybridization, polyploidism, and the correlation between ecological and morphological factors in the mutability of plant culture (they earned six Soviet patents on sugar beets); artificial selection of traits; natural and artificial parthenogenesis in bees; the genetic nature of adaptive properties of animals; and cancer research. Yet space limitations impeded this work: the institute had but 14,400 square feet for all its personnel and laboratories.[65]

By the beginning of the next decade, however, the biologists had embarked on a number of new research areas, no longer limited by the strains of insufficient space or staff, although the supply of modern equipment continued to lag. By now the institute consisted of twenty-two laboratories, a selection station, physiology and chemistry departments, and a total of 694 workers, of whom 247 were scientists (including 131 candidates and doctors of science). The total area of Akademgorodok's facilities was 125,000 square feet (of which 23,000 was for the selection station and 11,300 for the workshop), nearly a tenfold increase from the previous decade. Research centered on nine areas: cytology; molecular genetics; radiation and chemical mutagenesis; cytogenetics; genetics of malignancies; genetic foundations of evolution and selection (of agricultural plants and animals); ecological physiology; mathematical biology; and the complex study of man, including the study of cardiovascular and lower gastrointestinal tract diseases.[66]

Generally scientists focused on the laws and mechanisms of heredity and on the application of modern molecular biology techniques to animal husbandry, selection, and hybridization. By combining the techniques of the remote hybridization and cultures of cells, tissues, and organs, researchers succeeded in obtaining unique intergenetic cereal hybrids such as barley-rye, barley-wheat, wheat-rye, and wheat-couchgrass. Novosibirsk-67 spring wheat was developed by radiative mutagenesis and is now planted on up to 7.5 million acres.[67] New varieties of winter fodder rye and wheat were developed for various regions of Siberia. In cooperation with VASKhNIL, scientists bred sheep that successfully combined high meat and wool productivity with good adaptation to the climatic conditions of West Siberia. And the eternally green tomato, grown in hot houses, found life.

Under institute deputy director Rudolf Salganik and others, nuclease antivirus preparations were produced for the treatment of some serious viruses in man, bees, and silkworms. (Those in man were herpetic keratitis, herpes, infectious mononucleosis, and viral encephalitis.) Salganik, who headed the laboratory of nucleic acids, came to Novosibirsk from Kiev at Dubinin's request. He had given underground lectures on genetics at the Moscow Physical Technical Institute at Dolgoprudnyi. He enjoyed the atmosphere of Akademgorodok where the elite of genetics had gathered. After Dubinin was sacked, Salganik reluctantly moved into administration. This did not interfere in his work on nuclease antivirus preparations, however, nor his commitment to a philosophical seminar to promote the dialectical materialist approach to genetics.

Under Ratner and Liapunov, mathematical biology also expanded at the institute and at the university in the 1970s, and, like sociology and economics, was reborn in Akademgorodok. The first graduate students were accepted at Novosibirsk University in 1970. By the end of the next decade, research on modeling and computer methods in molecular biology and genetics had attracted the attention of researchers in a variety of fields: computer analysis of nucleotide sequences; computer analysis and modeling of protein structure; molecular evolution theory; molecular-genetic systems modeling; models of domestication of animals; molecular-genetic and molecular-biological data banks, including program packages for data analysis; and new computer information techniques, including expert systems, artificial intelligence systems, and knowledge bases; creation of a new experimental model of genetically conditioned arterial hypertension; hereditary predilection to catalepsy in rats: and an experimental model of schizophrenia.[68] Two problems arose, however. First, Academician Baev, a conservative, older scholar and chairman of the Russian genome initiative, seemed not to understand the need for mathematical modeling. To this day the institute has difficulty acquiring genome funds. Second, the Akademgorodok Computer Center, responsible to all of the city's twenty-one institutes, was unable to fulfill all the needs of the genetics institute.[69]

Another institute goal tied to the Brezhnev food program was to develop new breeds of cattle with higher slaughter weight, milk production, milk fat, and protein content. Because most Siberian milk is turned into butter, the issue of how to increase the butterfat in milk assumed great importance. The fat content of the milk of cattle in Siberia was generally low because of poorly thought-out crossbreeding carried out under Lysenko. In the early 1960s, in an effort to stem the attacks on his authority, Lysenko sought to raise milk production throughout the USSR in terms of overall yield and fat content by crossbreeding purebred Jersey bulls, obtained at high cost from the West, and other breeds such as East Frisian, Kostroma, and Kholmogory. Jersey cows have a very high butterfat content in their milk, often 5 to 6 percent, but their yield is significantly lower than many other breeds. The goal of crossbreeding under the circumstances between, say, Jerseys and any of the others would be to increase production with little loss in butterfat content. Knowledgeable animal husbandry specialists, armed with artificial insemination techniques, can breed such animals under controlled conditions.[70]

The problem is that although the offspring may have the desired productive qualities, their breeding value is diluted. Their introduction into pedigreed herds will destroy traits produced over hundreds of years of breeding. In this case, high butterfat milk lines would be weakened and eventually bred out. Lysenko hoped to counter the second generation deleterious effects by selecting cows of large stature, "feeding them copiously during gestation [so as to] force the embryo to maintain the desired high butterfat capabilities." But once again this was stressing environment over genes and was doomed to fail. Bulls in this line were used widely in state and collective farms, leading to a disastrous situation throughout the USSR, including Siberia.[71]

Beliaev's solution to this and other problems was to "industrialize" animal husbandry. Strains of beef and dairy cattle had to be upgraded through the management of stock bases and increased numbers of purebreds. This would be accomplished by creating massive breeding collective farms, factories, and artificial insemination stations. Examples included the crossbred sheep from Siberian genetic hybrids and a crossbreed of pigs with wild boars. (The publicity surrounding efforts to breed new animals generated curious queries from the masses. One group of kolkhozniks, experimenting with a new method of feeding pigs, were concerned that their ever larger animals might weigh too much for the pigs' skeletons. So they sent a skeleton to the Laboratory of Strength of Materials of the Railroad Institute in Novosibirsk for a series of tests on twisting, tension, and rupture strength of the bones.[72]) Beliaev was certain that scientific techniques, applied in the outdoor laboratories of the experimental stations, would play an even greater role in animal husbandry than hybridization of plants in experimental plots. The experimental stations would provide controlled conditions of feed and progenitors.[73] The result was the long-awaited "industrialization of animal husbandry."

Taking advantage of increased funding provided through the Brezhnev food program and agricultural investment, the institute established several new huge experimental agricultural stations and new animal preserves in the Buriat and Altai regions. These stations were up to 200,000 acres in size and were stocked with large numbers of animals. In January 1980, in an effort to spur the growth of Siberian agriculture, the Altai Experimental Station was founded at the Cherginskii Sovkhoz, which proved to be a model environment for the scientific activity of the institute. It had forests, mountains to 10,000 feet, and steppe, with moderate temperature and precipitation by Siberian standards. Efforts would focus on cattle, as well as zebu, bison, yaks and sheep, Iakut horses, wild goats, deer, and wild birds.[74] Beliaev hoped to expand research on cows, pigs, and sheep, on minks and foxes, and on the domestication of rare wild animals of Siberia to save them from industrial encroachment.[75]

The industrialization of animal husbandry was central to the burgeoning "Siberia program," which saw extensive development of all areas of the Siberian economy by the turn of the century. As chairman of the program's committee on "biological resources of agricultural production," Beliaev believed its biological research should focus on the study of those Siberian soils, plants, and animals with a potential for economic exploitation. Soil science, genetic engineering, hybridization, animal husbandry, and training of scientists would then serve as the foundation of modern Soviet agribusiness. With regard to soil science, research was directed toward increasing the fertility of soils, bringing highly alkaline Siberian land into the agricultural system, and taking measures to stem erosion and restore the land.

The Siberia program's second major task was to preserve the gene pool of both wild and cultivated forms of indigenous Siberian animals and plants. Beliaev recognized that three decades of Lysenkoism had set these efforts back a hundred years. He also knew that highly productive breeds of animals were more susceptible to illness and had a difficult time in the Siberian climate. But he was certain that the preservation and manipulation of the gene pool by the traditional method of hybridization and the more modern method of genetic engineering based on radiation and chemical mutagenesis could lead to brilliant results. He voiced deep concern about the encroachment of industrialization and water melioration projects on Siberia's natural resources. Like many Siberian scientists, he became an environmentalist of sorts. He hoped to see a large number of parks set aside to preserve the gene pool. Yet Beliaev and his ilk never hesitated to apply their science to "engineer" nature for human purposes. He never questioned the potential negative economic or social impact of modern agricultural techniques. All his methods, to be successful, required not just mechanization of agriculture but of animal husbandry, increasing the productivity of animals and plants, and driving animals, plants, and soils ever harder with the liberal application of antibiotics, hormones, fertilizers, and pesticides. Unfortunately the work of the Altai Experimental Station, in which Beliaev and the other applied biologists had set great hope, was hindered, like all things Soviet, by the system's inability to allocate adequate resources once the station had been approved. There was money for investment but not for maintenance or operation. Researchers often fed the animals bread as it was both more widely available and cheaper than feed. In addition, the station required 200 to 250 full-time scientific personnel for

maximum operation, which meant providing adequate housing and social services. As usual, little was done to make life comfortable in the countryside, and
attracting workers became increasingly difficult. Beliaev warned the authorities
that time was of the essence to ensure that experimental farms in the Altai
became a unique genome fund of cultivated and wild forms of plants and animals in the 1990s.[76]

WILL THERE BE GREEN TOMATOES IN THE 1990S?

On the eve of the next millennium, the research program of the Institute of
Cytology and Genetics stands poised to be among the leading biology centers
in Russia and Ukraine. The experimental basis for studying animal genetics is
found in a vivarium with its own nursery of more than 21,500 square feet. The
institute has twenty-five thousand mice, ten thousand laboratory rats (inbred
strains), fifteen hundred wild rats, five hundred field mice, and the largest
collection of drosophila in the country, as well as several large-scale experimental plots for applied agriculture, animal breeding, fur farming, and experimental feed and seed production.[77]

The institute faces an uncertain future, however, largely because of the political and economic crises Russia now faces. Russian agriculturists and Russian
biologists were caught off guard by the rapid disintegration of the Soviet
Union, and it is questionable whether they can work together with limited
resources to survive the ongoing crisis. As current director V. K. Shumnyi is
only too aware, the institute must provide salaries for a staff that has grown to
nearly a thousand individuals. [78] Yet like most Academy centers, it has had to
forego paying salaries for two to three months at a time and has virtually no
funds for research. Brain drain is also a danger. Nearly a hundred leading scientists currently work abroad, and although Shumnyi expects all of them to return, other employees are not so certain of that. As dean of the biology department at the university, Shumnyi feels confident that the institute will identify
ten to twelve students every year for research and graduate work in the institute. Students, however, are turning to business in droves, depriving the institute of its main source of personnel. The Moscow-based Institutes of Molecular
Biology, of the Gene, of Molecular Genetics, and of Developmental Biology are
seen by many western scientists as more promising centers of fundamental
research in the 1990s.

Institute directors have embraced a threefold philosophy to guide them
through the fallow 1990s. The first concerns the integration of all levels of
research to avoid specialization. Second, the institute's research will be integrated along a spectrum from genetics to population processes to evolutionary
processes. Work has currently begun to include the study of very small groups
of indigenous peoples of Siberia.[79]

The third aspect involves renewed efforts to turn scientific advances into
salable products. Shumnyi would prefer that his staff not be exhorted to adhere

to practical applications as they were under Communist Party direction, but rather allow the fruits of science to find their way into the production process naturally. He also recognizes that this is impossible today. In times of budget shortfall there is even greater accountability to the state. Indeed it is a fait accompli of applied research that pressure from small cooperatives and businesses which have sprung up around research institutes will garner the interest of potential funders while fundamental research suffers. An additional problem is that the institute's best young researchers are striking out on their own in new profit-making enterprises. On the other hand, Shumnyi recognizes the funds generated through contracts in areas vital to the health of the institute.[80]

The institute currently holds promise in a number of areas of biotechnology with direct applications in agribusiness, medicine, biotechnology, and molecular genetics. Its five major areas of research are molecular genetics, cytogenetics, plant genetics, animal genetics, and physiological genetics. Institute scientists' main efforts center on investigating the structure and function of the genome and understanding and elaborating on the laws of evolution and selection. In the Department of Molecular Genetics researchers study the mechanisms and control of gene expression, including molecular aspects of genomic recombination, transcription, and mutagenesis.[81] In cytogenetics, researchers apply methods widely used in all institute laboratories to solve problems of plant and animal genetics. In the Department of Cell Biology, work is directed on the organization and expression of tissue-specific genes and the organization and evolution of the genome of chironomids and drosophila. In the mutagenesis laboratory, scientists focus on abnormalities in the behavior of chromosomes in meiosis.[82]

Rudimentary work has been carried out with growth hormones, endocrine research, and antibiotics in feed. As in the United States, scientists recognize that these new production techniques increase stress on animals, making them susceptible to disease, but the use of antibiotics to combat this problem lags substantially behind the United States. Porcine and bovine somatotropin, however, are more likely to find application in Russia than in the United States. In the United States an interesting coalition of dairy-producing states, small-scale farmers, and social activists have banded together to oppose animal growth hormone. Some claim the hormones drive the animal too hard; others argue that the cost of hormones will drive the small dairy farmer out of business. The application of these hormones, however, may enable Russia to overcome its reliance on larger, underproducing herds for smaller, more efficient ones.

Plant and animal genetics continue to be promising areas of research because of modern genetic-engineering techniques. In many of the institute's laboratories the development of new strains of peas, alfalfa, barley, rye, and wheat of various lines produced through radiation mutagenesis proceeds.[83] The most important of these new plants include winter wheat, which can be planted in regions of Siberia where soil temperatures rarely exceed −7° because of snow cover, and even −20° or −25° in the southern steppe; a more productive Siberian barley that is good for fodder; winter fodder; rye; beans; corn, in cooperation with the Scientific Research Institute of Agriculture of the South-east (in Saratov); and Novosibirsk-67.[84]

In animal genetics, scientists are engaged in chromosome mapping, gene expression in ontogenesis, elaboration and application of new approaches to the genetics of quantitative characters of animal breeding, and the evolutionary genetics of animals.[85] Building on the work of Iurii Kerkis, researchers created a new breed of sheep that can be used for both meat and wool. Based on calculations conducted at the Medvedskii State Farm on roughly 300,000 sheep, this new breed will contribute 10 million rubles annually to Siberia's economy. Because it takes on the order of twenty years to develop a new breed of animal, any achievement of this nature in a thirty-year-old institute is reason to celebrate.

Finally, reflecting the abandonment of the xenophobic policies that characterized Soviet science until Gorbachev, the institute has embraced the "internationalization" of Russian science. As part of the Siberian division programs to establish new "international" centers of research to attract foreign interest and participation, the institute seeks to expand its horizons. Clearly the biologists in Akademgorodok cannot provide the best laboratory equipment or the comforts of home but they can offer western specialists an environment of unique geological, biological, and archaeological interest.[86]

Mikhail Lavrentev and his associates set out to create a scientific center in Siberia with a broad profile that would include modern biology. They naively assumed that the protection afforded this discipline by distance from Moscow and the umbrella of the physicomathematical sciences would enable Soviet biology to flower as it had in the 1920s. Lysenko, however, would not be stilled. Khrushchev, for all his faith in modern science, would not free scientists from the constraints of the Soviet social and political order. He insisted on allegiance to Lysenkoism, on practice over theory, and on party orthodoxy that still viewed genetics as a pseudoscience. The result was a discipline bound hand and foot by erroneous concepts, inadequate support, and poor instrumentation, facilities, and equipment.

As it turned out, the geneticists who gathered in Akademgorodok could not overcome the Lysenkoist legacy in a mere decade. In the second decade, their accomplishments were modest, largely limited to applications in plant selection and animal husbandry. Fortunately this dove-tailed nicely with the designs of institute director Dmitrii Beliaev and with the Brezhnev "food program."

Scientists were convinced that large-scale agribusiness was the way to go. It is not surprising that Soviet leaders also embraced big science as the solution to agricultural problems. But big agriculture required capital investment on a scale policy makers were unable to meet. The development of new breeds of animals, antibiotic feeds, and various animal growth hormones was too costly for fundamental Academy of Science institutes to handle. Basic research suffered. In the United States critics of agribusiness point to its high cost and energy intensiveness, its threat to the small farmer, and the fact that it produces bland and tasteless, if more packagable and less perishable foods. But agribusiness may be the appropriate path of development for large-scale, centralized Russian agriculture that traditionally underproduces all necessities.

For all the challenges they faced, the geneticists of the Soviet diaspora who gathered in Akademgorodok never lost faith in their abilities to tackle any problem. They believed that science was a panacea for the social and economic problems their nation faced. Having lived through decades of political control, they rejected any kind of regulation of their research. Left to their own devices, they will produce new medicines, plants, and animals whose benefits will outweigh the risks. As Beliaev once remarked, the great potential of genetics is "limited only by the most audacious dreams."[87]

Machines Can Think, but Can Humans?

ANDREI PETROVICH ERSHOV, a visionary and brilliant computer specialist, entered the debate in the 1950s about the future of thinking machines. A recent graduate of Moscow University, he was in the center of the cybernetics boom in the USSR. He jumped at the opportunity to move to Siberia, although he remained in Moscow until 1961 when the new Institute of Mathematics moved with its Computer Center to Akademgorodok. His many works include a number of elegant and versatile computer languages. His enthusiasm transfixed his staff associates and students alike. Ershov believed that the USSR had chosen the wrong path for computerization. He recognized that the emphasis on mainframes, imports, reverse engineering, and theft of technology would doom the Soviet Union to lag forever behind the West. Similarly, he thought that the Soviet stress on applications, on the use of computers as tools of management for central economic organs, was based on a narrow view of their potential. This view was destined to deprive Soviet society of the broad social receptivity necessary for the second industrial revolution—the computer revolution. Ershov fought to bring computer education into the schools and nearly single-handedly maintained scientific contacts with his western counterparts, the authorities having decided, in an act of xenophobic self-spite, that it was best to keep Soviet achievements strictly under wraps.[1]

Until his death in 1988 Ershov worked in the Computer Center in Akademgorodok. He and his colleagues—Mikhail Lavrentev, Aleksei Liapunov, Sergei Sobolev, and Gurii Marchuk among them—played a central role in the creation of modern computer science and technology in the USSR. From scientific dissident Liapunov to party scientist Marchuk, a more diverse group of scientists—both in terms of personalities and career patterns—would be hard to find. Together they struggled to raise computers to western levels in terms of both quality and quantity. In spite of Akademgorodok's fertile ground for such an achievement, systemic handicaps conspired to limit the effectiveness of their work. The handicaps included strict ideological supervision, division of labor between science and industry, and the failure to create the proper social environment for a modern computer culture.

Nevertheless the Computer Center had a number of positive attributes. First, mathematics and computation were the methodological glue in Akademgorodok. Second, Lavrentev himself was influential in early computer programming and was a powerful supporter within the Academy of Sciences and the government of those individuals who built the first Soviet computers. Third, Marchuk, the director of the center, was blessed by excellent contacts within the national scientific—and political—elite. He later became head of the

Siberian division of the Academy, then head of the State Committee for Science and Technology, and finally president of the Academy, always using his influence to help the Computer Center.

From the start, the founders of Akademgorodok saw the computer as crucial to their work. They planned to have a central computing facility accessible to all the institutes, perhaps through terminals or minicomputers connected by a powerful mainframe. This would encourage the pooling of data, sharing of resources, lowering of costs, and an interdisciplinary approach to research. All fields—from chemistry and physics to medicine to sociology and economics—would benefit from the modern electronic processing of information. Thinking machines with natural languages would soon be the rule. The most extreme proponents such as Sergei Sobolev had no doubt that the power of the machine was limitless. He believed that machines would be capable of full cognition, that they would translate, tally, compute, power artificial limbs, and supplant the worker through numerically controlled tools, even that they would think. In their belief in the promise of thinking machines, Ershov, Lavrentev, Sobolev, and Marchuk embraced the supremely rationalist tendency in Soviet science. Yet although Communist Party authorities embraced computer science and technology in word, in deed their fear of innovation and individual initiative, their failure to provide adequate support for fundamental research, and their desire to see computers applied for rather narrow, mechanical purposes were strong counterweights to the efforts of the computer scientists.

SOVIET CYBERNETICS AND COMPUTERS IN THE 1950s AND 1960s

Since the 1830s when the British inventor Charles Babbage invented a rudimentary mechanical digital computing machine, scientists have strived to build devices that can solve a wide range of problems by processing information in discrete form, problems that range from the mundane—the analysis and organization of data—to the sublime, even surreal—the creation of machines that can think. Those who adhere to a strongly mechanistic worldview and see the vital processes of living things as essentially the function of feedback mechanisms based on biochemical and physical laws also often argue that computers can take over many human activities. They see robots as freeing humans from nearly all forms of physical labor and even being capable of simple forms of thought through the application of Artificial Intelligence (AI).

The first electronic, programmable digital computer, the ENIAC (Electronic Numeric Integrator and Calculator), was built at the University of Pennsylvania in the period from 1943 to 1946. It weighed thirty tons, was as large as a two-car garage, and cost $500,000. It consisted of eighteen thousand vacuum tubes and, despite tube failures occurring every seven minutes on average, it was the most complicated and reliable electronic device built to that time. Today its functions could be handled by a tiny chip costing around $100.

Computers and microchips are everywhere: in automobiles, washing machines, VCRs, TVs, games, PCs, and in banking, commerce, and government. Perhaps half of all Americans are involved in the processing or analysis of information. In the USSR, on the other hand, the abacus was the major calculator used in virtually every establishment, even to complete transactions in electronics stores. Because of its reliance on copying western models, underproduction of domestic models, a clumsy division between fundamental and applied research, and political interference, the Soviet Union failed to develop a vital computer industry.

Paradoxically the computer occupied a hallowed place in Soviet society. The problems of declining labor productivity, the need to accelerate information processing, CAD/CAM (computer-assisted design/computer-assisted manufacture), and scientific research all seem at first glance to fall to the magic wand of the modern computer. Building on the pioneering work in linear programming of L. V. Kantorovich, Nobel Prize winner in economics in 1975, Soviet planners believed they could use computers and input-output analysis to solve the problems of bottlenecks, supplies, and prices that plagued their burgeoning centrally planned economy. Huge mainframes in Moscow would assist the party in its tasks of political rule and economic management. Others, like V. M. Glushkov, the director of the Institute of Cybernetics in Kiev, argued that computers could be programmed to mimic human thought up to a level of sophistication of the hypothetico-deductive method of reasoning. In a word, with its healthy scientism, Marxism, and belief in the power of the machine, the Soviet Union would appear to be a most fertile area for the development of computer science and technology. In keeping with past gigantomania, it is said that on display at the computer panorama at the Exhibition of the Achievements of the Economy (VDNKh) in Moscow is the world's largest microchip. Clearly computer science and technology were seen in the USSR as a panacea for a whole range of social and economic problems. Yet a country with vaunted achievements in space, nuclear physics, and chemistry, and a vital, even brilliant tradition in mathematics, failed to keep pace with the West in computers, indeed fell further and further behind, even in such an open environment as Akademgorodok.

For reasons described elsewhere—the absence of a computer culture; fear of hackers and samizdat ("self" or underground publication) leading to strict institutional controls over hardware and software; lack of coordination in the computer industry between competing ministries and the Academy of Sciences; unreliable equipment; inability to produce computers, disks, and peripherals at anything near the level of demand—the USSR entered the 1990s without the computer. Another problem, paradoxically, was the utter faith Soviet scholars put in computers as omniscient rational actors to solve a host of problems. They often overestimated the power of computers, applying them willy-nilly. They believed that computers would instantaneously arrive at the "one best decision" in areas requiring value judgments. They underestimated the social cost of unemployment associated with replacing skilled labor with numerically controlled machine tools. They never considered computer crime or the personal right to protect one's data. From schools to industry, computers did not have the impact in the Soviet Union that one would expect in a developed country.

The pervasive scientism of Soviet Marxism notwithstanding, the early history of computer science, technology, and cybernetics looked unpromising in the USSR.[2] A short-lived problem, but one that foreshadowed later challenges, was the condemnation of cybernetics in the early 1950s as "reactionary, bourgeois pseudoscience" by the guardians of official ideology. Cybernetics is the study of the common features of organisms and systems and their use of information to counter disorder (i.e., to fend off increasing entropy). Clearly cybernetics is not exactly computer science and technology. Indeed, the attacks on cybernetics excluded criticism of the technology itself. Yet the damage was done since cybernetics was closely tied to a number of fields—psychology, physiology, and genetics—which in Stalin's last years were subject to ideological restrictions. Although this view was quickly discredited, it gave ammunition to those who believed that machines would never "think." The attack on cybernetics led to long-term damage in the development of computer software and languages. It is unimaginable that a culture so fascinated with the potential of science to build communism, a culture whose achievements in the 1950s included the hydrogen bomb, nuclear power, tokamaks, and Sputnik, could dismiss the promise of cybernetics. Again, ironically, computer science and technology in the USSR fared best (compared to world standards) precisely when cybernetics was seen as a pseudoscience, and fell furthest behind when it was seen as a technoscientific panacea.

By 1958 cybernetics no longer lived underground. Unofficial seminars at Moscow University, which had kept the field alive, burst into public knowledge through the press. "This was a period," computer specialists Ershov and Shura-Bura wrote, "of unlimited optimism, a kind of computer euphoria—a childhood disease, like a pandemic, that enveloped all countries of the world at the time and led to the production and assimilation of computer technology."[3] Such popular science books as A. I. Kitov's *Electronic Computers* (Elektronnye tsifrovye mashiny, 1956) and I. A. Poletaev's *Signal* addressed the promise of computers and generated broad interest in them. Kitov's book was criticized for various political and scientific errors, but the book was well-intended, accessible, and successful.[4]

Poletaev was not as strong in the sciences as other computer scientists but was blessed with eloquence that made him a persuasive popularizer of the new technology. Since he had worked in the United States during World War II through the Lend-Lease Program, he was more western-looking than many and more receptive to the advent of the modern computer. Poletaev's *Signal* created a storm of interest among future programmers, hackers, and applied mathematicians in the USSR, including many at the Akademgorodok Computer Center. Poletaev was inspired to write *Signal* by Aksel Berg and Aleksei Liapunov. His goal in the book was "to give the reader a preliminary understanding of the genesis of the ideas of scientists on information and management" central to cybernetics. He saw virtually unlimited possibilities in the application of cybernetics and scientific management to machines and robots. "In essence," he wrote, "all production activity of man is the management of the forces of nature in his own biological and social interests." A large number of these activities—information search and retrieval, economic management, planning and production, medical diagnoses, and so on—could be undertaken by robots and automats. Was there anything machines could not master? Poletaev acknowledged that computers could not show emotion or reproduce, nor did they have an ego. But he believed that human intellect and artificial intelligence could be on the same level. The two differed primarily only in terms of the process of transmission and feedback of signals. This view had the advantage of freeing humans once and for all from the "fetishization of the human 'soul,' or 'souls.'" Poletaev included an indirect attack on the Lysenkoist opponents of genetics in the USSR when he likened the opponents of cybernetics to those nineteenth-century critics of Darwin's *Descent of Man*.[5] Many people believe that the article in *Komsomolskaia pravda* signed by "Engineer Iurii" in the early 1960s, which triggered discussion over the "two cultures" in the USSR, between the humanists and scientists, and considered which group had more to say about the human condition and the prognosis for humanity, was actually written by Poletaev. In all his work, especially during the Khrushchev thaw, Poletaev focused on the humanistic aspects of science and emphasized the need for academic freedom—a controversial topic, to say the least, in the USSR.

AKADEMGORODOK'S COMPUTER TROIKA

Lavrentev, Liapunov, and Sobolev stood poised to take advantage of Akademgorodok's vital intellectual environment to further the development of computer science and technology. As they had for genetics, physical scientists and mathematicians had had a long history of involvement with cybernetics and computers to protect the nascent science from misguided Stalinist criticism. They also personally advanced ideas for various early prototypes. Lavrentev may have contributed to the design of the first Soviet computers, the MESM and the BESM, through discussions with their chief designer and engineer, Sergei Alekseevich Lebedev. Lavrentev pushed the idea of the creation of powerful interdisciplinary scientific centers even before World War II.[6] In 1947 he gave a talk at a general meeting of the Academy of Sciences in Moscow on the thirtieth anniversary of the October Revolution. In his speech he drew attention to the lag of Soviet science in the area of electronic calculating machines. In Akademgorodok computers and mathematics were at the center of Lavrentev's vision. They would be the key to, and the symbol of, a scientific utopia where interdisciplinary education, research, and production came together.

Also in 1947, in Kiev, at the Institute of Electrical Engineering, a laboratory was established to design an electronic digital computer. Here Lebedev, a talented electrical engineer and head of the institute, advanced the notion of a machine like ENIAC. Lavrentev, who at the time was working on the theory of cumulative explosions, convinced the presidium of the Ukrainian Academy of Sciences to donate a largely bombed-out building just outside Kiev as a laboratory for Lebedev and his associates. In 1951 the small electronic calculating machine (*malaia elektronnaia schetnaia mashina*, or MESM) began to operate. Lavrentev departed Kiev in March 1950 to become director of the recently organized Institute of Precise Mechanics and Computer Engineering in Moscow, chiefly to give more prominence to digital computing. He enticed Lebedev to come to Moscow to head a department of digital computers. There, Lebedev's group created the BESM-1 and BESM-2 (*bol'shaia*, or large) computers. "When the machine was finished," Lebedev recalled, "it wasn't at all inferior to the newest American facility and was a real triumph of its creators' ideas."[7]

Around this time various Academy institutes and design bureaus began work on other machines intended for serial production, such as the "Strela" (Arrow) to be built at the Moscow Calculator-Analytic Machine Factory, and the "Ural." In Leningrad, in 1952, work on programming began in the Leningrad division of the Mathematical Institute under Kantorovich. Kantorovich, whose work in linear programming was essential to Akademgorodok economists (see chapter 6), began training "programmers" at the university in 1953. At this time Lavrentev gave up his directorship of the Institute of Precise Mechanics and Computer Engineering to Lebedev, and many of the staff left for the newly created Academy of Science Computer Center, including Andrei Ershov, a student at the time. In 1953 a Department of Programming was organized in the Steklov Institute of Mathematics under Liapunov and later Shura-Bura. At Moscow State University, Sobolev then organized a Department of Computer Mathematics to train the first generation of Soviet specialists. In 1955 Liapunov, Sobolev, and others organized a university computer center that worked on the M-2 computer in the laboratory of electric systems at the Institute of Energy.

In 1957 Sobolev joined Lavrentev in the effort to build Akademgorodok. He moved to Siberia as director of the Institute of Mathematics. Sobolev (1908–1988), a product of the Leningrad mathematics school, graduated from Leningrad University in 1929. He began work in the Seismological Institute of the Academy, then transferred in 1932 to the sector of differential equations of the leading Soviet mathematics center, the Steklov Mathematics Institute, where he worked on the dynamic theory of elasticity. He became a corresponding member of the Academy in 1933 and a full member in 1939. In 1940 he joined the party. When the Academy was relocated to Moscow in 1934, he moved there along with the Institute of Mathematics. He was a professor at Moscow University from 1935 to 1959, deputy director of the Steklov Institute, then director, and from 1944 to 1957 was the deputy director of the Kurchatov Institute for Atomic Energy. Sobolev had fantastic plans for a supercomputer capable of a billion operations per second based on parallel programming. He estimated that the project would consume twenty years. To demonstrate the promise of the computer Sobolev believed that some grandiose project like the deciphering of Mayan documents was needed.[8] He was also at the center of organizational efforts for Novosibirsk University. Sobolev was the point man on a series of committees concerned with Akademgorodok's physical plant. He questioned inadequate architectural plans and berated those who had the audacity to challenge his improvements. A specialist in computer science, he could not fathom designing a city of science without basic technology. Why were 20 percent of Akademgorodok apartments without toilets? The answer was that Gosstroi, the State Construction Committee, was needlessly cutting costs. Sobolev resented the assumption that an academician needed a bathroom but a worker could live without one. To the suggestion of V. A. Kirillin, head of the Central Committee Department of Science, that those without facilities bathe in the Ob Reservoir, Sobolev responded, "We will build to live in the 1960s." He was also critical of miserly plans for nursery schools and kindergartens.[9]

In the mid-fifties Sobolev wrote a series of articles on computers, many with Liapunov.[10] One of the most important joint articles grew out of a presentation to the October 1958 All-Union Convocation on Philosophical Questions of the Sciences in Moscow. This convocation represented a turning point in the relationship between philosophers of science, on the one hand, and scientists, on the other, who insisted on the right to determine the philosophical content of their research. The scientists resented the philosophers' interference in genetics, chemistry, quantum mechanics, and relativity theory, aspects of which the philosophers, egged on by Stalinist xenophobia, alleged were idealist. The scientists showed that modern science was fully compatible with Soviet dialectical materialism. At the conference Liapunov and Sobolev delivered a dry, even "instructional" presentation on cybernetics, showing its many applications, utility, and relationship to other sciences. They discussed how management, defense, automation, and mechanization required new mathematical tools to study increasingly complex processes.[11] Cybernetics was not in the least "idealist," they asserted, nor did it equate human thought with a mechanical machine. Sobolev and Liapunov criticized Lysenko for underestimating the centrality of cybernetic concepts in genetics. Cybernetics would merely help meet the challenge of the rapid processing and transmission of information in the modern communist society.[12]

Sobolev saw no limits or dangers in making science, most of all his beloved mathematics, central in the construction of communism. "We don't hear the dark prophesies of foreign fantasy-mongers about the coming reign of thinking mechanisms that destroy and overwhelm man," he proclaimed. "Cybernetics will move forward hand in hand with us, as a friend, giving machines more and more 'dirty work,' freeing up colossal reserves of human mental energy for higher creative activity. In the communist society the profession of mathematics will become one of the most widespread. We need to prepare for this now."[13]

These views toward cybernetics notwithstanding, unreasoned opposition remained for some time. Geometrist Iurii Reshetniak (b. 1939) read about Akademgorodok in an article Lavrentev wrote for *Pravda* in the summer of 1957. Reshetniak lived with his wife, their two children, and his parents in a 90-square-foot apartment in Leningrad. In 1954 his wife, a gifted astronomer, was given a palatial 160-square-foot apartment at Pulkovo, the historic observatory south of Leningrad. Reshetniak met Sobolev through Sobolev's daughter, who also worked at Pulkovo. The promise of a two-room, 520-square-foot apartment and Sobolev's conviction that astronomy would also find a place in Akademgorodok enticed the Reshetniaks to move to Siberia with their three children. The trip from Ukraine took four days by train. Reshetniak now lives in one of the Golden Valley cottages.

In 1959 Reshetniak flew to Moscow by way of Sverdlovsk, until quite recently a closed military city south of the Ural Mountains. He spent two days there waiting in line for a seat on Aeroflot, the Soviet airline. Had he not been detained he would have missed the article in *Vechernii Sverdlovsk* by one M. N. Rutkevich, future Academy member, condemning Sobolev and Liapunov for their idealism in cybernetics.[14] The article only further endeared Sobolev and Liapunov to Reshetniak, who, at Liapunov's request, ultimately lectured budding young cyberneticians on "mathematical analysis" at Novosibirsk University.

Aleksei Aleksandrov Liapunov (1911–1973) is considered to have founded programming as a scientific discipline in the USSR. He discovered his diverse interests while a student of N. N. Luzin, a brilliant mathematician who fell under sudden attack in 1935–36 for allegedly subjecting mathematics to foreign influences. As Luzin's student, Liapunov indeed strove to follow western developments closely and to embrace western traditions of academic freedom.[15] After studying with Luzin, Liapunov worked at the State Geophysical Institute, then the Steklov Institute of Mathematics, and finally in Akademgorodok. He studied mathematical biology, geology, astronomy, and the mathematical foundations of cybernetics from his early years in Moscow. Like Sobolev and Lavrentev, he contributed to the war effort, serving at the front. And, like Sobolev, he joined the Communist Party early on (1944). He was, however, a scientific iconoclast who loved to explore new ideas and share them with young students. He trained at least seven doctors and fifty candidates of science. He instilled his love of teaching in his students who later formed the core group of Akademgorodok computer specialists.[16]

Liapunov developed the fundamental concepts of programming while a member of a small group in the Department of Applied Mathematics in the Mathematics Institute in Moscow. He offered the first course on programming at Moscow University in 1952–53 in the Department of Mechanical Mathematics. The first half of the course closely followed a recently published instructions manual, one of the first programming manuals in the world, but because it was considered top secret, its impact was quite limited. A colleague of mine tried to track it down in Russian libraries, eventually convincing

Shura-Bura himself to lend him his personal copy for a few hours under Shura-Bura's watchful eyes in his own apartment.[17] Between semesters, Liapunov developed the basic approaches to what he called the operator method of programming.[18] The first algorithms for compilers of simple programs were then worked out on the Strela and BESM computers by Ershov, Shura-Bura, and others. The texts of these programs were set forth in the symbolics of Liapunov's operator method.

In the early 1950s, at Timofeeff-Ressovsky's "summer school," Liapunov delivered lectures on cybernetics to an audience of students and professors alike. Back in Moscow, Liapunov lectured in his apartment and then at Moscow University. Liapunov's Moscow seminar started off in November 1955 with a talk by Ershov on modeling of the process of simulation of conditional reflexes using EDSAC. Other speakers over the next two years included Poletaev on the promise of cybernetics, Kantorovich on mathematical methods in problems of economic planning, and Timofeeff-Ressovsky on "The Factors of Evolution."[19] His seminar became a semimonthly, interdisciplinary, all-Moscow affair. (The idea of the organism as a feedback mechanism accounts for the central place cybernetics holds for geneticists and physicists.) Even when Liapunov transferred to Akademgorodok, the Moscow seminar continued to meet, often during his frequent visits to that city.[20] With Aksel Berg, Liapunov published a series of articles which led to the founding of the journal *Problemy kibernetiki* (Problems of Cybernetics), almost forty volumes of which were published irregularly between 1958 and the 1980s.

Initially Sobolev's invitation to Akademgorodok had not been intended for Liapunov at all but for V. I. Zubov, who was blind. At the time Zubov, now a corresponding member of the Academy, was a mathematician at Leningrad University. Lavrentev vetoed the idea, however, believing that the Siberian climate and the absence of an infrastructure in Akademgorodok would be too much for the disabled Zubov to overcome.[21] So instead Sobolev invited Liapunov, whose controversial natural sciences seminar at Moscow State University was the talk of the scientific elite. Thus the gathering of highly select researchers needed to establish a modern computer facility in Siberia was complete. It remained to be seen if these researchers could make due with the poor equipment available to them in Akademgorodok.

THE DEVELOPMENT OF THE COMPUTER IN THE USSR

The Akademgorodok Computer Center, the major computer facility in all of Siberia, had a handful of Soviet-made computers.[22] These computers—the M-20, M-220, BESM-6, ES-1050, and others—were the best Russia had to offer but were inadequate to the task. They were slow in operating, had limited memory, and often crashed. Yet throughout the 1950s computer scientists made significant strides in the development of software and created at least eight new indigenous machines. Some of these were never serially produced but were designed for specific institutes; others were intermediate steps in the development of a larger research project; and still others were experimental. Some were vacuum-tube prototypes for the next generation of semiconductors, but most of them were semiconductors. While still in Moscow, Ershov, who had agreed to move to Akademgorodok, organized a small collective of programmers for the Siberian division. They worked with information received primarily from the United States and set to developing a Siberian programming language based on ALGOL-60 for the M-20 machine. Work on this project began in 1958 when Alan J. Perlis, of the Carnegie Institute of Technology in Pittsburgh, visited the USSR and brought with him the preliminary version of his publication. The result was that under Academy of Science leadership, a broad-based, vital computer program took root in Moscow, Leningrad, and Kiev, holding great promise for Siberia.

Just at this point, however, responsibility for computer development was usurped by the Ministry of Instrument Making, Automation Equipment, and Control Systems (Minpribor) and the Ministry of the Radioelectronics Industry (Minradioprom). At first these two ministries saw broad applications for computers, especially in the military, but also, along with Gosplan, the state planning authority, in economic development. Soon, however, they developed a limited vision of potential applications. They came to believe that copying western technology was preferable to spending billions of rubles competing with the West or following up on proposed models. Minradioprom was determined to "reverse engineer" the IBM 360 with its RIAD project, and Minpribor sought to emulate the Digital Equipment Corporation. Thus the Academy was "stripped of its central role as chief computer designer and forced to give up many of its research and production facilities." The shift in computer priorities from scientific toward industrial, military, and data processing applications for the control of personnel and economic policies ignored the specialized applications such as robotics so prominent in the West, ensured ministerial domination of computer development, and delayed the entry of the Academy of Sciences into promising specialized fields.[23] The nuclear and space programs required more powerful computers but were unwilling to underwrite fundamental research in the Academy.[24]

Scientists lamented the dependence of Soviet computing on the West. Granted, reverse engineering of the IBM 360 in the RIAD series would enable the USSR to make up for years of lag, as well as make use of billions of dollars of existing software, in one step. But this path threatened to make the USSR an "intellectual colony" of the West. One scientist, A. Kronrod, argued that it was "incorrect and even dangerous for the country" to bet on foreign hardware and software, that it was well worth the expense to develop indigenous technology and train qualified engineers. Although the USSR might lag in quantity in its "intellectual heavy industry," Kronrod believed that in quality it could rapidly achieve the level of the United States. To generate money for computer science and technology, he proposed piggybacking a prototype on the well-funded Kurchatov Institute research program in nuclear physics. Kronrod likened the task and the need for support to that which physicists and engineers had garnered in the mid-1940s when working on the atomic bomb.[25] However, the hierarchical ministerial system saw no need for rapid feedback, data sharing, or super machines. They remained firm in their decision to copy the West. The efforts of scientists to develop new programs in this area fell by the wayside until the early 1980s. What is more, favoritism toward each industry's own computers prevented the Soviet "consumer" or "market" from identifying meritorious computer designs and mobilizing resources for their production. An additional impediment to production were the usual shortages of manufacturing facilities and components. With the increasing velocity of the computer revolution, the USSR lagged ever further behind. Not all are convinced that the ministries deserved all the blame. One Akademgorodok computer scientist told me, "I know the guy responsible for the decision to copy western machines. I'd like to pound his skull." When I asked if he were in one of the ministries, he responded, "No, no, no, he was in the Academy, I'd grow weary quite soon if I had to pound all the heads necessary in the ministries."

INFORMATIZATION VERSUS COMPUTERIZATION

Difficulties in the organization, funding, and equipping of the Akademgorodok Computer Center persisted, reflecting systemwide problems in production and the battle between the Academy and the ministries for predominance in computer science and technology. The problems also mirrored a fundamental disagreement over the path the computer revolution ought to take in the Soviet Union. The dispute essentially was between those who favored "informatization" and those who favored "computerization." The former group believed in a second industrial revolution based on the efficient processing of information by a computer literate society armed with PCs and minicomputers. The latter group favored top-down planning of computer applications based only on the dissemination of mainframes.

Two leading academicians important to Akademgorodok's success embraced "informatization." Mikhail Lavrentev and Mstislav Keldysh, Academy president in the 1960s, repeatedly tried to impress on the party the importance of reestablishing Academy leadership in computer science and technology. At the very least they sought significantly greater funding for basic research. But the series of Central Committee meetings at which Brezhnev was advised of the Soviet Union's formidable lag in computer production did little to change his policies.[26] A series of top-secret commissions designed to consider the poor performance of Soviet computer science produced indictments of the Soviet system. They criticized limited applications, the emphasis on mainframes, the poor state of minis, dedicated lines, modems, terminals, and software. Training programs were also a target. Less than 20 percent of graduates of higher educational institutions had any experience with computers, and fewer than fifty thousand people had advanced programming training. The technological problems ran the gamut from mainframes to magnetic storage, from chips and drives to screens and keyboards. Inevitably the commissions called for redoubled emphasis on fundamental research.[27]

Late in the Brezhnev period Academy leadership exerted greater pressure to rejuvenate computer science and technology under its direction. In January 1978 A. A. Logunov, an Academy vice president, and B. N. Petrov, secretary of the Division of Mechanics and Management Processes of the Academy, wrote to Academy president A. P. Aleksandrov recommending that the Academy head a national program to create powerful computer systems. They criticized industry for actively avoiding new technologies and refusing to relinquish scientific policy making to the Academy.[28]

Only in the 1980s did the Brezhnev leadership relent and grant the Academy increased responsibilities in an effort to ensure Soviet participation in the computer revolution. A major step in this direction was the creation of a new Academy division in 1982, the Division of Informatics, Computer Technology, and Automation, inspired by an angry letter Andrei Ershov wrote to Academy President Aleksandrov (copies of which he sent to twenty-one other scientists). Evgenii Pavlovich Velikhov, a plasma physicist and head of the Kurchatov Institute for Atomic Energy, was appointed chairman of the new division. Velikhov, a product of the Soviet system, believed in computerization from the top down. His vision was based on his faith in government resolutions to bring about the production of millions of computers through the investment of billions of rubles. He promoted a fifteen-year national plan, approved by the Politburo and announced in January 1985, to introduce computer technology throughout industry and society. But as in the 1960s and 1970s the computer was merely a means to raise economic productivity by accelerating scientific and technological progress (so-called *uskorenie*, Gorbachev's May 1985 call for rapid modernization of the Soviet economy, a term that disappeared rapidly from the official lexicon), and production of computers never approached target goals. Like many other scientists of the Brezhnev era, Velikhov had a mechanical view of *uskorenie* that relied on an ossified, overly managed economy and R and D apparatus to achieve the diffusion of sunrise technologies through exhortation and slogans. Social receptivity rarely entered his mind as a necessary precondition for computerization; he merely saw increased production and distribution as the key.[29]

At least as significant as the lack of social receptivity and top-down management was the failure to reach production targets. The Soviet Union competently produced 16K chips but only belatedly manufactured 64K chips. In dynamic RAM chips, general purpose integrated circuits, drives, and so on, poor reliability was the pervading theme. Plans called for the production of nearly 400,000 PCs in 1989 and 560,000 in 1990. Less than half that number were produced by 1992. There were also significant problems with software. Granted, an informal army of hundreds of thousands toiled to produce software, and much of it was first-rate, but there was no commercial software. Each institution developed its own; thus software was repeatedly reinvented and much was of dubious quality, lacking even the support of its anonymous authors.

The Pushchino biomathematician G. R. Gromov and Akademgorodok scholar Andrei Ershov argued that Velikhov and others failed to address adequately the issue of social receptivity required of a true computer revolution.[30] It was not enough to publish reports, build computers, produce users' manuals, and spend billions of rubles. More vigorous steps were needed to disseminate PCs, spread applications throughout the consumer sector, support intellectual property rights, and promote professionalism among computer specialists. In a book entitled *National Information Resources* Gromov documented the information revolution in the West, particularly in the United States, and the need to go beyond mainframes to PCs in order to trigger an information revolution in the USSR.

Ershov echoed Velikhov's sentiment to put the Academy at the head of computerization, but he called for significantly increased levels of support for fundamental research. He believed that the impetus for the computer revolution must come from below, through the achievement of universal computer literacy and the creation of a computer culture. He saw "computerization" based on western models doomed to backwardness, divorced as it was from academic science and instead emphasizing the development of centralized computing facilities. The state's insistence on controlling print media, its fear of *samizdat* (self- or, here, desktop publishing), and its narrow vision of computer applications primarily to manage the economy ensured a narrow, mechanical embrace by society.

As a pioneer in artificial intelligence, Ershov recognized the importance of user-friendly operating systems and translators. He believed there were few limits to the computer's power and that in his lifetime computers might begin to think like humans. He therefore called for the "informatization" of society. To achieve this end, he strove to introduce computer education throughout the USSR. He fought to maintain ties with western specialists in his field. Xenophobic strictures on scientific contacts had deprived the Soviet Union of keeping current in its research. The computer culture required an open exchange of knowledge, nationally and internationally. Only in this way could the USSR take part in the computer revolution.

Ershov's notion of informatization went beyond the revolution in computer science and technology. As prescient as Daniel Bell's vision in *The Coming of Post-Industrial Society* (1973), Ershov's informatization was driven by recent advances in electronics technology, by the rapid growth of white-collar workers, and by the burgeoning of information generated in modern society. Automation paid little attention to the implications of computers for freedom of information, disestablishment of bureaucracy, and fundamental social change. Informatization, on the other hand, was consistent with early theories of the

self-organizing and self-regulating systems of cybernetics, offered greater potential for individual creativity, and was more consistent with what the Soviet
future realistically demanded to enter the twenty-first century.[31]

Velikhov's emphasis on technology was a concern for Ershov who believed
that research on programming and the training of qualified specialists was primary. Neither received enough attention from Minpribor and Minradioprom.
To double both the quantity and reliability of programs and programmers by
the turn of the century, a task made difficult by the development of new architectures and applications demanding new knowledge bases, languages, and
integral schemes, Ershov called for nothing short of universal computer literacy, starting with a "partnership" between schoolchildren and computers.[32] Of
course Ershov also understood that nothing less than decentralization of scientific and economic policy making was required to achieve informatization.

ERSHOV ARRIVES AT AKADEMGORODOK

Andrei Petrovich Ershov was the intellectual inspiration for computer science
in Siberia for its first thirty years—from its foundations in Moscow in 1958,
through the creation of the Computer Center in 1964, to the development of
fifth-generation computing in the early 1980s. He was a devoted family man,
a workaholic, and a poet. The following poem, written late in his life in 1983,
captures the essence of Akademgorodok:

THE PATH IN AKADEMGORODOK

For twenty years I've walked to work
Along a path broken through the forest.
If God has given me worries
I carry them with ease here.

This place is filled with all living things—
Birds, squirrels, grass, trees. . .
The course of life and the constancy of life—
These are the cherished words of that path.

A fairy made its home here not long ago.
The rays of her eyes pierce into your soul.
Her breath wafts into my face.
A quiet voice makes sounds of poetry.

But closest of all, the dearest of all
During the unbroken rush of quickly passing days
Is the passerby encountered not by coincidence
Who makes his way on this path of mine.

For twenty years, not having exchanged a word,
We meet each other with a quick glance.

At each encounter this glance
Gives my spirits a charge.

To live with people is the shield from every crisis.
Three families reign in my destiny:
At home is first, the second is at work.
The third is on this sunny path.[33]

As a mathematics student at Moscow University in the early 1950s Ershov
fell into the Sobolev-Liapunov crowd. His first public lecture on cybernetics
was given in Liapunov's seminar. He moved on to the Computer Center of
the Academy of Sciences in Moscow before deciding in 1958 to go to
Akademgorodok. He arrived there in 1961 and joined Sobolev and Liapunov in
lecturing on mathematical modeling and computers. The three of them were
the true popularizers of computing. They published widely, both original
studies and translations of American books on new computer languages, the
first of which appeared in 1958.[34] In July 1960, with two colleagues, Ershov
published a preliminary Russian version of ALGOL-60.[35] He recognized that
one had to be proficient in programming languages (PPS, PPMGU, FORTRAN,
and UNICOD) before developing algorithms for new programs. In 1961 he
published a volume of American programming languages for use on the IBM-
704, IBM-650, and Univac Scientific 1103A, "with which the Soviet reader is
virtually unfamiliar," including FORTRAN, UNICOD, SOAP-2, IT, FOR-
TRANSIT, and ALGOL.[36]

Work on computer operating systems, compilers, and languages required
healthy working relations among devoted scientists, mathematicians, and
computer users. Like the formal clubs of the physicists, such as "Under the
Integral" which served as a meeting place for Akademgorodok's scientists, informal clubs were crucial to the success of computing. The Coffee Club of
Cybernetics (Kofeinyi Klub Kibernetiki, jokingly referred to as the "KKK"),
founded during Akademgorodok's early days, provided such an environment
for the programmers, especially before the Computer Center received its own
facilities in 1965. Like many other formal and informal clubs in Akademgorodok, the coffee club met regularly to discuss issues both central and extraneous
to science. The atmosphere resembled that of the famous Vanderbilt "discussion" dinners at Harvard Medical School in the 1940s that led Norbert Wiener
and others to cybernetics.[37] The KKK had three honorary councils whose responsibility it was to call the meeting to order, then pass the gavel to the
speaker. The speaker had only one requirement, that he begin his talk with the
words "Respected nonempty set of thinking systems" (*Uvazhaemyi nepus-
toiemyi mnozhestvo mysliashchikh sistem*). According to one participant, the
KKK represented "collegiality to the extreme with almost everyone standing
at the blackboard, chalk in hand, and one person listening." The club met
weekly until 1968 when it was shut down by the authorities in the general
crackdown on academic freedom in Akademgorodok. In the early 1970s a
new coffee club, even more democratic than its predecessor, began to gather

within Ershov's sector. But this club centered on true coffee breaks, twice daily at 9:55 and 2:55, where, to be sure, scientific issues had a place but rarely was there a formal presentation, and one often heard the comment, "Let me tell you what you want to say." The authorities tried to prevent the club from functioning by putting up a blackboard and scheduling scientific talks, a ploy doomed to failure.[38]

One of the KKK's honorary councils was Andrei Aleksandrovich Bers. Tall and thin, sporting a long, unkempt beard, eyes bright and wild, fingers and teeth brown from smoking and coffee, he was always one of the first to arrive at the coffee club. Now in his fifties, and known to all Akademgorodok residents, he still enjoys his reputation as something of an eccentric. On his desk is the first edition of Akademgorodok's telephone book, not some newer version, and he prides himself on knowing by heart the numbers of all other owners of the first edition.

At the end of 1960 Ershov, yet to move to Siberia, attended a conference on cybernetics at the physics department of Moscow University. At one session an ungainly man with an unkempt beard took the podium and began to pontificate. Although his material was original without question, because of his arrogance his presentation was not very cogent and he withdrew from the stage confused, not knowing whether he had scored a victory or a defeat. Ershov was impressed, however, and when this man, Andrei Bers, soon to graduate from the Moscow Engineering Institute in computer programming, approached Ershov in the hall and expressed interest in moving to Akademgorodok, Ershov eagerly agreed to take him on. (When Ershov died, Bers would become the head of Ershov's Department of Informatics.)[39] Thus in April 1961 Bers arrived in Akademgorodok where he was determined to remain until his retirement.

The fall of 1961 brought great excitement to Akademgorodok: Liapunov, Ershov's mentor who, according to Ershov, had "secured the scientific status of cybernetics in the country," had transferred to the city of science. Upon his arrival Liapunov decided to subsume the entire programming group, which was working on the ALFA language for the M-20 computer, into the cybernetics department of the Mathematics Institute. Concern replaced joy when it was learned that some of the group, those Liapunov felt "were incapable of work," were not invited. Ershov disagreed with the idea, declined the transfer, and remained in the Computer Center with the group intact. But Bers could not avoid Liapunov's magical influence and left the ALFA group for cybernetics. This was a blunder as Liapunov then left the department, turning it over to Iurii Ivanovich Zhuravlev who intended to focus on mathematical biology. Nevertheless Bers maintained contact with Liapunov through the physics-mathematics boarding school where Bers became a legend for his ability to engender in young people a lasting interest in and enthusiasm for science. Bers was a product of the twentieth party congress, the Khrushchev thaw, the conquest of space, the Virgin Lands campaign, and of such folk singers as the Strugatskii brothers and Bulat Okudzhava. His dozens of students are now the key to informatization. They set the tone in the Computer Center, maintaining the historical optimism of the thaw but recognizing the challenges of the 1990s. Like Ershov, Bers always had grand expectations for informatization. Bers recalls articles and books from the early years of Akademgorodok with such titles as "Can Machines Think?" He says maliciously, "We've long know machines can think. The question is whether humans can think."

The KKK would have had much less material to consider had Ershov not fostered contacts with western specialists. He engaged bureaucrats in decade-long battles to attend conferences, to gain a sabbatical abroad, or to acquire current literature. He secured trips to Michigan, Pittsburgh, New York, and California and traveled to England to give a series of celebrated lectures later published in English.[40] He preserved hundreds of letters, written by him and by others in support of him, to Academy and party personnel requesting permission to accept invitations to go abroad. He also attempted to gain permission to invite U.S. scholars to visit Soviet facilities and begged Academy officials to allow the Computer Center to join officially in the international effort to develop ALGOL in the late 1950s. He wrote fusion specialist Lev Artsimovich, then secretary of the Academy's Division of Physical Mathematical Sciences; he wrote presidium staffers; he wrote Academy vice presidents two, three, even more times to maintain hard-won foreign contacts.

The authorities, however, were more concerned about secrecy than scientific progress. Several dozen U.S. computer specialists succeeded in making major research trips to the USSR in the fifties and sixties. Yet the authorities erected barriers to the dissemination of scientific information that restricted the flow of material. The intention may have been to prevent "anti-Soviet" propaganda from being read at home. The centralized abstracting, publishing, and censorship bodies that were established did much more than this, however. They slowed receipt of crucial western publications from months to years. Centralized control prevented Soviet scientists even from learning about their own colleagues' work. The All-Union Institute for Scientific and Technical Information (VINITI), created to handle the burden of rapid abstracting and distribution for a country of more than a million scientists, could not fulfill its duties. Ershov's expansive personal library is a testament to his determination to collect western literature. But he, too, ran into obstacles unknown to western specialists.

On one occasion Ershov enlisted Lavrentev's help to overcome controls on scientific literature. On October 10, 1967, Lavrentev wrote the deputy director of the Main Administration for Preservation of State Secrets in Print of the Council of Ministers, N. P. Zorin, asking permission for Ershov to bring reports on the AIST (Automated Information Station Software) language abroad in connection with his upcoming trip to Ann Arbor, Michigan. Ershov stood to benefit far more than the Americans who were well at work on third-generation computers and programming. The accounts of AIST had no "principally new technical ideas, were not inventions or patents, and did not contain details that would reveal the level of the technology of production of computers in the USSR." Computer Center director Gurii Marchuk added his voice to the request, trying to cut through layers of bureaucracy whose raison d'être was to erect obstacles.[41] Once again, needless secrecy had waylaid Soviet science. Whether Ershov was indeed able to carry the reports abroad is not clear.

In his last years Ershov was consumed by the effort to bring computers into the classroom. He believed there was no choice but to instill in schoolchildren the second literacy—computer programming. The conditions of scientific technological progress required informatization. "We don't live on another planet," he said, "and we can't be deaf to what is going on in other countries. The contact point between the computer and our everyday life is growing very rapidly. . . . Computerization of school education is a global process, a mass phenomenon, in which millions of people are involved." The process was irreversible; there was so much to do and so little time.[42] Ershov believed that only by bringing up a generation of schoolchildren on the computer could informatization be achieved. In the communist newspaper for elementary schools, *Pioneers' Pravda* (Pionirskaia pravda), he demystified the machine for children, comforting them that the computer, which would become central to their lives, was nothing frightening, indeed was a friend. In the national teachers' weekly he instructed teachers and students on how to make due with exercises for computer literacy using desktop calculators. He stressed the role of computers in modern research and the ease with which students could complete their projects.[43]

Based on his experience at Akademgorodok, Ershov was certain that computers could be introduced nationally. Several of the city's grade schools were equipped with special facilities. In 1984 AGAT computers, with a new program called "Shkol'nitsa," were introduced in Akademgorodok School No. 166. School No. 130 had modems and terminals connected to the Computer Center making anything possible. Students were instructed that "programming is not an end, but a means." The Academy of Pedagogical Sciences endorsed the goal of universal computer literacy in elementary schools. Educators and scientists set standards for computer literacy. A textbook, *The Fundamentals of Informatics and Computer Technology* (Osnovy informatiki i vychislitel'noi tekhniki) for ninth and tenth graders was prepared, syllabi were published, and films about computers were televised. A text for grade school appeared, G. A. Zvenigoroskii's *First Lessons in Programming* (Pervye uroki programmirovaniia). Then in March 1985 the Central Committee and Council of Ministers signed a resolution to raise the tempo of computerization in schools fivefold, so that by 1990 every tenth middle school would have a computer. "As the school reform goes," stated Politburo member Egor Ligachev on June 30, 1985, "so go all the affairs of the country." Even Minradioprom and Minpribor completed designs for mass production of school computers.[44] But hundreds of thousands of minis and PCs were needed, and AGAT computers, at several thousand rubles, were too costly for school budgets. Production never met demand. Teachers were poorly qualified, and texts were in short supply. Ershov believed this reflected a typical Soviet problem: although the computerization of schools was a revolution in education, it was a revolution from above. A school could not afford to wait for instructions from the Ministry of Education before starting work on its computers. Yet many schools waited years for instructions on how to use the machines, and applications and exercises were limited to those specifically spelled out in curricula provided from Moscow. Reliance on the West for software, a kind of "informational imperialism," made matters worse. Ershov suggested not waiting for every school to be equipped with computers but moving ahead with theoretical training using desktop calculators. Lest there were any doubting Ivans Ershov told Zamira Ibragimova, a famous Siberian journalist, "We Survived the ASU [Automated Management Systems, see below]. We will survive yet another universal education program."[45]

Together with efforts to create universal computer literacy, Ershov called for redoubled efforts to expand long-ignored fundamental research on computer architecture, languages, and programs for fourth- and fifth-generation parallel computer systems in the Computer Center. This required significant financial investment up front. Ershov believed that branch industrial research institutes were incapable of developing parallel systems since they were essentially design bureaus. Within the Academy research was poorly coordinated between institutes and personnel, and generally lacked staff and material support. Still Ershov was convinced that the Academy alone could provide adequate leadership for long-term "intensive development of fundamental and applied researches in the areas of [fifth-generation] architecture of computer systems and their mathematical apparatus." The Akademgorodok Computer Center, with its extensive research in the area of parallel programming, its long-term experience in the development of fourth-generation computers, and its western contacts and literature—all of which sits in the Ershov Memorial Library to this day—made it a logical choice to head up the program.[46]

That the research program of the Akademgorodok Computer center had grown from modest beginnings into one of the nation's most diverse program was compelling evidence that many of Ershov's recommendations were sound. But as the following brief history of the Computer Center indicates, such talents as Ershov, Bers, Liapunov, Marchuk, and others had to overcome too many challenges in their daily work, both political and scientific, to make significant headway on "informatization." Many of these obstacles to doing science were standard for Soviet research institutes, so it is not surprising that research on fifth-generation computing still lags.

THE COMPUTER CENTER'S RESEARCH PROGRAM, 1960–1990

From the far reaches of Moscow, Leningrad, Kiev, and Tbilisi, computer specialists met in Akademgorodok to staff a new Institute of Mathematics, home of the new Computer Center.[47] Before this time, in all of Siberia and the Far East, among more than 30 million people, there was only one full professor of mathematics (in Tomsk). This situation changed during the Academy's first elections for membership in its Siberian division when two mathematicians were promoted among the eight new academicians: I. N. Vekua, soon appointed rector of Novosibirsk University, and A. I. Maltsev. Soon Leonid Kantorovich, the creator of linear programming who headed up the new Department of Mathematical Economics in the Institute of Mathematics, was elected an academician. Together with Liapunov and Lavrentev, the presence of these other scientists lent legitimacy to the effort to develop cybernetics and related fields in Siberia.

Most staff members had to remain in Moscow until facilities were completed; others were forced to gather in two rooms at 20 Soviet Street in Novosibirsk, then on one floor in Iurii Rumer's Institute of Radiophysics and Electronics. Finally, in the fall of 1960 they moved to Akademgorodok although only one thoroughfare, Tereshkova Street, named after the first woman cosmonaut, was finished, and it was a quagmire. The first staff members occupied an apartment at 56 Morskoi Prospect, then a dormitory at Morskoi 2, then the Institute of Geology and Geophysics, then School No. 25, then even the fifth floor of the physics institute where the theoreticians were ensconced, before finally settling in their own building. Perhaps ten other organizations shared the half-finished facilities.[48] The computers themselves were actually located in the Institute of Geology and Geophysics. Enthusiasm and creativity had to carry the Computer Center until it accumulated adequate space and equipment.[49] In May 1963 the presidium of the Siberian division decided to make its Computer Center an independent institute which came about in 1964. And in 1965 it had its own building. Lavrentev appointed Gurii Marchuk its director.

The fierce competition for new facilities engrossed homeless institute directors each time an institute neared completion. The Computer Center was first denied its own building when the geneticists took over the building in what they considered a just (and bloodless) weekend coup. Then the medical researcher Meshalkin had designs on the next completed building. Meshalkin foresaw international fame resulting from the development of a human transplant and artificial organ program. He believed his biological research was far more important to Akademgorodok than cytology, genetics, or computers. But he took himself out of the running by violating Akademgorodok decorum: Behind Lavrentev's back he applied to the regional party committee for the building. As a result the computer specialists at long last won the traditional Akademgorodok jockeying for their own facilities. (Another institute was later built for Meshalkin—a distance from the other institutes. He became an outsider and had to find affiliation with the Academy of Medical Sciences.)

Even before receiving their first computers, the mathematicians and computer scientists moved ahead with their research. On March 5, 1960, a commission charged with getting the institute's first M-20 computer on line met to allocate computer time in 1960 and 1961 among competing research proposals. Khristianovich, Lavrentev, and Sobolev were among the decision makers. Cybernetics, mathematics, and programming were logically the big winners, receiving two-thirds of the total machine time; economics, biology, chemistry, and mechanics were the losers. Other Siberian division institutes were asked to inform the Computer Center by May 1 of their needs. The commission proposed that an interdisciplinary group be created to match needs with available time. It was already apparent that demand far exceeded availability so the commission proposed hiring an additional thirty to forty programmers, twenty technical support personnel, and fifteen mathematicians with a higher education—as well as requesting thirty additional apartments to house them all. To help provide such personnel, the university was asked to organize evening computer courses for recent graduates to begin in August 1960.[50] The requests for additional personnel and housing would not be met until years later; the first specialists in systematic and theoretical programming graduated from the mathematics department at the university only in 1964.

Each year scores of students from the fourth and fifth classes of the mathematics department at the university worked at the Computer Center, and by the mid-1980s some four hundred specialists had graduated with computer science degrees. Although these numbers are impressive from the standpoint of a scientific oasis in Siberia, in the United States tens of thousands of students graduate each year with computer science degress. Recognizing the dearth of specialists, the Siberian scientists founded a special school in 1979 for young programmers that attracted up to two hundred students annually.

Considering the persistent inadequacy of its facilities, it may have been a blessing that the Computer Center had so few students and scientists. At a party meeting in August 1960 under the chairmanship of Egor Ligachev, Sobolev noted that the mathematicians were "not in a position to use this unique instrument [the M-20] at full power . . . because workers without warning turn off our electrical energy and water." The operation of a diesel-powered cement mixer at the construction site next door which, to the mathematicians, seemed to operate according to the schedule of the M-20, also interfered with their work. Sobolev personally led the scientists in hand-to-hand combat with the builders in the machine hall! The victorious scientists cheered, "That's how the steel was tempered!"

Gurii Marchuk had the appropriate pedigree to be named the center's first director. His dissertation on nuclear reactors had been approved personally by A. P. Aleksandrov, himself director of two institutes in his lifetime and later president of the Academy of Sciences. Aleksandrov in fact considered Marchuk his protégé, as well as his friend. Marchuk's accomplishments were extensive: after serving as director of the Computer Center in Novosibirsk, he later became deputy chairman and then chairman of the Siberian division; chairman of the State Committee for Science and Technology (GKNT) when longtime Kremlin science veteran V. A. Kirillin retired; and finally, from 1986 to 1991, president of the Academy of Sciences. He presided over the "Siberia" and BAM (Baikal-Amur Mainline, the new trans-Siberian railroad) programs, and the development of ASUs (Automated Management Systems). Born in 1925 in Petro-Khersonets in Orenburg oblast, he entered the mathematical-mechanics department at Leningrad University during the war. He served in the army from 1943 to 1945 and graduated from Leningrad University in 1949. In 1952 he completed his candidate dissertation and began work at the Geophysical Institute. Like other promising physicists, he was drafted into the nuclear engineering establishment. He worked in Obninsk at the Physics Engineering Institute from 1953 to 1962. There he used mathematical techniques to refine calculations for reactor reactivity concerning neutron transport, research that led to his doctoral dissertation "Numerical Methods of Calculations for Nuclear Reactors" (1956). I. V. Kurchatov, father of the Soviet atomic bomb project, requested to see the dissertation in connection with the second Geneva conference on the peaceful atom. After skimming it, he called V. S. Emelianov, head of the Soviet Atomic Energy Commission as well as the Atomic Publishing House: "[Davochka]," he said, "This is Kur. Please see to it that Marchuk's book is published quickly." Without the interference of censors, the book appeared two months later, and in 1961 Marchuk received a Lenin prize for this work. In 1962 Sobolev and Lavrentev visited Marchuk in Obninsk and promised him election as a corresponding member of the Academy if he would accept the directorship of the Computer Center. He became the director in 1964 and in 1968 became an academician. His research from the 1960s on concerned short-term weather forecasting and the dynamics of the atmosphere and ocean, to which he later added immunological systems. In 1947 he became a member of the Communist Party and in 1981 a full member of the Central Committee.

A decent man, Marchuk found himself in a difficult position as he advanced through the ranks of the system. Some claim that Marchuk initiated a split between the Computer Center and the mathematics institute for personal reasons—to enhance his own authority and advance his career. This is not the case. The Computer Center simply could not grow within the confines of the institute. There were too many workers, and too many different directions of research. And the computers the researchers expected to receive would have to be stored until new facilities were found. Sobolev realized this and identified Gurii Marchuk as a promising candidate for the directorship. Lavrentev was initially opposed to Marchuk since Obninsk was a closed city, and Lavrentev believed that censorship was anathema to science. Eventually, however, he overcame his reluctance and approved the hiring of Marchuk. Marchuk had to show two faces: one for his fellow scientists, the other for party and administrative personnel. He began to be reworked by the system. "To be a good administrator," Aleksandr Nariniani, a specialist in artificial intelligence, said, "you must be a manipulator."[51] Although Marchuk's colleagues respected him as an administrator, many resented his devotion to the party.

The difficulty of being both a scientist and an administrator in the Brezhnev era is illustrated by Marchuk's appointment to the State Committee for Science and Technology. When two right-wing members of the Politburo, Kirilenko and Suslov, approved of Marchuk's appointment, Marchuk attempted to decline. But in a veiled threat, they told him that if he wanted to see the continued flowering of the Siberian division, he would be in a better position to help from

Moscow. Unlike the relationship between Lavrentev and Khrushchev, that be-tween Brezhnev and Marchuk was distant. When Brezhnev traveled to Siberia in 1972 to discuss the new trans-Siberian railroad there was little beyond for-malities between the two, and when Marchuk was in Moscow as head of GKNT Brezhnev never asked to see him. Because of his access to higher circles, Marchuk gained a reputation as a conservative.

During his tenure as chair of the Siberian division of the Academy of Sci-ences, its apparatus grew manyfold. Perhaps bureaucracy was forced on the Siberian scientists by the central party apparatus in the wake of Czechoslovakia and the general crackdown on academic freedom in 1968, but Marchuk was without question more inaccessible than Lavrentev, a quality that did not serve him well. Although certainly much of the blame for the failure to reform Soviet scientific administration can be laid at his door, nonetheless he displayed human qualities and exceptional patience at Akademgorodok. He gave com-plete independence to laboratory directors; he supported his staff in the face of undue party pressure; and he approved the election of A. S. Nariniani, one who was not afraid to speak his mind on matters such as housing and foreign contacts, as chairman of the institute's union committee (Mestkom), a position usually given, for obvious reasons, to a politically reliable senior scientist. Nariniani became disenchanted with the union committee of the presidium. In an open meeting he accused it of hoarding institute funds and doing nothing to promote better housing or cultural conditions in Akademgorodok. He called for a change in leadership. Rather than dressing him down, or even firing him, Marchuk merely said, "You should have talked with me about your concerns first." Marchuk also went to bat to seek permission for Nariniani to go to France as a member of a delegation in 1974, when Nariniani was not a party member.[52]

In the spirit of Akademgorodok Marchuk supported the infamous "Fakel" project in the 1960s before it too was shut down by the central party apparatus. Fakel was one of the few Soviet small-scale, profit-oriented, self-financed en-terprises to exist since the 1920s when all economic organizations were subju-gated to state power. In the early days of Akademgorodok, there was no need to secure permission to register as a social organization. The Council of Young Scientists, and even the Komsomol, were active in Akademgorodok social cafés. Under the direction of N. N. Ianenko, now an academician, such young scientists as Andrei Bers in the Computer Center suggested producing modest software packages for local industry to help assimilate modern production processes. Fakel was the middleman that fulfilled contract orders and made money on Computer Center equipment. The scientists worked overtime and charged a fee. But the Ministry of Finance refused to accept funds so the local Komsomol opened a bank account into which millions of rubles flowed. The national press spoke highly of Fakel. With some of the proceeds, Fakel paid for festivals, bards, and social cafés. Other institutes in Akademgorodok and then institutes in other cities duplicated Fakel. Like the production facilities at the

Institute of Nuclear Physics, this was an independent enterprise that owed its existence to the unique privileges of Akademgorodok. But the central party apparatus became concerned. The Ministry of Finance and the national Komsomol organization were frightened by individual initiative. Perhaps recalling Ilf's and Petrov's tale of the Soviet "bourzhui" in the novel *The Twelve Chairs* (1925), they saw Fakel as being run by hucksters. They arrested some programmers and closed down Fakel's bank accounts. Lavrentev and Marchuk appealed to Kirillin at the Central Committee, to the Ministry of Finance, and to the Komsomol to reverse the decision. But the central authorities saw the Fakel experiment as nascent capitalism and refused to budge.

When the Computer Center finally received a second-generation M-220 with semiconductors instead of tubes, research took a surprising turn for the worse. The disassembly of the M-20 and difficulties in bringing the M-220 on line drastically curtailed computer time for months. The one BESM-6 designated for the institute could not pick up the slack, since it lacked such mathematical apparatus as compilers and systems software.[53] When a machine is designed in the West, engineers, programmers, and architects work side by side. In the USSR, on the other hand, the programmers are given the hardware components, required to assemble machines on the spot, and then write the languages to run them on.

Because of its confusing array of tubes and the absence of factory support, repairing the M-20 was just as difficult as repairing later-generation machines like the M-220 or BESM-6. A veteran of the first generation "categorically disagreed" that second-generation computers were more complex. "The fact that our engineers could shout out where the problem was when getting information just from the control room spoke to their detailed knowledge of the machine, not its simplicity," he recalled. When the system crashed in 1966, nearly three days elapsed before the engineers found the source of the problem. Diagnostic oscillographs constantly remained switched on so they would not burn out. This speaks volumes about the pressure to keep the clumsy M-20 operating "normally" in the face of regular breakdowns. The engineers succeeded in keeping the machines operating twenty to twenty-one hours a day, but lamented that a BESM-6 would still have been almost incomparably more productive operating only ten hours a day.

Although these difficulties are not atypical for any first-generation machines, Soviet or otherwise, an early Computer Center annual report described the extent of the problem: "[The M-20] is characterized by low reliability and dependence on temperature fluctuations, and requires a large support staff to guarantee operation. The peripherals of the M-20 do not satisfy the needs of the programmers (there are no large format printers, graphics, etc.)." The Siberian division had requested an M-220 in 1966 and received it in 1967 from Gosplan. In anticipation of the M-220, the M-20 was shipped to Vladivostok, the Far East branch of the Siberian division. The M-220 arrived late, resulting in the center's "computation power [being] cut in two." That the

M-220 arrived without any systems software also hampered computing throughout the Siberian division and delayed the creation of the operating language, AIST. In November 1967 a BESM-6 was put into operation, but its software was laden with bugs. Since most problems required four to five minutes of computing time during which the BESM-6 could not operate at full power, the computer specialists decided that it was not worth using the machine. As if this were not enough, the machinery needed to make punch cards worked poorly, although some of the problems with the punch cards resulted from human error. One pensive mathematician placed a deck of them at the punch card input, having forgotten to remove them from her shirt. The technicians were confused for a long time as to why the punch cards were not being accepted. All in all, neither software nor hardware corresponded to the volume or content of research in the Computer Center. Clearly more computers and peripherals were needed.[54]

Efforts to secure modern equipment were complicated by ministerial barriers of various flavors, all of which proved "untasty" to the computer scientists. The major problem was the physical and bureaucratic separation between hardware and software. According to Gosplan directives, in exchange for software for the M-222 (a modernized version of the M-220 needed to create the AIST-0 software), the Kazan Computer Factory of Minradioprom was to produce auxiliary equipment and eventually an M-222 for the center. But at the end of September 1967 Kazan held up delivery of the equipment, which already was paid for, claiming it represented an overfulfillment of Kazan's M-220 production plan. To settle the disagreement Lavrentev had to call on the deputy chairman of Gosplan.[55] The lack of an "800 number" and support staff plague computer operations to this day.

In an attempt to circumvent this type of problem, the first of two BESM-6 computers for the center was built with Computer Center engineers on site at the factory in Moscow. There was reason for optimism in July 1975 when the center received its first ES-1050, a new third-generation machine, from the Penza Computer Factory.[56] A Hewlett Packard time-sharing system was connected to the M-222 to operate forty-eight terminals throughout the Siberian division. Equipment problems notwithstanding, the 1960s were considered "the good old days," a time when informal contacts were important. According to one senior specialist: "There was a higher percentage of interesting people. When a research center is new it is always filled with a large number of active individuals."

In its early years Computer Center research focused on four main areas: complex models of ocean and atmospheric physics for applications such as weather forecasting; economic cybernetics with research on management systems and economic problems; the application of cybernetics and modeling to mathematical problems in geophysics, nuclear physics, the mechanics of complex media, and chemistry, for example, the modeling of processes of zone refining, and on the quantum chemistry of surfaces and valent electrons; and computer science and technology, with work on principles of construction of self-regulating systems, automated programming, and the creation of new algorithmic languages.[57]

Marchuk organized the Department of Dynamic Meteorology in 1964 to study the first area, essentially an area of applied mathematics. Work focused on the modeling of atmospheric and oceanic processes; weather forecasting and climate theory; and research on the huge circulating systems of the Pacific Ocean. Researchers elaborated a theory of radiation transfer and the mechanics of continuous media. They studied the interaction of radiation, heat, atmosphere, and ocean through various algorithms, statistical methods, and hydrodynamics. They developed a weather forecasting program for small regions on the BESM-6, which finally came on line in the late 1960s. The first major study, completed in 1967, was Marchuk's *Numerical Methods of Weather Forecasting* (Chislennye metody prognoza pogody).[58] In the department's first twenty years more than fifty candidate and seven doctoral dissertations were defended, and more than twenty monographs and a thousand articles were published. When Marchuk moved to Moscow a number of staffers moved with him. They formed an independent institute when Marchuk stepped down as Academy president in 1991.

Work in cybernetics was supported by special government research programs of the State Committee for Science and Technology and the Council of Ministers. The goal was to develop "new . . . methods [for constructing] highly productive informational-computational systems based on optimal organization of computer processing and the joint effort of linked computers." This meant that Ershov and his colleagues were occupied with the creation of a series of languages, compilers, and systems software. These included the SIGMA system for the solution of logical problems on the M-20 and M-220; BETA for logical problems; development of Automated Managements Systems, or ASUs; EPSILON, by Ershov and Pottosin, an operating system for the M-20; the ALFA-BESM system for other computer centers; even payroll software. Bers ultimately created the software program—RUBIN—for the writers and editors at *Pravda*. From the start Ershov, Nariniani, and Kotov were interested in the theory of mathematical machines and programming, especially work on parallel processes. They wrote special programs to assist specialists in other institutes, for example, "The Application of Cybernetics and Computers in Chemical Research" for the Institute of Catalysis. Computer scientists developed "Spektr" for use on the BESM-6, Minsk-32, and M-220 for molecular spectroscopy to identify chemical compounds by their spectral, structural, and physical characteristics for the Institute of Organic Chemistry.[59]

The Department of Informatics, which Ershov founded from a group in the programming department, grew out of the language design efforts of several groups, including those who had worked on the ALFA translator for the M-20 and M-220. The scientists set out to solve a challenging problem, namely, to create a compiler that automatically built programs equal in quality to those of the algorithmic languages on which they were based. ALFA, a variant of ALGOL-60 for use on the M-20 and later the M-220, was one of the first

Soviet compilers and was considered highly effective by its authors. It used 4,096 bits of operating memory without disks, display, text editors, and even without a printer.

As historian of Soviet science Greg Crowe points out, in many respects Soviets were far more computer literate than many Americans. The average computer science student at the university was a much better programmer and engineer than his American counterpart. One man Crowe met spent a month reducing the size of a layover program for Microsoft Word that would produce Cyrillic so that it would fit into 17K instead of about 60K. Why the emphasis on program size? Because he did not have enough twenty-five-cent floppy discs to hold the bigger version. This was a strange way to achieve efficiency, but his program was much tighter than similar western ones. Very few Americans actually "program" their PCs. Not so for the Russians—perhaps three-quarters of them wrote their own programs.[60]

Although Ershov, Pottosin, and G. I. Kozhukhin wrote ALFA in a short time, to them the process seemed endless. No one wanted to write by hand any more, so an automated system of programming—EPSILON—had to be created. An overriding problem for the programmers, among others, was that by 1963 ALFA had grown to some thirty thousand commands and would not fit in the memory of the M-20. Experiencing delays in machines, memory, and debugging for two to three months at a time, Kozhukhin succeeded in dividing the software into two smaller, complementary programs. The project culminated in ALFA-6. Other software included the BETA system; the automatic information station, or AIST, which permitted the collective solution of the most diverse problems in dialogue with an M-220 using the language SETL; and successful debugging of the EPSILON program for the BESM-6.[61] Soviet specialists consider the achievements of the Novosibirsk school of informatics in graphics, operating systems, data banks, and software as pioneering. By 1980 the informatics group had produced more than twenty different systems with more than a million commands, whose various versions had been applied in a host of minis and micros and were introduced in dozens of enterprises, affecting the economy to the tune of millions of rubles.

Two problems hounded the Computer Center's productivity. First, equipment problems had only been solved superficially by 1970. At that time the only machines in full operation were two M-220s and a BESM-6, a Minsk-22 for work on AIST-0, and an Ural-14D, which was intended for the AIST-1 system but was plodding and unreliable. Another BESM-6 and M-220 had arrived but were not yet operational. Two graphics machines and several magnetic disks from France had been acquired. Efforts to create a series of data bases and to provide more computer time for other institutes continued. However, scientists were hampered by a lack of external memory—an absence of tapes, diskettes, as well as peripherals. Getting ALGOL-type languages to work presented special difficulties, particularly with regard to keyboards. Certainly the major computer facility in all of Siberia should not have been plagued by such problems.

The second obstacle interfering with the Computer Center's efforts was that work on applications hindered fundamental research. On the positive side, applied science generated welcome income and demonstrated in the minds of central planners the efficacy of the institute's program for Siberian development. In 1970 alone the ALFA system was installed in fifty organizations, ALGIBR in seven, EPSILON in five, and SIGMA in a handful of others. In October 1970 the West Siberian Administration of Gidrometsluzhba (the national weather service) began to make weather forecasts with Computer Center software. Although payroll software systems were introduced in ten Moscow organizations and fifteen others in Leningrad, Kazan, Saratov, Alma Ata, and Barnaul, they signified an empty promise since workers continued to receive payment in cash, not checks. Instead, staffers had to wait in two lines for their pay—one line to fill out the paperwork for receipts, the other to exchange the receipts for rubles. Other applications included a program to increase productivity 2.5–3.0 times of an experimental reactor at the Novomoskovskii Chemical Combine, an advance that "exceeds the best foreign achievements." In one ten-month period in 1970, contract work exceeded 2.1 million rubles. In 1975 the Computer Center earned more than 2 million rubles—64,000 above plan. Scientific laboratories generated more than 700,000 rubles of this income, including 140,000 in the Department of Informatics (under Ershov) and 190,000 in the Department of Physics of the Atmosphere.[62]

The down side to the emphasis on applications was that as the Computer Center expanded, the percentage of computer time going to Siberian economic organizations increased. In 1967 the center allocated roughly two-fifths of computing time to its own needs and two-fifths to other institutes of the Siberian division, with the remainder going to regional economic organizations and enterprises. By 1970, of 19,065 hours of computer time, 5,749 (30.2 percent) went to institutes of the division, 4,972 (26.1 percent) to other enterprises and economic organizations, and 8,344 (43.7 percent) to the Computer Center.[63] In 1977 more than fifteen hundred programmers representing 115 different organizations used the center computers through various time-sharing programs and widely placed terminals. The mainframes that the other Siberian institutes had acquired hardly worked any better than those in the Computer Center and were often hand-me-downs.

In the 1970s research in computational and applied mathematics grew more complex and diverse. Work on meteorology and the physics of the ocean and atmosphere continued to be the central focus. Detailed forecasts could now be made for up to three days, a pedestrian accomplishment in view of their poor reliability and rare broadcast to the populace. Researchers developed a system to analyze clinical biochemical data for pediatric hepatitis. Work on graphics, operating systems, data banks, ASUs, and other software applications continued. Other algorithms and mathematical models led to applications for other Akademgorodok institutes in such diverse areas as explosion welding, the theory of nonstationary kinetic equations, the filtration of two-phased liquids, global atmospheric and oceanographic processes, and high energy physics.[64]

Lavrentev's son, Mikhail, now director of the Institute of Mathematics and a corresponding member of the Academy, and Anatolii Semenovich Alekseev, currently director of the Computer Center, worked on mathematical problems of geophysics and geology. The history of their work dates to May 1959 at 20 Soviet Street in Novosibirsk. Sobolev, the young candidate of mathematical sciences, Mikail Lavrentev, Jr., and V. A. Tsetsokho discussed the application of mathematics in geological and geophysical problems. The small group agreed that Lavrentev would develop an analytical apparatus for mathematical problems in geology and geophysics. Soon Alekseev arrived from Leningrad and joined Lavrentev in this research. They worked crowded in a tiny room. Since the M-20 was not operational, they undertook calculations on the FE-LIKS arithmometer (a Soviet adding machine). They developed algorithms, wrote programs in code for the low-power Strela, and occasionally flew to Moscow to run programs on modern computers. In late 1960s the department grew rapidly, hiring recent graduates from the university. Research expanded into seismoholography, mathematical theory of problems of photometry, and conditional correctional problems.[65]

The golden years of this research were the mid-1970s. The scientists developed mathematical systems for understanding the structure of the earth's crust and upper mantle on the heterogeneous structure of layers, permafrost, and rockshelves. The research was applied for mineral, ore, and especially oil and gas seismological prospecting in west and east Siberia, in the Pamir, Baikal, and Kurilo-Kamchatka regions. The computer data were supplemented by data from airplane and satellite reconnaissance. In the rich Tiumen fields, ninety oil and gas reserves were identified by techniques developed in the institute. The techniques in question, however, cannot be used for surveying the nonstructural deposits of Tiumen, deep paleozoic complexes of West Siberia, and other sites owing to the complex three-dimensional structure at significant depths, and traprock which accelerates the speed of dispersion of the waves. So in connection with the rapid growth in the availability of mapping data produced by satellites (which showed changes in vegetative patterns, courses of diseases, forest fires, agriculture, and so on), Marchuk and Alekseev proposed the creation of regional Geophysical Information Processing Centers. The centers were based on a multimachine complex united through general disk memory of three BESM-6 computers and a mini M-6000 with fifty program modules. Alekseev and his colleagues have long sought new equipment and processes, including seismic stations, instantaneous processing of data through satellites communications, on-site analysis with smaller microcomputers, and the expansion of the regional centers. More than once the scientists requested funding for an airborne laboratory.

In the 1980s the profile of the Computer Center remained stable. In general, as befitting a research institute under Brezhnevite science policy, the central focus was on "complex" national economic development projects like "Siberia" that had a rather immediate impact on economic performance. Work on mathematical models, software, algorithms, conditionally correct and inverse problems, and graphics were intended, in the words of an official publication, to enable "the practical solution, in the shortest time possible, of a broad range of problems of paramount importance to the national economy." Because of inadequate hardware throughout Siberia and the Far East, the Computer Center took on a number of tasks which in the West would be filled by on-site mainframes and microcomputers: it processed economic data that was shared by territorial industrial complexes; it functioned as a regional automated geophysical data processing center; and it continued to be the major source of computer time for the entire Siberian division.[66]

The physics of the atmosphere, oceans, and climate change remained central. Scientists developed one- and three-day weather forecasting programs, and five to seven-day and fourteen-day planetary circulation forecasts. They applied methods of statistical modeling to multivariate problems in radiation transfer theory for application in atmospheric optics that took into consideration radiation balance in clouds and cloud brightness, signal-to-noise ratio in optical probing of the ocean, and the effect of ocean temperature anomalies on the thermal conditions of the atmosphere. Mathematical and computational methods were applied to other fields of research: the development of algorithms for modeling of nuclear reactors and radiation safety; statistical modeling methods for gamma and neutron scattering in oil, gas, and other mineral prospecting; the theory of conditionally correct problems in mathematical physics; geophysical and geological prospecting of the crust and ocean bed in the search for commercial minerals and in the study of volcanoes and earthquakes. Finally, Ershov, Nariniani, and others made some headway in artificial intelligence.

Research on global meteorological processes inevitably led to studies of anthropogenic effects on the environment, including the greenhouse effect. Scientists noted in the mid-1960s the ecological impact of human economic activity through discharges of thermal energy and waste products into the atmosphere, water, and land. Only in the 1970s were they able to accumulate sufficient data for analysis. But statistical extrapolation was insufficient. Scientists saw the need to create mathematical models to evaluate "the influence of human social and productive activity on the environment." They argued that mathematical modeling would "in the near future play a central role" in technology assessment and environmental impact statements for all industrial and construction projects. They created a model for the microclimate of industrial regions that took anthropogenic factors into account. The model was applied in Novosibirsk, Sofiia, Kansk-Achinsk, and elsewhere. Yet this research was inhibited by policy makers who did not wish to see more evidence of the devastating impact of Soviet development programs on the environment.

The quantitative indexes of this vital research program reflect its quality (their numbingly relentless upward climb notwithstanding). The institute grew rapidly in its first three decades in terms of staff, publications, and facilities. In 1967 center staff consisted of 2 corresponding members of the Academy, 34 doctors and candidates of science, and 433 staffers, with 21 graduate students. By 1974 there were 2 full and 4 corresponding members, 24 doctors and 88 candidates of science, and 200 other scientific workers without higher degrees with more than 600 employees. Fifty-nine—roughly one-tenth—were Communist Party members. By the mid-1980s, of a staff of roughly 1,000, there were 449 scientists including 2 academicians and 1 corresponding member, 27 doctors and 156 candidates of science. The university provided a steady stream of students for the center, with more than 100 each year doing work at the Computer Center. The scientists published scores of monographs and thousands of articles.[67]

The Computer Center owned an HP-2000, three BESM-6 computers with common external memory (disks); an ES-1052; and an M-6000. The BESM-6 computers were used in the batch processing mode by more than a thousand programmers from various institutes of the Siberian division and other organizations, and handled an average of a thousand jobs a day from ninety terminals throughout Akademgorodok institutes and schools, using ALGOL, ALPHA-6, FORTRAN, BESM-H, ASTRA, and MACROEPSILON.[68] In 1984 the Computer Center generated nearly 2.5 million rubles through thirty-four contracts providing computer graphics, instructional systems, modeling, and data processing services to other Academy, higher educational, and ministerial institutes. The center acquired IBM-clone ES-1052 and ES-1060 computers, a Burroughs-6700, and a number of minicomputers. Excluding minis and micros (of which there are more than sixty), the total productivity of the Computer Center reached 15 million operations a second in 1987. In the mid-1980s the ES-1052 and ES-1060 were exchanged for an ES-1055M, -1061, and -1061M. The 1055M and 1061M came on line in 1986, and the ES-1060 and ES-1052 were given to other Siberian organizations.[69]

AUTOMATED MANAGEMENT SYSTEMS

The quantitative growth of the Computer Center masked a national problem: narrow, mechanical applications hampered the computer revolution. Computers were part of the management science revolution that swept the USSR under Khrushchev and Brezhnev. In the most striking vision of their power, terminals in all economic units of the country—enterprises, stores, collective and state farms—would be connected to massive mainframes in the central government. Instantaneously, the mainframe would make adjustments for inputs, outputs, and prices, ensuring the most rational production and distribution of goods and services. Planners in the center would then gain complete power over the production process from their comfortable chairs in Moscow. Efforts to introduce this national system commenced on the local and regional level through the Automated Management System (Avtomaticheskaia Sistema Upravleniia), or ASU.[70] ASUs, like the Taylorist Scientific Organization of Labor (Nauchnaia organizatsiia truda, or NOT; see chapter 6), reflected a belief in the complete rationality of modern science and technology. The Akademgorodok computer effort was tied to the national development of ASUs.

In 1964 the Computer Center concluded a contract with the Barnaul Radio Factory for the introduction of an ASU as part of its research "in the area of the theory and practice of automated management of industrial enterprises." The Barnaul Radio Factory ASU was connected with inventory control and statistical analysis of production processes, norms, and quality control.[71] Located in Barnaul, some 125 miles south of Akademgorodok, the radio factory produced a wide variety of electronics, including some for the military.

Responsibility for the ASU program fell to the laboratory of economic information systems (later the Department of ASUs) under Igor Maksimovich Bobko. Bobko was another Soviet success story, a devoted communist born in a family of village intelligentsia who became a laureate of the grand Soviet Council of Ministers prize. He received higher education in both engineering and mathematics. After serving as a naval officer, the native Siberian from Ust-Sosnovo returned home. He was appointed a junior scientist in the Akademgorodok Institute of Mathematics. A shy man, Bobko preferred working alone and assumed a heavy load. He quickly rose to the top, becoming a doctor of technological science, a professor, head of the ASU department, and first secretary of the Computer Center's party committee.

Bobko embarked on the creation of ASUs when the majority of computers in enterprises did not work for more than two hours a day. Factory managers did not know what they were for or how to use them. The first ASUs were failures, serving at best as electronic bookkeepers. Enthusiasm waned. There was, however, a "catalyzer of progress," one Boris Vladimirovich Doktorov, the director of the Barnaul Radio Factory. Doktorov tired of asking Novosibirsk economists and mathematicians for assistance in introducing modern management techniques in his factory. He contacted the Institute of Cybernetics in Kiev. The institute director, Academician V. M. Glushkov, advised Doktorov to contact Bobko, Marchuk, and Aganbegian in Akademgorodok.

Bobko had long dreamed of ASUs. He had asked Novosibirsk enterprises to serve as guinea pigs in working out the kinks. None accepted. The investment in time and money and the uncertainties associated with developing a new technology troubled most managers. Others saw no reason to abandon tried-and-true techniques. Doktorov and Bobko made a perfect match. Like Bobko, Doktorov was not afraid of the difficult fine-tuning required for the first ASUs. Unlike new machine tools, an ASU had to be adapted to the responsibilities and work style of managers, workers, and scientists alike. Feedback mechanisms were needed. The ASU in the Barnaul Radio Factory resembled an automated Taylorist system, where each worker at each station received a norms card spelling out all parameters of the required technology, work, pace, and production. Mathematicians and economists then designed dozens of algorithms, solving hundreds of problems. The goal was to create an elegant system that mirrored production information from the shop-floor level, the production department, the factory system to the managers.[72] Although production increased in Barnaul, the improvement was not dramatic and may even have been the result of scientists' and managers' heightened attention, as much as the new ASU.

ASUs were tried out elsewhere as a prelude to a national system. Using the Lvov Television Factory and the Barnaul Radio Factory as prototypes, ASUs were produced for a series of research institutes, enterprises, and design bureaus, which were often closed military establishments. An ASU for the government of the Soviet region of Akademgorodok was also introduced. The greatest success was the Sigma ASU, based on third-generation computers, which was installed at more than sixty enterprises, including Elektromashina in Cheliabinsk.[73]

Natasha Pritvits, an accomplished scholar and also first academic secretary of the Siberian division and later its press secretary, described the general rapture with early ASUs: "There will be more and more such automated enterprises—the cells of the future unified system," she wrote after the twenty-fourth party congress in 1971. A series of automated enterprises would be united with regionally based computer centers in Akademgorodok, Novosibirsk, Irkutsk, Krasnoiarsk, and Iakutsk. This complex of shared-use computers would work "like a single organism," Pritvits claimed. To link all the communication cells the Computer Center provided AIST, which permitted workers "not to bow before the machine with each problem and not to wait while it completes each task" but to interact directly with an entire complex of computers from a terminal in each institute. Pritvits believed the ASU would become universal—like the telephone, electricity, and central heating.[74]

Yet the early ASUs, as mentioned above, were merely glorified inventory control systems. In one, an architect utilized information taken from sociological surveys in Novosibirsk microregions to make recommendations for improvements in housing and services. Another system, called "Health Care" (*Zdravookhranie*), modestly increased the diagnostic and prophylactic capability of medical institutions and tracked the health patterns of the region's inhabitants with an eye toward understanding lost labor time. Still another ASU, "Capital Construction and the Utilization of the Territory," was intended to streamline resource use and construction. Finally, a series of ASUs were utilized by government and party organizations to improve information storage, retrieval, and processing.[75] Knowing the level of Soviet medical care, the quality and aesthetics of housing, and the efficiency of Soviet bureaucracy, it is doubtful these ASUs contributed anything to the health, comfort, or representation of the inhabitants. The ASUs were designed to cut costs through judgments of "efficiency." As substitutes for rational economic managers they paid little attention to "equity" or "justice."

Marchuk saw great reserves of economic potential in the computer but also significant problems with the ASU. "If in the next five to ten years we succeed in radically perfecting planning and management at all 'levels,' using computer technology and economic mathematical methods, the productivity of labor will grow sharply," he said. ASUs had the necessary speed but were not used properly owing to the "psychological inertia"—the lack of interest—of many managers. Another problem was the belief that computers were autonomous, that the acquisition of one was half the battle to develop an ASU. Many managers were passive. Others overloaded small computers with complex problems. This led the Computer Center to arrive at the idea of time-shared, centrally located computers. The enterprise bought machine time, not the machine. ASUs were brought on line by the branch principle, that is, one ASU per branch of economic activity. This led to duplication, however, since there could be as many ASUs as branches. More crucial, significant technological barriers slowed assimilation of ASUs, for example, the absence of dedicated lines for the transmission of information, not to mention a shortfall in computers. There was also a huge lag in the training of young specialists. None of this, however, prevented the Communist Party from embracing the promise of the computer for economic management at the twenty-fourth party congress in 1971.[76] During the next three five-year plans, the party attempted to hasten the introduction of ASUs but without success.

One reason the computer revolution failed to take hold in the USSR was the nearly exclusive emphasis on economic applications for central control and planning. Computer scientists and economists with visions of rational planning on a national scale, planners fascinated with heavy industry and increasing productivity, and others of their ilk drunk with the power of the rational computer dominated the ASU program. In the Brezhnev years central planners were consumed by the desire to apply input-output analysis universally, as well as to put a terminal in every enterprise and a mainframe in every ministry. This was to be a pyramid of perceptive computers connected to omniscient mainframes. The belief was that if enough computer power were available, the giant mathematical models about which Soviet economists dreamed might lead to optimal central planning. Yet the ASU was intended to ensure the primacy of planners' preferences over the manager, and the last thing the enterprise manager wanted was more accurate and timely information in planners' hands, for this tied his in knots.

ARTIFICIAL INTELLIGENCE IN AKADEMGORODOK

The vitality of Akademgorodok's core group of computer specialists—Ershov, Kotov, Nariniani, and others—made the Artificial Intelligence program in the city of science one of the most original in the USSR. These scientists had come a long way from the 1950s when their claims for the promise of a thinking machine provoked hostile criticism. Now their research focused on theoretical programming, programming languages and systems, man-machine dialogues in natural language, processing of texts, and educational informatics. In their effort to develop high-level ALGOL-type languages and to elaborate a general theory of parallel programming, the computer specialists set out to optimize

ALPHA, ALGIBR, ALPHA-6, EPSILON, EPSILON-MB, SETL, and the AIST-0 programming systems.[77]

Igor Vasilievich Pottosin, director of the Institute of Informatics Systems which split off from the Computer Center in 1990, was closely involved in these efforts. A native of Tomsk, Pottosin stopped in Akademgorodok on his way home from Moscow in 1959. Sobolev recommended that he meet with Ershov, who asked him on the spot to set up a programming department in the Institute of Mathematics. At first Pottosin's employer in Moscow would not let him go, but the Siberian division had a special order from the Council of Ministers to "requisition" personnel in just such cases and eventually it had its way. Since the Siberian Institute of Mathematics had no machines, Pottosin and the others remained in Moscow for a year anyway. Three groups in the country developed compilers: Shura-Bura's, Korolev's, and Ershov's with Pottosin and others who created ALFA. The Ershov group focused less on architecture than on how to operate the machine itself. It was also involved in a joint effort with Minradioprom's Kazan Computer Factory to build a computer. Over the next fifteen years the group produced a series of effective translators and programs. In the early 1980s this group tried again to develop indigenous technology under Vadim Kotov and worked to achieve parallel programming in the modular asynchronous reconfigurable systems project, or MARS.

It was hard to get money for any project, especially one involving uncertain payback. On one occasion physics students at the university proposed building a promising new computer prototype. Pottosin turned to Kotov, who in turn gained Marchuk's support. The idea was to have the university pay the physics students with funds from the Computer Center, but the university refused to support the project, claiming the students were out to cheat the center. So instead Marchuk's son, Pottosin, and others each took on a few student assistants at the center, and in a couple of years a prototype was designed. But branch industries refused to consider the project. The Computer Center specialists soon tired of the charade, convinced that Soviet industry would never help them defray costs or provide special equipment without having arms twisted by the central party apparatus, so their dream to one day establish their own industry was relegated to a fantasy.

In 1964 Ershov shared his hope for AI in an article entitled "Man and Machine," published in the weekly Akademgorodok newspaper. Like Allen Newell and Herbert Simon at the Rand Corporation, Ershov predicted with confidence that in the near future computers could be made to understand human instructions in common, informal languages and to respond in a way that showed understanding. To achieve this goal the computer had to have the ability to comprehend. Ershov acknowledged that it was no simple task to make all basic grammatical constructions in a language such as Russian machine-readable. No real dialogue might be possible for two reasons. First, the machine might not "understand" the input as intended. Second, making the task clear to the machine might be too time-consuming and tedious. To achieve true understanding, Ershov believed, the computer had to be capable not just of storing and analyzing information but of refining accumulated information, making deductions and analogies, and asking appropriate questions. The process had to be two-way during which the computer's level of understanding approached that of the programmer. All this had to be done in a nonformal language both to verify the computer's real "comprehension" and to ensure its accessibility to a broad body of users. When Ershov first arrived in Akademgorodok he had no doubt that scientists would succeed at this task and that it was bound to be a lengthy, labor-intensive study. Nearly forty years of research into artificial intelligence throughout the world has demonstrated just how complex, and some would say impossible, this problem is.[78]

One of the leading architects of Soviet artificial intelligence, Aleksandr Semenovich Nariniani, continued Ershov's Novosibirsk tradition. A handsome, fit, middle-aged man with graying hair and a soft, wide smile, Nariniani now divides his time between Akademgorodok and Moscow, where his roots and family are. His father, Semen, was a famous writer connected with Boris Pasternak and others in the writers' colony of Peredelkino outside Moscow. His ties to this literary community ensured an affinity for language when AI became his true avocation. Although he intended to become a mathematician, or perhaps an architect, he found these studies too abstract. Then he attended a Moscow seminar on cybernetics which convinced him to attend Moscow Physical Technical Institute, an institute of Minsredmash, where a Department of Cybernetics had recently opened. Minsredmash saw value in computers for building hydrogen bombs and intended to foster the rapid development of cybernetics. In 1957 Nariniani read about the founding of Akademgorodok, and in 1960 several of his friends went there. They told him of the freedom, the nightlife, the nude bathing at Ob Reservoir "seminars." He decided to join them there for a year and take courses while working on his senior thesis. Although it was difficult to get away from a closed Minsredmash institute, Sobolev was able to arrange for a quick exit and Nariniani was given an assignment in the Department of Parallel Computer Technology in the Institute of Mathematics.

For ten years Nariniani lived in a dormitory as few buildings were finished. Although cramped quarters and poor food were the rule, Nariniani met outstanding talents and intellectual outcasts and encountered Soviet high culture. Here Andrei Bers, considered a social misfit in Moscow for his outspoken manner, could develop to his fullest potential. Here people would gather to exchange polemics and anecdotes in the smallest apartments, where the only place to sit was on bookshelves, and at the coffee clubs. This was a time of debate—over C. P. Snow's two cultures, over who was the greater authority on human nature and the truth, physicists or lyricists (*fiziki i liriki*). They met at "Under the Integral" and "The Smile" (*Ulybka*, also known as the "Crooked Smile" because of its reputedly bad food). The youngest of the scientists and the most irreverent met at the Club of the Gay and Resourceful (*Klub Veselykh i Nakhodchivykh*, or KVN), which was reborn with perestroika and glasnost. "When people visited," Nariniani said, "they stood immobile, their mouths and eyes gaping. In the dorm we had our own 'Under the Integral' where dance, pierogi, card games, and discussions were a good synthesis." Some of the more traditional scientists fretted about the cards and the women, concerned that the place would be closed down.[79] For the nascent computer specialists, the debates centered on whether machines could ever be made to think. In *Signal*, which Nariniani, Bers, and Ershov read and discussed, Poletaev expressed his firm belief that machines could indeed become intelligent. For Nariniani, as for many others, the question about thinking machines was like asking which culture should be dominant, for so many issues remained to be resolved. One candidate of science said it was like asking, "Is immortality possible?" For, what is immortality, and what does it mean "to think"?

In the early 1970s Nariniani joined a section of the Computer Center concerned with AI and known as "Intellect." A laboratory group was formed in 1972 and the laboratory itself, under Ershov's Department of Informatics, was established in 1978, with Nariniani at the helm. This was a period of great activity, resulting in the creation of such languages as SETL, VOSTOK, and DIALOG. In its first years the laboratory had two main goals: man-machine interaction in natural language and the development of the SETL-BESM programming system. Work on a formal model of a natural language quickly led to the basic principles as well as the elaboration of experimental systems that modeled understanding of communication. The system VOSTOK-0 was capable of accepting information that represented a semantic network of predicates of the first order. The system included a model of time with a mechanism of logical conclusion and a thesaurus, as well as an active interface based on simple natural language. Work began on VOSTOK-1, which included a hierarchical semantic network, encyclopedic information, and a model of temporal and spatial relations. DIALOG was created for the automatic processing of text in natural language. The laboratory earned thousands of rubles annually from regional party organizations, national ministries, and the university. It maintained active contacts with other institutions at home and abroad, including those in Czechoslovakia, East Germany, France, and the United States.[80]

Like Ershov, Nariniani regretted the extent to which emphasis on applied problems waylaid the development of computer science in the USSR. He eloquently defended research on AI. The success of systems that operate on natural languages depends largely on the level of development of the formal model of the language. The concept of the nature and scale of the model constantly evolves. Computer scientists in AI—specialists in logic, linguistics, and semiotics—focus on the improvement of the formal means of representation of the model, the theory of language (linguistics itself), and the expansion of the formal description of the language. According to Nariniani, the dynamics of the development of these languages and problems associated with them are such that often the model "succeeds in aging morally long before its realization." This "early death" is explained first by their "complex of ineffectiveness." The unsolvable conflict between fullness and naturalness of the model, on the one hand, and the demand of effectiveness of the corresponding machines, on the other hand, has always been solved in favor of the later. The path that emphasizes the computer is no longer fruitful. Emphasis must be placed on the choice of the proper ideology of a formal model of languages. "It is time," Nariniani asserted, "to separate the development of the model of language, as fundamental research that is directed by the problem's 'internal' laws of development, from its applied, 'current' realizations."[81] Some AI critics say that the failure to create intelligent machines comes from a complete misunderstanding of what "intelligence" is to begin with and that Nariniani's entire approach to the problem was therefore doomed before he even began.

For Nariniani, the joy of working on AI in the 1970s and 1980s did little to replace the intellectual excitement and esoteric debate of the early years. The party had cracked down on "high cultural" discussion after the podpisanty affair. The second wave of scientific settlers to Akademgorodok consisted of gray, if competent personalities. The social clubs like "Under the Integral" disappeared. In AI, systematic interference of ministerial and military interests in computer science caused many specialists to loose faith. Only after repeated prodding from Ershov in Novosibirsk and other scientists in Moscow, Leningrad, Kiev, Tartu and Talinn had the scientific establishment belatedly recognized AI as a genuine discipline. In August 1975 members of an Academy of Sciences Council on Semiotics of the Cybernetics Council discussed the problems of man-machine interaction. There was concern, however, that AI's strong theoretical basis had not been reflected in practice. The AI effort was small-scale, poorly organized, and underfunded, with inadequate access to powerful machines or avenues for publication. The scientists called for redoubled efforts to create systems of formal representation of thought; the development of dialogue systems that utilized large dictionaries, semantic systems based on natural languages, and new languages; and continued semiotic study of complex systems in AI and applied cybernetics. Academy specialists established a national section on AI in the Scientific Council on Cybernetics only in 1978. The AI section was chaired by V. M. Glushkov, head of the Kiev-based Institute of Cybernetics, with Ershov and eleven other leading specialists on its board. AI projects included DIALOG, under Ershov; "Situatsiia," under D. A. Pospelov; "Konstruktor"; "Intellect of Robots"; "Bank"; and several commissions on "Methods of Searching and Decision Making," "Psychophysiological and Psycholinguistic Aspects of AI," and "Methods of Mathematical Logic in the Area of AI."[82]

Maintaining their bold leadership in Soviet and international AI efforts, Ershov, Kotov, and Nariniani proposed to create a fifth-generation artificial intelligence, something no group in the world has yet achieved. They tried to involve Marchuk in their work. In the fall of 1983 the Academy of Sciences held a special session attended by representatives of Minradioprom, Minpribor, and the military. The participants agreed to form a government council to work on programming, architecture, and intelligence (language) on several levels. Ultimately, however, the "scientific mafia" decided it was better—cheaper—to buy and copy fifth-generation science and technology. Academician Seminikhin, the "black genius of computer technology," said, "Let's just copy. It's easier to be second than first." Even in the face of this opposition Marchuk created a "temporary creative collective," essentially an institute without walls, made up of Moscovites, Siberians, and Estonians, with two brigades in Novosibirsk of 150 to 200 people, with hard currency and Australian machines; within three years they had developed a prototype. "We were clearly ahead," Nariniani claimed. But the technocrats again decided to copy Japanese and western efforts. The loss of support slowed research. Only in 1988 were Akademgorodok Computer Center specialists able to publish their first substantive collection of articles on AI.[83]

As with AI, so the Soviet supercomputer effort lagged considerably owing to bureaucratic and political problems. But without a major change in national policy it was unlikely that the three or four major research centers that were needed to develop supercomputers capable of 1 to 10 billion operations per second would ever be created. A promising supercomputer effort commenced under Vadim Kotov of the Akademgorodok Computer Center. Already in 1967 a new laboratory was established under Ershov to explore the theory of computer processes.[84] In 1975 Ershov was succeeded by Kotov, who thought he had found a way to achieve supercomputer capacity using a parallel processing system. He might thereby avoid the systemic handicaps on the development of Soviet computer science and technology—unreliable, slow chips, inadequate support for fundamental research, and ministerial domination. Kotov, who resigned as director of the Akademgorodok Institute of Informatics Systems for a position at Hewlett Packard in Palo Alto, California, is a specialist on the theory of concurrent systems and processes, concurrent software, and architecture. When the Japanese launched their fifth-generation project in the early 1980s, there was a broad response from Soviet computer specialists. They sought government support for a large-scale, long-term, and well-funded program. Others, like Kotov, recognized that the government would not be interested in such a project because of declining economic growth rates and general avoidance of long-term technological investments. Moreover, Soviet leaders rarely relied on informatics or AI; centralized decision making discouraged government interest in such decentralized, user-friendly technological systems. Finally, the growing Soviet lag in computer technology dampened enthusiasm for this program.

Two approaches in fifth-generation computing, concurrent architectures and intelligent softwares, several of which were based in Akademgorodok, managed to get off the ground. Kotov initiated one of them, START, in 1985.[85] A central component of START was the modular asynchronous reconfigurable systems concept, or MARS, "to experiment with new architectural concepts, concurrent languages, software, and methodologies." MARS was developed in response to the fact that contemporary computer programs were being applied to run increasingly complex systems—ocean liners, power stations, rockets, jets, and military hardware—but the development costs had grown disproportionately large, there were limits to productivity, and the programs were frustratingly time-consuming to debut. The major difficulty was that the speed of transmission and processing of signals "within" the computer approaches the limit of the speed of light. The only way beyond this limit was to break a problem down into a series of logically independent subproblems and solve them simultaneously. MARS was intended to achieve simultaneous parallel processing of different fragments of one and the same problem, so-called parallel programming.

The START collective was hampered by bureaucratic impediments in paying and hiring workers.[86] This did not preclude further development of MARS. MARS-M involved research on "Cray-like supercomputers with few powerful pipeline processors for 'vector crunching.'" MARS-T focused on the development of "multiprocessors with a larger number of simple scalar (micro)-processors." After the START project was completed, several participants organized small companies and software houses to generate financing for a second research stage and produce commercial projects for interested industrial parties. What remains of Kotov's group in Akademgorodok is seeking, with Kotov's assistance, foreign partners who have time, equipment, and money to support their efforts.[87]

Gorbachev himself considered the development of supercomputers as central to the revitalization of Soviet industry. When he addressed the twentieth national Komsomol congress on April 16, 1987, he drew attention to the work on supercomputers by young scientists at the Lebedev Institute of Precise Mechanics and Computer Engineering, the Minsk enterprise "Integral," and the Akademgorodok Computer Center. Supercomputers based on parallel processing and capable of billions of operations per second were crucial to manage the economy, automate planning and modeling, undertake research with global implications on climatological change and space, and accelerate technological progress, Gorbachev told the young communists. The project "Elbrus," under specialists in the Lebedev Institute, and "Mars-Kronos-Start," under Vadim Kotov in the Computer Center, were indications that research on supercomputers was moving ahead, albeit in fits and starts.[88]

COMPUTER SCIENCE ENTERS THE 1990s

By 1990, with the exclusion of the USSR and Eastern Europe, perhaps $19 billion had been spent throughout the world on the development of artificial intelligence, with $12 billion spent in the United States alone. The United States had nineteen AI journals, and more than a hundred universities offered courses on AI. In the USSR, on the other hand, the choking grasp of Minpribor, Minradioprom, and conservative Academy leadership prevented the field from flowering. There was a dearth of highly qualified individuals, not one journal for the entire field, rarely a professional conference, and virtually no university courses. There were few textbooks, and contact with the West was limited. In the West linguists, engineers, psychologists, mathematicians, and programmers shared interest in the field, while in the USSR they hardly knew one another. Indeed the field was contracting, and training had slowed to a stand-still. Anything short of the general state of the art in the West—long-range, well-funded fundamental research, training and retraining, the creation of cur-ricula, funding for attendance at conferences at home and abroad, and special-ized journals—meant further lag for the USSR.[89] The result of ideological, po-litical, and economic factors was that the USSR missed what has been called the second industrial revolution, that is, the electromagnetic storage, retrieval, communication, and manipulation of quantitative data, the efforts of scholars in Kiev, Leningrad, Moscow, and Novosibirsk notwithstanding. Its citizens had remarkably limited access to modern telecommunications with dedicated lines, electronic mail, satellites, cable TV, and cellular radios and telephones.

Under Brezhnev, computers were instructed to have a major role in bringing about automation in industrial processes, but few envisaged some of the other more individualized applications so prominent now in the West. Computers were used in banks, enterprises, research institutes and hospitals, statistical administrations and central party organs, but rarely in the home or for "con-sumer" purposes. Contributing to the problem was the domination of the computer field by industrial ministries with narrow economic interests. The paradigm of the Brezhnev years—science and technology for economic growth—found its obvious manifestation in the Computer Center in the devel-opment of ASUs.

In the United States the word *hacker* provokes an image of a young man with straggly hair, smoking, drinking lots of black coffee, and working late into the night at the computer. The Soviets have a counterpart. But in the late Brezhnev years the authorities tried to impose order on these peripatetic programmers, tried to get them to conform to Soviet norms, even tried to force them to work on a nine-to-five schedule by creating an internal pass system at the Computer Center and threatening "persons who violate the pass and internal regime" with "disciplinary measures." No one was allowed to work after midnight or on holidays and weekends without special permission, and even regular workers had to turn in their passes while on extended leave. Those who lost passes were put through a rigorous inquisition—and in some cases punished—before a new pass was issued. An exception was made for "members and candidate members of the party, members of the government, secretaries of the Obkom, Gorkom, and Raikom of the party, the chairman of the Siberian division . . . his deputies, staff of the apparatus of the presidium . . . and individuals accompa-nying them" who could "enter the Computer Center . . . on the basis of their identity cards." Finally, materials of value could be removed from the premises only with special passes with an exact description of the item(s) and person(s) involved. Watchmen were given permission to inspect all property.[90] The major intent of the rising tide of ideological control was to prevent samizdat publishing. Of course Bers, Nariniani, and the others had better things to do than pay strict attention to the letter of the law.

Computer specialists in Akademgorodok labored mightily to overcome the systemic problems that plagued Soviet computing. They overcame poorly equipped facilities, the emphasis on applied research at the expense of fundamental, and the impediments of conservative ministerial domination of science policy. Third-generation computers with limited memory and power, of course, were the primary obstacles scientists faced in their daily activities. Yet the Computer Center had the ingredients for success: qualified, foresighted, enthusiastic personnel, a collegial environment, and, what to the authorities must have been infuriating, the ability to go outside accepted paths to augment their facilities and gain funding for their programs. Through a combination of creativity, ingenuity, and stubbornness, the computer specialists achieved most of their goals. Indeed the story of computer science and technology in Akademgorodok is primarily one of achievements. It stood at the forefront of Soviet research in meteorology, seismology, and other areas of geophysics; mathematical modeling; economics; applications for other disciplines; innovations in software; and the first steps toward fifth-generation architectures and parallel processing. Lacking supercomputers, programmers developed skills in other areas: languages, semiotics, architectures, mathematical modelling, and so on.

The future of the Computer Center, however, remains uncertain. Only after perestroika did programming become a valued and lucrative profession, with software viewed as a commodity to be traded commercially. Laws to protect authorship by according rights of intellectual property are only now being established. Many talented researchers have set up their own computing cooperatives and seldom come to work. Financial shortfalls in the Academy have made life bleak. Nariniani sees the uncertainty surrounding the institute as symptomatic of the decay of Russia. His solution? He wants to create a "free intellectual zone" outside Moscow where *fiziki* and *liriki* can build up and export Russia's intellectual potential—its science and art. Nariniani worries that world civilization will enter the twenty-first century without Russia. The political and economic crises enveloping Russia today has focused attention only on moving ahead, not on the direction to take. Nariniani disagrees with those who say, "Stop the experiments, the special path of Russia is a myth, let us become like the others." But who are "the others"—Brazil? The United States? Sweden? Italy? Japan? Other people see "market mechanisms" as a panacea. For Nariniani, the "market" is not the answer. Why this fascination with entering the world market, he asks, when there is little in the way of finished goods Russia can sell? He recognizes that there is a limit to the export of natural resources. One marketable commodity remains, however, and that is intellect based on a millennium-old culture. Since the revolution in science, education, and culture, great strides have been made. Nariniani is deeply concerned in saving this culture before a brain drain destroys its foundations. Like Japan, which rose in the period from 1970 to 1990 to world leadership in computers and automobiles, Russia can triumph in its industry of intellect.[91]

When Siberia enters the nineties computer science may lead the way. A large number of young, talented programmers and applied mathematicians have set out to establish their own consulting groups, think-tanks, and software firms.

The Computer Center has tried to remain on top of the burgeoning computer industry. It has split into two, more manageable institutes, although some say it is still unwieldy. With Marchuk's support, in April 1990 the Institute of Informatics Systems was created, primarily out of three other departments—the Department of Informatics, which consisted of two laboratories, experimental informatics and artificial intelligence (Nariniani); the Department of Programming, with laboratories of systems programming (Pottosin) and a group on theories and methods of translations; and Kotov's Department of Computation Structures and Processes, consisting of five laboratories.[92] The new Institute of Informatics Systems carefully guards its allegiance to fundamental research. Three-quarters of its funds come from the Academy of Science and only one-quarter from contracts.

The Computer Center continues its main research directions: the improvement of mathematical models related to the physics of the atmosphere and ocean, geophysics, mechanics, chemistry and physics; the development of universal software; methods of numerical statistical modeling for solving multivariate problems of the theory of radiation transfer; the theory of conditionally correct problems in mathematical physics; and automated control systems. The center provides roughly 40 percent of its machine time to various institutes of the Siberian division, and perhaps 15 percent to enterprises. The qualifications of personnel at the Computer Center remains high. Many scientists have gone abroad, however, to start their own businesses, or they use the facilities for their own purposes.[93] The hope is that western investment, privatization, and a programming culture will come together to rejuvenate the Computer Center. But old habits die hard, money is tight, and Siberians are averse to taking risks, so many wonder if the opportunities are not better abroad.

Siberian Scientists and the Engineers of Nature

SOVIET SCIENCE was big science, applied science, science to tame nature and have it serve the state. Owing to the advent of superpower competition, the growing hubris of technical specialists, and the absence of public opposition, Soviet scientists and engineers tackled increasingly grandiose projects in the postwar years, with little regard for the environment or for public health. These "engineers of nature" tapped virtually every aspect of Siberian natural resources, equating conservation with "intensive utilization." The projects acquired great "technological momentum," that is, the ability to take on a life of their own, to prolong their own existence long after their usefulness had passed.

Under Stalin, nature's engineers pursued transformationist visions in canal, dam, and irrigation construction that rivaled the efforts of the Army Corps of Engineers and the Bureau of Reclamation in the United States. Khrushchev and Brezhnev provided additional impetus to these projects, Khrushchev through his support of Siberian development and various agricultural projects and Brezhnev through his "southern strategy" to produce grain in Ukraine and southern regions of Russia, areas with virtually all the meteorologically desirable characteristics for grain except water. Brezhnev's southern strategy therefore required the building of thousands of miles of irrigation canals.

Siberian scientists became active in environmental politics over the irrationality and gigantomania of two Siberian projects involving water resources. The first involved the construction of paper mills on the shores of Lake Baikal; the second concerned the diversion of the flow of Siberian rivers thousands of miles to the west.

In 1954 the state committee of the cellulose, paper, and woodworking industry of the USSR learned that the United States had begun producing a super-strong rayon cord. Members of the committee and representatives of the defense industry recognized that the cord was important for advanced aviation design. They determined to build a plant to meet the perceived U.S. threat and decided that Lake Baikal, the largest freshwater lake in the world, with a low mineral content perfect for various manufacturing purposes, was the ideal location. Over the next twenty years, despite Siberian scientists' efforts to delay or halt construction, two paper mills were built on Baikal's shores, joining dozens of other industrial and agricultural concerns, all threatening the beauty and wildlife of the region.

In 1958 N. A. Grigorovich, an engineer from Mosgidep, the Moscow division of the All-Union Hydroelectric Design Institute, proposed using thirty thousand tons of explosives to turn Lake Baikal into a seemingly endless source of water for downstream Angara River hydropower stations. Grigorovich was disturbed by the imperfect, narrow outflow of Baikal at the source of the Angara. He proposed tapping Lake Baikal and the water that had backed up behind the Irkutsk hydroelectric station, often called the Irkutsk Sea, and known by some locals as "Baikal Bay." The force of the detonation (thirty kilotons compared to Hiroshima's twenty) would make the aperture of Baikal twenty-five meters deeper and a hundred meters wider for a distance of ten kilometers. Grigorovich calculated that the millions of gallons released would create billions of kilowatt hours of electricity.[1]

The Siberian Rivers Diversion Project was even more extravagant. It would have exceeded the pyramids as a monument to modern engineering prowess. All the major Siberian rivers—the Ob, Irtysh, Enisei, and Lena—flow from the south to the north into the Arctic Circle. Engineers planned to divert the flow through a series of canals, locks, and new riverways, replete with hydropower stations and irrigation networks to distribute water to arid Soviet Central Asia. Some far-sighted dreamers even suggested using small, so-called peaceful thermonuclear devices to melt snow, lift soil, move mountains, and dig trenches, all in the service of socialist geological engineering. No one seems to have thought of the irreversible global impact such diversion would have had.

Environmentalism, waylaid by Stalinism and hindered by Khrushchev's allegiance to Lysenkoism, owes much to Siberian scientists for its resurrection. The scientists joined local citizens and Russian writers in the fight against big projects to hatch environmentalism in the post-Stalin USSR. The rise of the environmental movement in Siberia showed that interest groups whose views differed from the dominant party and government hierarchy would inevitably form in a modern, increasingly well-educated society. Here an unlikely amalgam of free-thinking scientists from various disciplines, social and political backgrounds, and cities—A. A. Trofimuk from Akademgorodok, G. I. Galazii and M. M. Kozhov from Irkutsk, and others from Moscow—joined leading Russian village prose writers to protest decisions made in Moscow. Their battle to save Baikal and the Siberian rivers reveals the extent to which economic plans overwhelmed scientific objections in communist Russia, even in the face of irrefutable scientific evidence.

The mismatch between Soviet environmental protection laws and practices is the subject of legion tales. Rivers were dammed to capacity for electrical power generation and irrigation, so that evaporation increased, the flow turned to a trickle, and brackish waters invaded delta areas formerly rich in fish and wildlife. Mining, drilling, and other extraction processes were forced ahead without regard for fragile tundra and taiga. Haphazard, if not intentional disposal of hazardous waste led to ecological disaster seldom encountered in the western world.

Many Soviet and western observers assumed that a communist government with central economic planning would be far more sensitive to environmental issues than a pluralist system with a market economy. In the former, social ownership for the common good would protect resources from profligate use. In the latter, capitalists would compete for scarce resources and strive to maximize profit at any cost. Environmental concerns, so the argument went, would take a second seat. As it turned out, the Soviet developmental model was far more costly to the environment. Because of the desiderata of investment in heavy industry, the need to increase production and fulfill plans at any cost, and the Marxist labor theory of value which fails to attach significant costs to air, water, or land, the environment was seen solely as something to be tamed by man's will. Environmentalism existed in legal statutes, not in practice. It did not help that environmentalists were seen as "bourgeois theorists" whose purpose clearly was to "wreck" economic plans through the establishment of nature preserves, research stations, and university departments that were intended to foster the rational use of the nation's great natural resources.

By 1960 most of the European USSR was subjected to human geological engineering on a vast scale. Rivers were tamed for power and irrigation. Chemical pesticides and industrial pollutants were dumped into drinking water. Ore was extracted without regard for the scars left behind. Only Siberia's climate and geographic isolation had prevented the Communist Party from unleashing economists, planners, and geologists on its natural resources. To a certain degree Khrushchev's policies tempered this breakneck economic development. But at the same twentieth party congress where he loosed the de-Stalinization thaw with his "secret speech," Khrushchev also set Siberian development in motion by pointing to its rich natural resources and their role in building communism.[2] It was left to such scientists as Andrei Alekseevich Trofimuk from Akademgorodok to urge caution in Siberian development.

Trofimuk, a short, bespectacled, somewhat frail man now in his eighties, arrived in Akademgorodok in 1957. A man of peasant roots who was given the opportunity to receive specialized higher technical education by Soviet power and channel his political aspirations into party work, he was a most unlikely opponent of extensive Siberian research development. As director of the Institute of Geology and Geophysics from 1957 on, he willingly embraced dogma about the party's leading role in economic development. He believed that scientists ought to rework nature to suit man's needs. He favored ocean shelf oil exploration and widespread construction of nuclear power stations. Frequently he spoke up at party economic planning meetings to criticize the pace of development of Siberian oil and gas resources. A geophysicist by training, Trofimuk made his career by identifying oil fields that some dismissed as unproductive. To their later regret, Trofimuk and others advocated a technique of pumping water into wells to force oil out. Although increasing production in the short run, the technique ruins the site and virtually precludes recovery of oil after the initial surge in production. But in one area, the Baikal question, Trofimuk rejected party gospel. Siberian scientists, native or immigrant like

Trofimuk, joined together in a thirty-year battle to save Siberian lakes and rivers from the designs of the engineers of nature.

In 1958 Trofimuk entered the increasingly divisive debates about the future of Baikal and maintained a visible profile until his recent retirement. How he got to Baikal—both physically and intellectually—is one of those wonders of Soviet history. He was born in Belarus in 1911, and his mother died when he was quite young. His peasant father fled with him eastward to avoid war, and they arrived in western Siberia in 1918. His curiosity, diligence, and proper social origins secured him a place in the geology department of Kazan University from 1929 to 1937. He defended his candidate of science degree in 1938 on "Oil Bearing Limestone of Ishimbaev" in the Volga-Ural region. In 1944–45 he moved to Romania at the behest of the Soviet government to reestablish oil production in lands taken from the Germans. In 1949 he finished his doctoral work on "Oil-bearing Paleozoic Deposits of Bashkiria [now Bashkortostan]." He completed his higher education at the Evening University of Marxism-Leninism of the Ufa City party committee, but it is a mystery as to what Marx and Lenin had to add to oil discovery and recovery. In 1950 he moved to Moscow as chief geologist of the main fossil fuel administration of the Ministry of Oil Industry. In 1953 he became deputy director and, in 1955, director of the All-Union Oil-Gas Scientific Research Institute. His achievements in oil prospecting were considered sufficient accomplishments to earn him a the status of corresponding member of the Academy. In 1958 he became an academician owing to significantly easier entrance qualifications worked out by Lavrentev for the Siberian division.

From 1958 to 1962 Trofimuk divided his responsibilities between Moscow and Novosibirsk, but more and more of his energy was devoted to the organization of the Siberian division's Institute of Geology and Geophysics. Symptomatic of the big science of the Brezhnev era, by 1975 his institute employed more than a thousand workers, including six academicians, five corresponding members, forty doctors of science, and two hundred candidates of science. Trofimuk was a member of several government commissions with responsibility for the development of the oil and gas industry including SOPS, several other Academy of Science councils, and a number of editorial boards. He also served as deputy chairman of the presidium of the Siberian division, and, from the 1930s on, as a member of local and regional soviets and party committees, eventually as a member of the Supreme Soviet of the RSFSR and the Novosibirsk party committee. That he had time to continue looking for oil and to publish is surprising. But, like many academicians, he was able to sign his name to research conducted in his institute, and in the fifteen years between 1960 and 1975 he published four times as many articles than he had in the preceding quarter of a century.

Considering Trofimuk's allegiance to party principles and his support of technologies with a potentially negative environmental impact, it is difficult to understand his vocal defense of Lake Baikal. Yet, in the words of his colleagues, he turned into a "wild animal" in his efforts to protect the pristine lake. He was shocked by the cavalier attitudes of Soviet engineers and could not fathom planners' insistence that the "self-cleaning ability of Baikal, its natural bio-filter" would protect the lake from pollution. But he also participated in environmental folly when he suggested building a pipeline to the Irkut River to transport paper mill wastes out of the Baikal basin. "Perhaps, for once," he later admitted, "industrial leaders were right, recognizing the cost and fantasy, if not the absurdity, of 'diversion of wastes.'" [3]

SIBERIA FACES THE SHOVEL AND AX

Russian scientists have been studying Siberian resources with an eye to tapping them since before the turn of the century. A vast region spanning as many as eight time zones and three climates, Siberia has ninety percent of the USSR's hydroelectric potential; half its fossil fuels, including the richest oil and coal regions; major deposits of rare metals, gold, and platinum; four-fifths of Russia's freshwater and nearly half its freshwater fish. The Commission for the Study of Natural Productive Forces (Kommissia po izucheniiu estestvennykh proizvoditel'nykh sil), or KEPS, an organization under the umbrella of the Imperial Academy of Sciences, was formed to explore Russia's great natural wealth during World War I. The war had cut Tsarist Russia off from its traditional suppliers of chemicals and ores for gunpowder, iron, and the like. In the 1920s KEPS gave way to SOPS, the Academy and governmental council responsible for the development of the oil and gas industry. Its first chairman was oil specialist I. M. Gubkin, whom Trofimuk considered his mentor, and then V. L. Komarov, president of the Academy of Sciences from 1936 to 1945. When German armies advanced during World War II, conquering most of the Soviet Union's industrial potential, SOPS focused administrative energies on the rapid development of the Ural region to provide resources for the front.

As soon as postwar reconstruction was concluded, the Communist Party turned to the development of Siberia. In 1947 the first postwar conference on the development of the productive forces of East Siberia was held in Irkutsk, a large Siberian town on the Angara River, seventy kilometers downstream from Lake Baikal. Irkutsk, the center of the East Siberian branch of the Academy of Sciences, home of an active chapter of the Russian Geographical Society, and the future center of a burgeoning hydroelectric power network, was the logical choice for the conference. Indeed, although SOPS offices were located officially in Moscow, the true center of activity was in Irkutsk. The Communist Party developed policies to encourage settlers to move from the European industrial and population centers to Irkutsk, Omsk, Tomsk, Krasnoiarsk, and Novosibirsk. Between 1913 and 1961 industrial production of Irkutsk oblast, where Lake Baikal is located, grew sixty-one times; two-thirds of that growth came in the 1950s alone.[4] But systematic study of Siberia commenced after

Khrushchev's 1956 party congress speech when Akademgorodok was added to the mix of industrial and agricultural ministries involved in Siberian management. Party officials asked its scientists to assist in bringing Siberia fully within the grasp of the nature creators.

No resource escaped the fertile minds of the engineers of nature, from the artificial regulation of the snow cover to the new field of "ice technology." The transformation of nature itself was the goal—the taming of tundra, taiga, forests, lakes, swamps, and wetlands to meet "the demands of the economy."[5] Dams were the major tool. As early as May 1920 engineer A. A. Velner completed a report for the State Electrification Agency, GOELRO, in which he described the Angara River's great potential for hydropower. He admitted that the scale of the effort presented significant obstacles, "but at the same time [one cannot] forget about the three million horsepower of cheap water power the production of which is technically possible."[6] In the 1930s limnologists continued to study the Angara in preparation for the construction of a series of hydropower stations. Since 1950 huge hydroelectric stations have been built on all of Siberia's major rivers, including one just five miles from Akademgorodok.

Engineers were thrilled by the unanticipated payoffs of their projects. One scholar, for example, was pleased to discover that the hydropower stations destroyed the small, nearly invisible blood-sucking mites (*gnusy*, in Russian) that infect the taiga of Siberia. Their presence had occupied scientists for most of the 1950s and early 1960s. All-union conferences met to discuss the *gnusy*. Attempts to exterminate them with DDT, first in aerosol powered by huge fans with jetlike engines, then dropped by crop dusters on rivers, and finally dissolved in the Angara itself, had failed. Then specialists discovered that the insects were far more susceptible to changes in the flow and level of rivers. The villain here was Lake Baikal, which contributed to the even flow and clean water of the Angara which the insects loved. Thus "the triumph of Soviet energetics" was a dam that interrupted flow and temperature, killed many midges outright, delayed egg laying and the formation of maggots, and eventually got rid of them.[7] What, then, could possibly be wrong with a paper mill?

At the beginning of July 1954 F. D. Varaksin, the Minister of the Paper Industry (Minbumprom, later the Ministry of Forestry and Paper Products, Minlesbumprom, used throughout interchangeably) invited B. A. Smirnov, a Leningrad engineer, to select a site for a cellulose plant near Baikal. Smirnov and his commission quickly selected a spot, from among fifteen potential sites, on the Angara River, just seventeen miles above Irkutsk. In 1956 local authorities, fearing pollution of city water supplies and also planning to build a party pension at that location, objected to the site so Smirnov's group chose instead a small river on the southwest side of Baikal, where the town, Baikalsk, and the paper mill were eventually built. The minister of the forestry industry had proposed an alternative near the Bratsk hydrostation and a new forestry complex, but this would have required the construction of a water purification station the likes of which the world had never seen. Of course the economic planners promised to build purification systems that would return water to Baikal even cleaner than it had been. If this were truly possible, couldn't the paper industry have built its mills anywhere, first producing clean water for its purposes? Brushing all questions aside a Gosplan commission approved Smirnov's choice, and limited construction began in 1958.[8]

A major conference in August 1958 on "the development of the productive forces in East Siberia," sponsored by SOPS chairman V. S. Nemchinov, was pivotal in drawing scientists' attention to the Baikal paper plant and set the tone for Siberian development. Caught up in the Khrushchevian fervor to transform Siberia into an important economic region, the conference spoke to the goal of surpassing the United States in per capita industrial production by the end of the sixth five-year plan in 1965. The conference participants, representing the Academy, Gosplan, the Council of Ministers, and various other ministries, discussed the major role the Academy would play in the study of geological formations, ore mining, and the extraction of fossil fuels; ferrous metallurgy; machine building; the construction industry; the chemical industry; forestry; agriculture; and transportation, including thousands of kilometers of new roads and railroads.[9]

Nemchinov described transforming the forests into "rayon for the textile industry, cord of high quality for tires, paper and boxes . . . and other kinds of chemical products."[10] Other participants discussed the development of Siberia's great natural resources, for example, its coal and hydroelectric potential, in similarly transformationist terms. The successful construction of the Bratsk, Irkutsk, and Kasnoiarsk hydroelectric stations convinced many that efforts should be redoubled to finish the Eniseisk, Ust Ilimsk, and Siansk hydroelectric stations (known in Russian by the acronym GES) along the Angaro-Eniseisk Cascade.[11] Enjoying Sputnik's recent conquest of space, and seeming to have the endorsement of scientists, the engineers of nature launched an all-out assault on Siberia's rivers, forests, and lakes. Lavrentev himself, although a budding environmentalist, gave impetus to the onslaught with comments he made on the conference that were prominently displayed in *Pravda*. The conference, he said, "had made a great impression on him. It showed once again how great and responsible was the role of scientists in the unmasking of the fantastic wealth of Siberian minerals and in the search for more rational ways and means for their utilization for the benefit of the Motherland."[12]

But the conference ended on a tense note after Grigorovich proposed a project too bold even for Soviet scientists. After the construction of the Irkutsk GES, Baikal had risen one meter, destroying much of the shoreline. Grigorovich now audaciously proposed dropping the level of the pristine lake five meters by widening the flow at the source of the Angara. Using a crude kind of cost-benefit analysis, Mosgidep engineers had determined that one could tap Baikal's power by opening the mouth and producing a hundred billion kilowatt hours of electricity over time. But what would happen after all this water had flowed out? Would the engineers use more explosives to lower Baikal further?

Mikhail Mikhailovich Kozhov, a leading specialist on Baikal mollusks and fish, could take no more. He strode to the podium to attack Grigorovich. Kozhov had taken a circuitous root to scientific prominence and to this conference. Born in a small village to a poor peasant family in 1890, he was self-taught, receiving a teaching certificate in 1913. He was mobilized in 1914, trained as an officer, and served at the front in Poland during World War I. When he returned to the Irkutsk region in 1918, he was caught up in the civil war and intervention in Siberia by the Czechs, like the protagonists in Sholokhov's novel *And Quiet Flows the Don*. Having been forced to serve by the Whites, he deserted three months later, in March 1919, and made his way across Lake Baikal to Krasnoiarsk where he was captured and again mobilized, this time by the Reds.

When he was finally discharged in 1921, Kozhov returned to school. The Russian university was in turmoil. Hundreds of demobilized soldiers and students from *rabfaky*, special workers' schools, who could barely read, write, or do simple fractions, flooded the schools. Those without the proper social or party background found their entry to the university barred. Faculty and administrations resigned en masse rather than face the disorder of civil war, revolutionary entrance requirements, shortages of books, and so on. Chaos reigned at Irkutsk University. The leading university faculty—eleven professors, sixteen teachers, and seven assistants—had recently walked off the job. At age thirty, having been taught in life's school, Kozhov filled the vacuum, quickly acquiring authority among remaining faculty and students alike and serving several administrative positions. Upon graduating in 1925, he entered the newly opened Biological-Geographical Scientific Research Institute (BGNII) of Irkutsk University. He studied the flora and fauna of streams around the Angara River and Lake Baikal, focusing on sponges and mollusks. The result was several publications in *Izvestiia vostochno-sibirskogo otdela russkogo geograficheskogo obshchestva* (Proceedings of the Far Eastern Division of the Russian Geographical Society). Kozhov became docent in 1931, professor in 1932, and later director of BGNII. After Stalin's government made the awarding of degrees again acceptable, Kozhov received his candidate of science degree in 1935 without defense, and in 1937 he finished his doctoral dissertation on "Mollusks of Lake Baikal."

Kozhov expanded his interests to the origin and the horizontal and vertical distribution of the fauna of Baikal. He became convinced of the indigenous origin of many of the species in the Baikal basin. The result was the volume *The Animal World of Lake Baikal* (Irkutsk, 1947). He then turned his attention to the area's fishing industry, organizing several expeditions to other lakes in East Siberia of similar tectonic structure. He completed three more fundamental monographs: *Freshwaters of Eastern Siberia* (Irkutsk, 1950), *The Fish and Fish Industry of Lake Baikal* (Irkutsk, 1958), and *The Biology of Lake Baikal* (Moscow, 1962).

Throughout the 1940s Kozhov defended his beloved mollusks and fish, in particular the omul, a salmonlike fish indigenous to Baikal, which were threatened by overfishing using motorized drift nets. He openly challenged the party's exhortation to local fish factories "to overfulfill yearly plans in six months."[13] Representatives of the district party committee and fishing trusts called him a *predel'shchik*—one who sets limits and is obviously not concerned with plan fulfillment. This made him a "wrecker"—a charge that led, during the purges, to imprisonment or even execution. But he stood his ground, adamantly criticizing fish industry workers who were concerned only with fulfilling the plan, arguing that "measures to increase the fish reserves" also needed to be part of the plan. To the dishonest attempt to attribute the decline in fish population to new migration patterns, he responded, "This matter has nothing to do with changes in fish behavior."[14]

Kozhov was well-equipped to take on economic and political organs of power. He had unassailable scientific knowledge, and his career path of war, civil war, and Soviet power had made him fearless. So at the critical Irkutsk conference in August 1958 he defended nature from economic desiderata. In his official presentation, entitled "Basic Problems of Research in the Fish Industry of Eastern Siberia," he raised the specter of overfishing. But he could not allow engineer Grigorovich to go unscathed:

> I did not intend to speak, but now I cannot keep quiet. Excuse me for speaking without preparation. Not only are fishermen against [the detonation]. The Sovnarkhozy [regional economic organs] are also against it. Baikal is a unique gift of nature. This is the deepest lake in the world . . . but for the animal world of Baikal the surface waters have the greatest significance. Reducing Baikal by five meters will expose the entire shoreline circumference, an area of 100,000 hectares, and dry up the fish spawning grounds. . . . Schools of Baikal fish will all leave the area. . . . Grigorovich calculated that a hundred billion kilowatt hours meant a gain of two billion rubles. But the Sovnarkhozy also know how to calculate. The losses would also be two billion. Yes, fishermen are against it, and they declare this openly. We do not have the right to destroy the harmony and beauty of this unique gift of nature.[15]

Grigorovich thanked Kozhov for his impromptu speech but criticized his alleged emotionalism: "You value Baikal as one would value the beautiful eyes of a woman. But we energeticists—we do not consider aesthetic beauty." And he continued: "We are not enemies of Baikal. We want Baikal to be utilized not only so people will fall in love with it but so that it gives the country the maximum it can give." But the tide had turned. Speaker after speaker attacked the economics, aesthetics, and grandiosity of Grigorovich's project. From the floor came calls to "Close the aperture!" Conference participants voted to reject Grigorovich's project and the paper mill, and to turn Baikal into a *zapovednik* (national park and nature preserve) for a ten- to fifteen-kilometer-radius around the lake.[16] But that was not to be. On the initiative of Gosplan, Minbumprom, and the Ministry of Defense, construction on the paper mills on Baikal's shores commenced.

Paper Mills and Pure Water

I arrived in Irkutsk on a windy, cold December morning at half past four. Owing to a typical but inexplicable oversight, no one from the Limnological Institute of the Siberian division, the major center for research on Baikal, met me. Indeed A. M. Grachev, the director of the institute who had agreed to meet me, had unexpectedly gone out of town and failed to inform his staff or me. I made my way to the hotel, took a room with a view of the Angara River, slept a few hours, grabbed breakfast, and called the institute. The scientific secretary kindly arranged several meetings for me and gave me an invaluable file of articles gleaned from the Soviet press that provided a historical perspective on the Baikal question. She introduced me to scientists, frustrated from decade-long battles with paper mills, who described their research and the institute's program. I took a crash course in the basic ecology and geography of Baikal with Dr. Tatiana Khodzher, a specialist on air pollution in the Baikal basin.

As Grachev had taken the Limnological Institute's only two vehicles on an expedition, I talked a taxi driver, Zhenya Sokolov, into driving me the seventy kilometers to Baikal for 350 rubles ($3 at the time)—about one month's salary. The road was icy but clear. Zhenya and I talked politics and history on the way to the lake but became reverentially silent on our arrival. Clearly the gem of Siberia, snow-covered mountains on all sides, the lake had crystal clear water that appears blue, green, or gray depending on the clouds and the sun, with three-foot waves whipped by a thunderous wind.

The first written accounts of the lake date back to the seventeenth century. D. G. Messershmidt, a naturalist invited by Peter the Great to Russia's newly established Academy of Sciences, conducted the first scientific observations of the lake in 1724. He included maps and charts that were published only in 1936 from the archive of the Academy library in Leningrad. Modern, hydrobiological research began in the 1860s under B. I. Dybowski, a Polish nobleman, who measured variations in the level and temperature of Baikal. A. V. Voznesenskii, director of the Irkutsk Geophysical Laboratory, part of an observatory movement that swept the Russian empire somewhat later than that in the United States in the nineteenth century, then organized eleven hydrometeorological stations in the lighthouses of Baikal between 1896 and 1901.

In 1916, most likely as part of KEPS, the Academy of Sciences created a Baikal Commission under Academician N. V. Nasonov. The Academy organized an expedition using a cutter named the "Seagull" (*Chaika*). Gleb Iurevich Vereshagin from St. Petersburg then established the Baikal Limnological Station with the support of the Russian Geological Society. Most research focused on the flora and fauna of the basin. So that the station could acquire modern equipment and several boats, several applied studies were conducted to provide additional income; these included the planning of a hydroelectric station on the Angara before World War II and an analysis of ice thickness and strength during the early 1940s, a critical issue in helping planners design the

Lake Lagoda supply route to ease the Leningrad blockade. A modest two-story contemporary building was erected on the shoreline that today serves as a museum and bookstore. By the end of the 1950s the station had a hundred employees. By 1961, when it was renamed the Limnological Institute of the Siberian division, scientists had completed more than seventeen hundred studies, five hundred of which were published. The Limnological Institute moved to permanent quarters in Irkutsk proper in the 1980s.[17] I earned the deep respect of my hosts for eating lunch in the public cafeteria across the way from the institute. Apparently most westerners shy away from the spartan mass-produced food of questionable origin and texture.

Research in the Limnological Institute focuses on the unique flora, fauna, and other natural resources of the lake; on its hydrometeorology, hydrobiology, and geology, including the epilimnion; and on the biology of the shoreline, currents, and chemistry of the water. Scientists have determined, for example, that all salts are distributed evenly throughout the lake, while organic materials vary by depth and their proximity to shore. Most important for the rayon cord desired by the defense industry was that silicon in Baikal water is 1.5 mg/l, as opposed to 5 to 10 mg/l in other freshwater bodies.

As mentioned above, Baikal is the largest freshwater lake in the world. Its surface area is 31,500 km^2, averaging a width of 40 km and a length of 636 km. The lake is 900 m in depth on average, with a maximum depth of 1,700 m (about 1 mile). Its water volume—23,000 km^3—comprises 20 percent of the world's freshwater. The Selenga and Barguzin Rivers feed the lake, with the Angara providing the outflow. Roughly fifteen hundred kinds of animals inhabit the lake and eleven hundred plants, of which two-thirds are endemic. Baikal fauna are relict, consisting largely of ancient ocean and freshwater forms that have been partly transformed during their long life in the lake. There are fifty-two kinds of fish from twelve different families, of which twenty-seven are endemic. Some sixty thousand seals populate the lake; I did not see any. Seeding Baikal fish to other bodies has had some success, but not in seeding fish in Baikal itself. About a third of the fish have economic significance: omul, white fish, sturgeon, salmon, ide (carp), roach, and pike. Baikal sturgeon live fifty to sixty years and reach 100–130 kg in weight and 1.5 meters in length. Overfishing has been the rule, with commercial fishing prohibited periodically. Some scientists claimed that Chernobyl fallout killed five to seven thousand Baikal seals in 1987. Whether or not this is the case, the Baikal ecosystem is fragile, sensitive to slight changes, and industrial pollution upsets the balance. So it is all the more surprising that the Ministry of Forestry and Paper Products was able to have its way to build such environmentally unsound mills on its shores.

The entire Baikal basin came under assault from Soviet industry. The Baikal Paper and Cellulose Combine (Baikal'skii tselliuloznyi i bumazhnoi kombinat), or BTsBK, began operating in 1967, tossing heavy metals, sulfur, phenols, and other pollutants into the water daily. A brother plant, the Selenginsk Pulp and Carton Mill started up in 1973. And in the 1980s the new trans-

Siberian railroad, BAM, crossed the northern end of the lake, destroying forest, accelerating erosion, and blighting the scenery. The BTsBK uses 400,000 m³ of water per day, the Selenginsk TsK 100,000 m³, the Angarskii Oil Refinery over a million cubic meters per day, and the Ust-Ilimsk TsBK a little less. Heavy metals—especially lead and tungsten—flow into the atmosphere; phenol and formaldehyde from Ulan-Ude and Selenginsk flow into the water. Chlorine- and phosphorous-based pesticides and heavy metals from as far away as Mongolia enter the lake through the Selenga and Barguzin Rivers. Acid rain, inversions of sulfur and nitrogen oxides, and increased automobile use also affect Baikal.[18] How could such environmentally destructive factories be built and operated in the face of such clear evidence of pollution and three decades of scientific opposition?

Like elsewhere in the Soviet Union, Siberian ecology labored under the influence of Stalinists who viewed the study of ecosystems as the first step toward their eventual transformation into viable production units of the national economy. In 1957 B. G. Ioganzen—a biologist at Tomsk University, a specialist in forestry, and considered to be one of the biggest Stalinist nature transformers of all, truly an enemy of protected territories—called for the formation in the Siberian division of an Institute of Ecology with a specialized journal to provide the scientific basis for putting Siberian national resources in the hands of planners. Ioganzen's *Fundamentals of Ecology* (Osnovy ekologii) (Tomsk, 1959), a university text, was the subject of a special session in Novosibirsk organized by the Biological Institute of the Siberian division, attended by sixty specialists in all areas.[19] A commission on environmental protection of the Siberian division of the Academy of Sciences soon was established. It had two major functions that were compatible with the aims of rapid economic development: first, to supplement the work of SOPS in classifying and cataloging mineral, plant, and animal resources; second, to evaluate the impact of water melioration and hydroelectric power stations on river flow, groundwater, shorelines, flora and fauna, swamps, forests, fields, and meadows, both "from the point of view of environmental protection and scientific nature utilization."[20] In this atmosphere engineering, construction, and scientific organizations linked up with industrial and agricultural ministries and acquired substantial technological momentum.

In opposition to scientists like Ioganzen stood Trofimuk, Kozhov, and others who opposed the construction of Baikal paper mills. When economic planners and industrial interests began their assault on Lake Baikal in earnest, Siberian scientists, in November 1959, wrote an open epistle to the party, the government, and society calling for the preservation of Baikal. They criticized development plans. They raised the specter of pollution. As a reflection of increased openness during the de-Stalinization thaw, the Soviet press carried a debate that pitted the concerns of these scientists against those of the developers. Headlines for and against economic development such as "Big Chemistry for the Baikal Region" and "Preservation of Baikal or Pollution?" were played off against each other in the Soviet press.

By the time of the Fourth All-Union Conference on the Environment (Novosibirsk, September 1961), a public schism opened between the engineers of nature and scientists who proposed cautious utilization of Siberia's unique natural gifts. The chairman of the Siberian division's Commission on Environmental Protection, G. V. Krylov, delivered the keynote address at this conference on the "problem of resource conservation in Siberia and the Far East and tasks of scientific institutions." Like other traditional Soviet conservationists, Krylov gave lip service to the conservation of Siberian resources but called for their "rational use." L. K. Shaposhnikov, the scientific secretary of Gosplan's Commission on Environmental Protection, discussed the development of nonpolluting industrial technologies and proposed the organization of environmental science in schools and universities to achieve these ends.

But Siberian environmentalists came to do battle against "conservation as usual." They were certain that "rational use," which in the Soviet sense relied on technological solutions to technological problems, preserved nothing. The hydrobiologist Grigorii Galazii, the main defender of Baikal and the leading scientist at the Limnological Institute, detailed the anthropogenic assault on Lake Baikal. Other environmentalists proposed the creation of nature preserves throughout Siberia to protect water, fish, and other natural resources.[21]

Researchers at the Institute of Hydrodynamics, the Institute of Thermal Physics, and the Computer Center joined those in the Limnological Institute to dispute specious Minbumprom projections that pollution would be local and small scale. To garner authority to move ahead with construction, the Ministry of Paper's engineering firm responsible for the design of the BTsBK, Sibgiprobum, had predicted that pollution would never exceed 0.7 km^2 owing to the "calm," "self-cleaning" waters of Baikal. Akademgorodok scientists used computer diagnostics to demonstrate that constantly circulating water, assisted by wind currents, dispersed industrial wastewater along the southern coast of Baikal a minimum of 15–20 km^2 and perhaps 40 km^2. (Later surveys by the state hydrological and state hydrochemical institutes, both of which were subordinate to "Gidrometsluzhba," the Council of Ministers Environmental Protection Agency, determined that pollution would spread 200–250 km^2, that is, three thousand times greater than Sibgiprobum guesses. Only massive additions of clean water to the wastes, which exceeded plant requirements, could limit local concentrations.[22])

Galazii raised the stakes in the debate with a letter published in *Komsomol'skaia pravda* (the Communist Youth League daily). He revealed the design parameters of the BTsBK which, according to the Moscow Design Institute of the paper industry (called Giprobum), would result in the planned discharge of heavy metals such as zinc, sulfates, and other poisons into the lake. He proposed building the mill instead near the Bratsk GES, downstream from Baikal. He called for the creation of a standing coordinating committee under the jurisdiction of the Siberian division of the Academy of Sciences with the right of control over all engineering organizations.[23] But the question had already become not where to build the mill but how to operate it "properly," since the authorities had rejected other sites and the military had convinced party leaders of the need for rayon cord for their jets.[24]

Unfortunately for Lake Baikal, the growing body of scientific evidence that demonstrated the dangers of pollution failed to slow construction. Even scientists of the All-Union Scientific Research Institute of Paper of Minbumprom acknowledged that an emergency discharge of phenol would release concentrations of 25 mg/l, whereas the allowable amount was 0.001 mg/l. If this happened, ten percent of Baikal waters could be poisoned in one day. In practice, violations of the norm were the rule. In the first month of operation of the BTsBK, the Baikal Basin Inspectorate of the Ministry of Water Resources and Melioration detected twelve violations of norms for phenol and sulfurous compounds. Despite opposition from the scientific community, from local authorities, and from the fishing and water industries, the Baikal mill pollutes to this day.[25]

Throughout the early 1960s a series of intergovernmental technology assessment groups investigated the construction site. All came away deeply concerned. Government action did little to allay their fears. In May 1960 the Council of Ministers of the RSFSR passed a special resolution about ways to preserve Baikal through the proper use of forest, flora and fauna, and fish resources. In December 1961 an intergovernmental commission of the presidium of the East Siberian Filial of the Siberian division of the Academy reported on the pollution control equipment of the Baikal mill. It concluded that the equipment in no way "guaranteed complete cleaning of the wastewater and the protection of Baikal water from pollution."[26] In April 1962 an independent group of scientists including Mikhail Lavrentev presented evidence to the Council of Ministers of seismic activity and the danger of earthquakes in the Baikal region. They noted at least two major earthquakes in the previous century—one in 1862, the other in 1959—with an epicenter under Baikal. The group also questioned the availability of pulp and other raw materials because of problems of transportation and weather. Academy of Sciences president M. V. Keldysh intervened to reiterate to the government the Academy's extreme discomfiture with the construction of the paper mills in an active seismic region. He called for the creation of an independent monitoring station, long-term testing of all pollution control equipment, and the search for a way to dump the wastes outside the Baikal basin.[27]

Perhaps because they were dealing with an especially vulnerable site scientists now adopted a NIMBY ("not in my backyard") attitude toward the pollution of Baikal. Even Galazii joined in advocating the construction of a thirty-five mile pipeline to transport paper mill wastes outside the Baikal basin into the Irkut River which flowed into the Angara much further downstream. A February 1965 Academy study concluded that the pipeline was cost-effective.[28] The pipeline would run south and west through the mountains, and then north into the Irkut. The State Committee of the Gas and Piping Industries was instructed to produce blueprints, with Minsredmash, the nuclear weapons ministry, to supply pumping stations.[29] In 1973 Galazii and K. K. Votintsev, another leading hydrobiologist, published a manuscript showing that industrial effluents from the two paper mills were highly toxic and mutagenic, and both endorsed the pipeline ploy.[30] Of course the construction of a pipeline through a mountainous region proved to be more difficult and costly than initially thought, and it was never completed.

Scientists of the Siberian division were often torn between the pressure to complete research that supported plans of various economic organizations and the need to question honestly the environmental viability of those plans. In as much as contract research with those organizations generated income for their institutes, they often erred in support of development when scientific uncertainties remained in their conclusions. This was clear in the case of the Irkut waste pipeline. The Irkutsk-based Institute of the Earth's Crust was established in 1962 to study the structure, makeup, and dynamics of the earth's crust through seismological, magnetometric, and gravimetric methods. The institute built more than twenty seismic stations in the Irkutsk, Buriat, and Chita regions of the USSR. Its research was central to plans for construction of the Bratsk and Ust-Ilimsk hydropower stations on the Angaro-Eniseisk cascade. The primary customers of its scientific research were design institutes and geological planning organizations like Giproproekt, Tsvetmetproekt, and Giprobum for "big science" projects. Still, M. M. Odintsov, the director of the institute, in a report to the presidium of the Siberian division, refused to endorse the paper mill plans precisely because of earthquake concerns, although admitting that the pipeline project could be completed successfully.[31]

B. A. Smirnov, the main engineer of Sibgiprobum in charge of site selection and mill construction, somehow rejected Odintsov's conclusions of seismic activity. Having few friends among Siberian scientists to support his contentions, he wrote Gosplan, the minister of Minbumprom, and Gosstroi, the state construction agency, with a request to "verify who is right, if need be by asking the Institute of the Earth's Crust in Irkutsk to conduct further research." He concluded that Odintsov's findings "were established not by any kind of scientific investigation, but by a uniform desire not to permit the construction of a paper factory on Baikal, to preserve the lake in its virginal state."[32] Smirnov had Siberian geologist V. P. Solonenko testify to the seismic safety of the area. According to Odintsov, however, Solonenko's work was "unqualified, tendentious, and has the goal of verifying [the designs of] the engineering organization instead of facts."

Other Siberian division institutes got into the act of adjudicating the science of economic development. The Institute of Forestry, which was transferred from Moscow to Krasnoiarsk in the late 1950s, conducted studies of forests, soils, and relief with an eye toward meeting the growing demands for wood and pulp production. Although this research program was largely directed toward "intensification" of cutting practices in the most "scientific" manner possible, the scientists' studies of the Baikal basin between 1961 and 1970 must have bothered Minlesbumprom personnel. They concluded that the forests of the basin were an important "filter" for Baikal water, were a major source of soil stability in the rocky, sandy soils nearby, preventing excessive mud and silt from running into the lake. They suggested leaving Baikal alone and instead focusing on the taiga regions further away.[33]

In this environment of contentious scientific debate, of disputes between engineers and environmentalists, Andrei Trofimuk grew increasingly concerned about the inexorable march forward of BTsBK. He made a series of presentations at Academy of Sciences meetings. He wrote letters to the Communist Party Central Committee. He fought Minlesbumprom tooth and nail. In the winter of 1965 he chaired an Academy commission, including Nikolai Vorontsov, academic secretary of the biological sciences of the Siberian division, Odintsov, Galazii, and others, on the BTsBK question. Their investigation revealed a "series of mistakes" from site selection to surveys to "the consequences for Baikal. Eliminating all the mistakes is already impossible, but the danger of pollution may be averted." Pollution control equipment remained to be approved. Early startup had to be avoided: a year of operation without equipment would be comparable to thirty years with it. Yet pollution control equipment would not be in order for at least four more years. Trofimuk's source for this assertion was a newspaper article written by the chief engineer of Baikalstroi which included norms for the operation of paper mills.[34]

On April 9, 1965, Trofimuk wrote General Secretary Leonid Brezhnev directly about the Baikal and Selengsk paper combines. Trofimuk pointed to the series of special Academy commissions that demonstrated the impropriety of the paper plants. He called for either relocating the mill to Gusinoe Lake in Buriatia or building the Irkut pipeline. Trofimuk condemned the staffs of the paper mills: "Comrade Orlov, having long ago lost the honor and conscience of a communist, systematically deceives the government and Central Committee with his assurances that all questions of construction and operation of the [plants] have been solved correctly . . . Defending the honor of his uniform prevents him from objectively evaluating and accepting the suggestions of the Academy of Sciences." The cost of pollution would far exceed any benefit society might derive from the mill. The Central Committee had to intervene immediately.[35] Trofimuk also discussed these issues in an open letter published in *Literaturnaia gazeta* in which he accused the paper mill proponents of spreading a "fog of bureaucratic optimism" along the shores of Baikal.[36]

A year later Trofimuk again wrote Brezhnev, saying that deteriorating conditions "force me to turn to you a second time on the question of the irreversible pollution of Lake Baikal by wastewater from enterprises of the cellulose industry." Trofimuk described a 1966 investigation by Gosplan, the State Committee for Science and Technology, and the presidium of the Academy, which concluded that the assessment of paper industry technology was "based on false information in terms of the quality of wastewater discharged into Baikal and its influence on the fauna and flora." He accused Minlesbumprom employees of misrepresentation since they could not prevent pollution nor produce high-quality cellulose. Trofimuk implored Brezhnev, "Considering the great significance of the question of the irreversible pollution of our unique Lake Baikal, I ask you to take up this question in the Politburo."[37] Brezhnev, who had come to see Siberian development as a monument to his rule, refused to act.

Grigorii Galazii joined Trofimuk in the defense of Baikal and in tireless lobbying for policy changes. He differed from his allies in his uncompromising and high moral stance, rejecting any development at all on his sacred Baikal. This led him into conflict with his allies—biologists, geophysicists, and chemists of various subdisciplines who urged the application of science to "minimize" the environmental effects of development. For example, Valentin Koptiug, who succeeded Lavrentev and Marchuk as chairman of the Siberian division, considered Galazii's strategy "on the whole correct" but found his approach "pretentious and demonstrative." Galazii's absolutist position, however, gave him unassailable independence during the Brezhnev years, enabling him to take on the supporters of BTsBK and "ministerial science" nearly single-handedly. He believed, quite simply, that "even distilled water tossed into the lake would then and there change the habitat of the lake's organisms for the worse."[38]

Now director of the Baikal Ecological Museum, Galazii served as director of the Limnological Institute for thirty-four years. Born in 1922 in Kharkiv in Ukraine, he moved to Siberia and graduated from Irkutsk University in 1947. He completed his graduate work in Leningrad and then moved to the Kolsk Peninsula for his first hydrobiological investigations. In 1953 he was asked to study what would happen to Baikal were a dam to be built on the Angara. Using trees, many of which live five hundred years, and geological formations as his data, he found that in the past five thousand years the amplitude of Baikal varied no more than three meters. His conclusion? A dam would have a profound, negative impact by causing great change in the water level. Galazii by now had fallen in love with Baikal and spent the next forty years studying the local ecosystem in great detail. His familiarity with the lake led him to conclude that economic planners must have looked at a map, shut their eyes, and picked a spot without any sense of Baikal's uniqueness and fragility.

Galazii first gained prominence by signing an article with Kozhov and others that was published in *Literaturnaia gazeta* in 1958. The article attacked Grigorovich's plan to widen the Baikal outlet. Galazii tried to orchestrate criticism of the BTsBK by his public writings. It was hard to make any progress when the government's Environmental Protection Agency was in cahoots with the economic organs, and when articles were censored. Fortunately he had the assistance of such scientists as Lavrentev, Trofimuk, Ianshin, and Vorontsov. "But the planning organ listened instead to the opinion of Academician N. M. Zhavoronkov," Galazii recalled. "Zhavoronkov and others wrote complaints about us to the Central Committee and everywhere else they could. They likened our efforts to save Baikal to complicity with imperialism. It wasn't easy to defend ourselves from that slander."[39]

In a series of monographs and articles published in the 1960s and 1970s Galazii shared his study of all aspects of Baikal life: its flora and fauna, geology, hydrology, and meteorology. In *Baikal and the Problems of Clean Water in Sibe-*

ria, he documented the drastic decline in the number of freshwater fish—sturgeon, white salmon, and white fish—in Siberian lakes and rivers because of unencumbered industrial expansion. By 1968 Baikal produced only 500,000–600,000 km of omul per year. The reason was clear: 10 million cubic meters of insufficiently treated wastewater from chemical, oil, metallurgical, and paper companies were dumped into the lake annually. Pollution control equipment was designed to remove only easily oxidized organic compounds, so that never more than 90 percent, and often 40 to 60 percent of the organic materials remained. Concentrations often exceeded norms 250 times. Salts were untreated. The natural concentration of chlorides in Baikal is 0.6 mg/l, whereas in wastewater it often exceeded 150 mg/l. Sulfate concentrations grew five times greater than normal. The pH of discharges ranged widely from 2.7 to 11.5. Making matters even worse, the wastes were malodorous.[40]

The major goal of Limnological Institute research became the development of the theoretical basis to measure the impact of industrial wastes on microbes, algae, plankton, mollusks, and fish to determine maximum allowable concentrations (PDKs). Galazii called the PDK a "poor consolation." "In general, an entire science that orients economics toward the achievement of norms of PDK is eyewash," he said. "You can stay within accepted norms and regularly release into Baikal hundreds of tons of oil, tens of thousands of tons of mercury, huge quantities of heavy metals, and other matter."[41]

As in any scientific controversy certain members of the community see their colleagues as being too averse to risk, of favoring small furry animals over the well-being of humans. Academician Nikolai Mikhailovich Zhavoronkov played that role in this cellulose passion play. His career mirrors that of the so-called party-scientist of Brezhnev's command-administrative system. Zhavoronkov (b. 1907), a specialist in organic chemistry and chemical technology, graduated from Mendeleev Moscow Chemical-Technological Institute in 1930. From 1948 to 1962 he was rector of the institute, after serving briefly as director of the Karpov Institute of Physical Chemistry. He joined the Communist Party in 1939. A corresponding member of the Academy in 1953, he became a full member in 1962, at the same time that he was appointed director of the Institute of General and Inorganic Chemistry and scientific secretary of the Academy's Division of Physical Chemistry and Technology of Organic Materials. Serving on several scientific commissions for Gosplan in the 1960s and 1970s, Zhavoronkov always sided with Minlesbumprom on the safety of BTsBK. From these administrative posts, he dismissed any doubts about the central role that chemistry would play in building communism. He saw chemistry as a panacea for Soviet agriculture and industry.[42] Lacking a Russian counterpart to Rachel Carson's *Silent Spring* (1962), he faced no one who might question the improper use of pesticides, herbicides, and fertilizers, who might suggest that rayon cord was being produced at excessive environmental cost, who might even challenge his technological bent in general.

Zhavoronkov's actions spoke volumes about his position on the role of big projects in the Soviet system. In 1966 Gosplan appointed him director of an intergovernmental commission to investigate Baikal. This commission found nothing wrong either with the site or the pollution control technology, conceding only that pulp ought to be harvested from outside the Baikal zone.[43] Next he was involved in a commission to determine the PDKs of discharges. The commission had close ties to the paper and defense industries. It freely changed the PDK as needed. Commission members used the specious argument that the ability of science to measure substances in increasingly minute quantities—to parts per billion level—and not the ability to measure risk itself created a greater sense of danger about pollutants than warranted. Indeed Zhavoronkov, the minister of the paper industry, the director of the paper industry's Institute of Ecological Toxicology, A. M. Beim, O. M. Kozhova, and the president of the Academy, A. P. Aleksandrov, penned a letter to the Council of Ministers concluding that discharges simply were not harmful. Fifteen years of research had shown Baikal to be "self-cleaning" and the surrounding forests in fine shape. Most of the pollution of Lake Baikal, they asserted, came from sources other than BTsBK.[44]

In his effort to support big chemistry projects in Siberia, Zhavoronkov was joined by A. M. Beim of the Minlesbumprom Institute of Ecological Toxicology. A conservative and closed-minded individual, Beim defended the paper mills for the paper industry. He used smokescreens familiar to western readers who have studied acid rain and the greenhouse effect. He strove to postpone action by highlighting "scientific uncertainty," calling for more study, and stressing the importance of economic growth versus ecology. Beim wore the mantle of his employer with pride. He sought to discredit as emotional the efforts of such scientists as Galazii, and earned their enmity. Never was there any sense of urgency in Beim's bland, pro-technology "party language" descriptions of the goals of the Institute of Ecological Toxicology. His research was intended to develop "scientific recommendations for the rational utilization and conservation of natural resources of the Baikal basin." For Beim, science was a central factor in economic development. "Scientific workers of Siberia carry a special responsibility for the solution of problems of the accelerated development of the region which are an integral aspect of the economic strategy of the party," he wrote.[45]

The Institute of Ecological Toxicology was established initially as a separate laboratory in Petrozavodsk University under contract from Minlesbumprom. When Beim, a candidate of science, was appointed director, he quickly assembled twenty biologists and chemists as staff. The university soon moved these facilities to the paper mill site, establishing an "independent" laboratory that was linked through contract research with Academy institutes. Beim's institute was instructed "to introduce a regional system for environmental monitoring of Baikal," based on "the elaboration of scientifically based PDKs." The institute acquired a wide geographic range of responsibilities, so that the efforts of its staff to develop "methods of quality control and ecological norms" were stretched thin along the Baikal region, the Angara, Amur, and Volga Rivers, the Baltic, Karelia, Kazakhstan, Ukraine, and even Sakhalin Peninsula.[46]

In spite of its broad profile and contacts with academic laboratories, the institute fell strictly under the jurisdiction of Minlesbumprom, which was apparent from the papers on which its reports were printed. Beim's coworkers presented evidence that all was in order with the BTsBK in *The Baikal Cellulose Worker* (Baikal'skii tselliuloznik), the official mouthpiece of the paper mill, the local party committee of Baikalsk, and the institute. The articles defended the design and operation of purification technology. They testified to the health of the local workers and their families. All claims, past, present, and future to the contrary, including a shocking radio report that questioned the health of Baikalsk children, were characterized as "disinformation." Mill officials hid under a smokescreen: their data on air and water pollution were classified.

Beim was not the only local scientist who defended the paper mill. Little did Mikhail Kozhov know that his daughter would fall under the spell of cellulose. Olga Mikhailovna Kozhova, a chain-smoking, coffee-drinking, energetic individual, is director of the Irkutsk University's Biological Scientific Research Institute. She maintains that she consistently defended Baikal from industrial encroachment. The record shows otherwise. She served on commissions with Zhavoronkov that supported Minlesbumprom's designs on Baikal. Her research exudes faith in the notion that nature can be improved, for example, through "acclimatization" or seeding of species of fish in Baikal to raise overall fish harvesting productivity. On one occasion, she cited a ten- to hundredfold growth in waste-eating bacteria in Lake Baikal in response to discharges as evidence for the lake's "self-cleaning" capacity. Granted, the bacteria kept pollution local. But Kozhova was not disturbed by the loss of up to 50 percent of all plankton by those very same discharges, perhaps because the loss was limited to several square kilometers and the surviving plankton had not lost the capacity to produce organic matter for the fish. Unlike her father, she found no evidence of overfishing of omul. Rather, she counted this as an irrational fear based on data that showed instead the dynamic character of fish populations.[47]

In a way Kozhova maintained independence from the BTsBK controversy. The university, being poor, far from the center, and small, was not viewed as a threat to the paper industry, its institutes, and factories. The Limnological Institute and Galazii always remained in the crosshairs of Minlesbumprom scientific sniping. But Kozhova defended the Institute of Ecological Toxicology from charges of being captured by the industry it served. Data that seemed to indicate the impact of pollution she attributed to existing temporary fluctuations of a hydrometeorological or hydrobiological nature, or to scientific uncertainties. Although acknowledging the influence of BTsBK on local forests, she was convinced that such anthropogenic factors as atmospheric warming, ship traffic, tourism, and BAM had contributed just as much to deforestation. It remained for scientists armed with the facts to point the way for engineers to bring safe technology into production.[48]

Kozhova's views are rather typical among Soviet scientists. She combines faith in progress with a strong belief in the possibility to avoid its negative impact through technological solutions. She believes that scientists can anticipate and avoid the anthropomorphic costs of progress. An interdisciplinary approach such as that provided by ecological toxicology is the key. Ecological toxicology permits us to determine "what concentrations and what elements are safe for humans, animals, and plants, and for their populations and communities; what the mechanisms are of their detoxification, not only in the organism but in the population, the community, and the ecosystem as a whole." Kozhova sounded one warning: interdisciplinary approaches such as those embodied in ecological toxicology were required to ensure conservation of nature but had not penetrated Soviet planning. Intensive development of agriculture and industry were impossible without them.[49] Zhavoronkov, Beim, and Kozhova never accepted for a moment the argument that he who pays the piper calls the tune.

BAIKAL MEETS THE BULLDOZER

Brezhnev turned a deaf ear to Trofimuk's and Galazii's pleas. To defend the paper mills Gosplan and Minlesbumprom marshaled the evidence of Zhavoronkov, Beim, and Kozhova. Although Gosplan approved the operation of the paper mills in June 1966, several provisos were required: that the pollution control equipment clean at least 95 percent of the wastes; that the equipment work twenty-four hours a day; and that tree-cutting guidelines in the Baikal region be strictly observed. Gosplan ordered monitoring stations installed to verify compliance and required Minlesbumprom to development more modern, less polluting equipment in short order. In reality, Gosplan forced the pace of construction and rejected any concession to environmentalism, for example, by building a reservoir to hold wastewater.[50]

During the first four months of operation alone, the BTsBK violated norms a hundred times and in 1967 did so virtually every day. The Council of Ministers had no choice but to urge strict compliance through a resolution issued on January 1969, "On measures for the Conservation and Rational Utilization of Natural Resources of Complexes of the Baikal Lake Basin." Within eighteen months Nikolai Vorontsov delivered a report to the Academy presidium criticizing the failure to fulfill the resolution. Vorontsov, Odintsov, Galazii, and others had visited the Baikal region and had examined data and reports. Their worst fears were confirmed: "The leaders of Minlesbumprom and bureaucracies subordinate to it cannot guarantee safe operation of [BTsBK]." The report detailed extensive erosion and pollution with norms for dangerous wastes of all sorts exceeded by four, ten, up to ninety-eight times. The Vorontsov commission recommended expansion of the staff and responsibilities of the Limnological Institute with the assistance of other Siberian division organizations, and the implementation of a long-range plan, put together by Trofimuk and others, to ensure compliance with all laws. Minlesbumprom would be ordered to clean up the local rivers, reclaim forest, and adopt erosion prevention measures.[51]

In November 1970 a Gosstroi commission visited the site. Even a commission from Gosstroi, an organization whose very raison d'être was large-scale projects, reported that if all the problems associated with the site had been properly analyzed at the start, the site would never have been chosen. In as much as the factory was already built, however, efforts had to be focused on making discharges safe for the plant and animal life of Baikal. In a word, nothing was done to complete work on pollution control equipment, prevent accidental discharges, or regulate logging.

The 1970s and 1980s witnessed a charade: norms were established and then revised, ignored by Minlesbumprom with slight of hand, and then revised again. In June 1971 the Central Committee and Council of Ministers passed yet another resolution, "On Supplementary Measures to Guarantee Rational Use and Conservation of the Natural Wealth of Baikal." Pollution control equipment was to be upgraded and operated properly, the maximum levels of pollutants to be determined. The Siberian division thereupon produced a seven-volume work, the result of the research of fifty different organizations, which established maximum allowable concentrations, and this was then forwarded with recommendations to the Academy, the State Committee for Science and Technology, Gosplan, Minlesbumprom, Minvodkhoz (the Ministry of Water Melioration), Minrybkhoz (the Ministry of the Fish Industry), and Minzdrav (the Ministry of Health). The plan was never fulfilled.[52]

Boris Nikolaevich Laskorin, who served on at least five different government and Academy committees on environmental issues, often as chairman, knew the history of Baikal well. The first commission, chaired by Laskorin and appointed by the State Committee for Science and Technology (GKNT), met five times from January to April 1972 with Trofimuk and Liapunov as members. It determined that pollution was the least of BTsBK's problems. More important, the Council of Ministers learned, was that the combine could not produce cord to specifications.[53]

Faced with a pollution fait accompli, Academy scientists strove to set PDKs based on irrefutable hydrochemical analysis to prevent the destruction of Baikal. Trofimuk himself tried his hand at creating norms. In January 1975 the Siberian division approached the Central Committee with a scientific study that argued in no uncertain terms about the need to adopt radical measures. But the party leadership instructed GKNT to carry out further study. GKNT appointed an investigative commission over the objections of Marchuk and Trofimuk from which Baikal activists were summarily excluded. Surprisingly, the commission admitted that industrial discharges were quite dangerous for the endemic species of Baikal. However, it ordered the Siberian division to identify the influence of specific compounds and to determine the norms for each with a prognosis to the year 2000. Trofimuk was incensed. He knew the study would take years during which time pollution would continue. Instead of forcing action, the commission had tied the Siberian scientists' hands.[54]

Trofimuk would not rest. In his position as chairman of the scientific council of the Siberian division on problems of Baikal, Trofimuk asked the Institute of Economics in December 1978 to evaluate the economic costs of the "assimilation" of Baikal. In October 1979 economist Abel Aganbegian responded with a thirty-eight-page report. The report outlined how air and water pollution from industry, private cars, inadequate wastewater treatment facilities, erosion from overharvesting of forest, and profligate use of fertilizers, pesticides, and herbicides on collective farms damaged Baikal flora, fauna, and the fishing industry, almost beyond recovery. The institute singled out Minlesbumprom and the paper mills as the prime culprits. Aganbegian's report called for unconditional fulfillment of the resolutions of the Central Committee and Council of Ministers; a sharp increase in capital investment in nature conservation measures, noting that even "minimal investments" would have a positive influence; cessation of clear-cutting near Baikal; and the creation of a park in the Baikal basin.[55] But Trofimuk could do little more than keep a watchful eye on the efforts to develop Siberia of Beim, Zhavoronkov, and their dozens of allies throughout the paper industry and government.

Another commission under Laskorin reported its findings in 1982. The report indicted Minlesbumprom activity in every respect. Minlesbumprom "constantly advanced newer plans and programs which as a rule, for various reasons, were not fulfilled, and the deadline for the realization of these measures was pushed back." The damage to Baikal, the destruction of local forest, the transport of millions of cubic meters of wood for hundreds of kilometers, all these problems were "the consequence of a series of errors committed by Minlesbumprom." Worse still, Minlesbumprom had yet to master the technology of cellulose cord production.[56]

No matter how hard they had tried to prevent the construction of the plant or to delay its operation, Siberian scientists recognized by the early 1970s that the worst had occurred. The paper mills operated at low capacity, manufactured a poor product, and polluted heavily without remorse. Minlesbumprom falsely claimed that only by 1984 had the Academy successfully set norms. Every step of Trofimuk, Galazii, Vorontsov, and the others, every report, every letter was met by requests for more information, promises of action without substance, and naked lies. They now faced down Siberian river diversion. In this case they were successful, but only because of the great cost involved, too great even for the communist engineers of nature.

THE SIBERIAN RIVERS DIVERSION PROJECT

Plans to divert flow from Siberian rivers—the Ob, Enisei, and Irtysh—to Central Asia, and northern European rivers into the Volga basin, gathered momentum in the late 1960s.[57] Diversion was intended to provide an open faucet for agriculture and to save the Aral, Azov, and Caspian Seas from rapidly dropping water levels. In all nearly two hundred scientific research institutes, enterprises, and scientific production organizations floated on the waters of this project. An academy institute, the Institute of Water Problems (Institut vod-

TABLE 5.1
Number of Institutes Involved in Research and Planning of Territorial
Water Redistribution Projects, by Organization

Organization	No.	Organization	No.
Academy of Sciences	10	Minrybkhoz SSSR	13
Siberian Division	6	Mingeo UzSSR	5
Karelian Branch AN SSSR	4	Minvuz	4
Minenergo SSSR	11	Minvuz RSFSR	10
Goskomgidromet SSSR	30	Minzdrav SSSR	3
Komi Branch AN SSSR	2	Minzdrav RSFSR	13
AN Azerbaidzhan SSR	2	Minzdrav KazSSR	3
Minvodkhoz SSSR	10	Gostroi SSSR	3
Minvodkhoz RSFSR	2	Gosleskhoz SSSR	2
Gosplan (SOPS)	1	Minselkhoz SSSR	4
VASKhNIL	3		

Source: G. V. Voropaev and D. Ia. Ratkovich, *Problema territorial'nogo pere-raspredeleniia vodnykh resursov* (Moscow: Institute of Water Problems AN SSSR, 1985), pp. 488–497; published in a limited run of a thousand copies and classified.

nykh problem), or IVP, was designated as "head" institute for the project. Most of the other organizations involved in the planning and research for diversion were subordinate to economic ministries (agriculture, fishery, forestry, and geology) with a vested interest in rapid progress. For the Siberian diversion project, design work was assigned to Soiuzgiprovodkhoz (of the Ministry of Land Reclamation and Water Resources, or Minvodkhoz), and for the northern diversion project, to Gidroproekt (of the Ministry of Electrification and Energetics, or Minenergo). IVP had the leading role in research for both projects. Such poetic sounding Central Asian engineering organizations as Sredazgidroproekt and Sredazgiprovodkhlopok joined on as full partners (see Table 5.1). Nature, beware!

As with many other projects intended to tap Siberian resources, water diversion had roots in the Tsarist era. The engineer Ia. G. Demchenko, who was inspired by French success in building the Suez Canal, saw diversion as a means of solving recurrent Russian famines and improving transport. Foreshadowing objections of twentieth-century critics, Demchenko's detractors in the Russian Geographical Society were deeply concerned that diversion would flood vast tracts of land and generate significant climatic change.

In connection with Bolshevik fantasies to transform nature from a capricious force into a rational tool of socialist construction, the Soviet Union supported diversion from the 1920s on, partly linked to plans of the State Electrification Agency (GOELRO) to build hydropower stations on every conceivable site. Already in 1933, at a special session of the Academy, Gleb Krzhizhanovskii, head of GOELRO, and others discussed a partial diversion of waters from northern rivers into the Caspian.[58] The Leningrad-based institute

Gidroenergoproekt, whose personnel later were actively involved in diversion, offered a Siberian rivers variant in the 1940s to advance irrigation in Central Asia and Kazakhstan.

The Siberian Rivers Diversion Project is an intriguing case study of Soviet scientific politics. As Michael Bressler has shown, in diversion we see the replacement of Khrushchevian-style politics, with its impetuous lurch from one large-scale project to another, by long-term, rational, and incremental policy making under Brezhnev, although in both cases the projects remained ambitious. Under Brezhnev the major difference is that state-sponsored technocrats took broad Communist Party goals and turned them into concrete policies with tremendous institutional momentum and support not only among technocrats but among various interest groups throughout the USSR. For example, Central Asian party officials and water resource ministries wholeheartedly supported the project as crucial to future economic development.

Soviet water resource managers believed that they alone could overcome decades of poorly designed irrigation and dam projects advanced by their colleagues in agricultural and hydropower ministries. Irrigation and dams had increased evaporation, salinity, and pollution, and destroyed productive farm land, submerging it under water. As many as 2,600 villages, 165 cities, more than 7.5 million hectares of land—3.1 million of these agricultural land and 3.0 million forest—were flooded by Soviet dams.[59] In the United States 70 percent of farm land can be cultivated without irrigation, whereas in the former USSR engineers claimed that only 1 percent could. Soviet agriculturalists had drained European rivers and lakes of flow. Industry took its share, then dumped filthy wastewater back into the system. Hydroelectricians dammed the rest.

With the guidance of scientists at IVP, the engineers of nature would overcome those man-made problems with man-made solutions. They planned to add dozens of cubic kilometers of water annually to Central Asian rivers and agriculture from Siberian ones to fight ecological devastation caused by profligate use of water resources. They justified their own devastation through detailed empirical study. Only a unique combination of Russian writers, Siberian scientists, and local officials managed to waylay the diversion projects.

Those who favored diversion were encouraged by three factors. One was the grand history of water melioration projects, beginning with the Roman aqueducts. In the USSR, too, canals built during Stalin's time seemed to demonstrate the feasibility of such vast, costly projects as the Volga-Moscow canal. In all, thirty-six major canals were built in the Soviet period at an increasingly rapid pace. Seven were built before World War II, seven from 1946 to 1960, and twenty-two over the next twenty years.[60] As of 1983, 115 km³ of water was redistributed by canals annually, most of it locally, and three-quarters in Central Asia.[61] Siberian diversion was huge, even by Soviet standards, for it promised to take 60 percent as much water as all the other existing projects combined.

The grand history seemed to ignore the great human costs of these canals, for example, the infamous Belomorsk-Baltic shipping canal that was built with slave labor and accompanied by tens of thousands of deaths. Granted, the deaths were more a function of the cruelties of the Stalin era in general than of science run amok. Yet the tragic epidemiological consequences (e.g., high infant mortality) of poor irrigation practices that accompanied a number of diversion canals indicated just how little regard the planners and technocrats had for Soviet citizens.

The second factor heightening scientists' enthusiasm for diversion was that they intended to divert—by their own calculations—very small amounts. From all of Siberia, the engineers of nature planned to divert annually just 60 km^3 from a total of 1,500 km^3 (4 percent), including 27 km^3 from the Ob (with a flow of 400 km^3, nearly 7 percent). In the European USSR the scientists had their eyes on 40 km^3 from a total annual flow of 600 km^3, or less than 7 percent, of which the Pechora would contribute 18 km^3 of its annual 130 km^3 flow (nearly 14 percent).[62]

Finally diverters pointed out that the Soviet Union had a total annual river flow volume of 4,750 km^3, second in the world only to Brazil. There were another 40,000 km^3 in ice flows, ice, lakes, and swamps. Most of the river volume (93 percent) originated on Soviet territory. As usual, however, Russian geography played a mean trick on planners' preferences: 80 percent of the demand for water was in regions with only 20 percent of the resource. Four Siberian rivers (the Enisei, Lena, Ob, and Amur) contributed a third of the total volume. By 1960, owing to agricultural, industrial, and electrical needs, water demand far exceeded supply on the Volga, Dnepr, and other rivers; the Caspian, Azov, and Aral Seas had dropped tens of meters. Their waters were increasingly salty and polluted. The fishing industry had been destroyed. Scientists from the Institute of Water Problems concluded that scientific "management" of water resources was the only rational way out.[63]

The Institute of Water Problems actively pursued the general policy outlines of Minvodkhoz. Minvodkhoz had been created from the State Committee for Irrigated Farming and Water Resources in 1965 to implement policies, including diversion, to improve the performance of Soviet agriculture. By the fall of 1968 Brezhnev delivered a report before the Central Committee asking that Minvodkhoz, Minselkhoz (Ministry of Agriculture), and Gosplan work up long-range reclamation plans with a diversion component. Northern Russian rivers would be diverted to the Volga basin, Siberian rivers to the Syrdar'ya and Amudar'ya Rivers which had been drained as the major sources of the now dying Aral Sea for Central Asian cotton. When disastrous droughts destroyed the ninth five-year plan (1971–75) for agriculture, the Institute of Water Problems was in a position to seize the initiative and press for more financial and personnel resources for diversion.

Founded in 1968, the Institute of Water Problems ambitiously charted a course through Russia's rivers. In 1972 its first director, A. N. Voznesenskii, published a provocative article in *Vodnye resursy* (Water resources) that proposed diversion projects to overcome the serious mismatch between geographical demand and the availability of water.[64] Armed with a firm belief in the efficacy of their tools, IVP scientists moved ahead in planning with the certainty that they would avoid environmental degradation. Studying every aspect of the "project of the century," as it came to be known in leading government circles, they quickly accumulated data on every conceivable variable. They authored dozens of environmental impact statements on climate, land, and water conditions, producing detailed projections to the end of the century.

Grigorii Vasilievich Voropaev, the bête noir of environmentalists, wholeheartedly embraced Voznesenskii's vision and devoted his career to water melioration. Before joining the staff of IVP in 1969, Voropaev studied the hydrobiology and economics of the Aral Sea and Central Asia. In 1971 he became director of a special laboratory of water melioration, then director of another laboratory studying the economic problems of water use, and finally deputy director and director of IVP. He called for complex, interbranch solutions to the water problem, as opposed to what he alleged were heavy-handed, shortsighted water resource programs developed by such bureaucracies as the Ministry of Agriculture. Voropaev came to believe that the so-called southern strategy of the Brezhnev years to cultivate more sunny arid Central Asian land would result in further irrational use of scarce resources. Voropaev saw diversion as the only way to save his beloved Caspian Sea while simultaneously supporting agriculture and industry in the European USSR.

IVP analysis indicated that the level of the Caspian Sea, already almost fatally low, would continue to drop throughout the century owing to increasing industrial and agricultural use. (For reasons that remain unclear, the level of the Caspian is now actually rising.) Contrary to popular belief, the major cause of the drop was not the Fort Shevchenko BN-350 breeder reactor that desalinates 120,000 m^3 of water per day, although the Shevchenko oil and atomic industry indeed pollutes heavily. Voropaev grew convinced that Central Asia would "blow away within twenty to thirty years" if nothing were done. He therefore proposed a twofold program to move the southern strategy along. First, he suggested changing the patterns of water use, that is, introducing new technologies and economic incentives to cut down on its use. The second measure was to tap Siberian water. "We have a large number of regions with a great deal of water. Too much in some cases," Voropaev stated, "thousands of cubic kilometers of swamps in western Siberia [that] are not part of the water exchange." Like U.S. developers who run roughshod over the birds that nest in the wetlands, Voropaev believed that up to 100 million hectares of land that he considered "unproductive" should be drained, reclaimed, and put to use as farms.[65] By 1985 diverters in Voropaev's institute completed two major studies and sixty classified and unclassified reports. They concluded that the benefits of diversion far exceeded environmental and economic costs.

Examining the Ob River that flows past Akademgorodok, for example, to determine what the impact of diversion would be, scientists examined weather, geography, and geology; water chemistry, flow, sources, and tributaries; temperature, pollution, and salt levels; flora and fauna; ice cover; and so on. No water molecule escaped their calculations. Their studies led to the following conclusions: diversion of 60 km³ per year would influence microclimate in some areas. Brackish waters, normally 70° to 70.5° north latitude, might penetrate the Ob, perhaps as far south as 68°, a distance of about 250 km. Bays and inlets would become increasingly saline. Ice cover patterns would be altered. Slight increases in winter humidity would accompany higher temperatures. A loss of heat in downstream waters would lead to the onset of ice two to five days earlier. In spite of the litany of detail, the scientists concluded that diversion from the Ob of up to 100 km³ per year would have no impact on the climate of the Arctic.[66]

IVP specialists anticipated significant difficulties in the construction of the water diversion system using old river beds, valleys, and canals for water diversion. Regarding new canals, the change in seasons was the major problem. The onset of winter would reduce flow. Ice would interfere with pumps, reducing their efficiency 40 to 60 percent. Fall and spring freezes and thaws would deform the shorelines, although this could be mediated by limiting average flow in those seasonal periods to around 1 mile per hour. According to the All-Union Institute of Water Geography (VNIIVODGEO), canals also had the disadvantage of significant filtration loss of 2 to 4 percent of total flow, with contamination and loss of ground water a strong possibility. Of course there would also be a significant impact on the recipient lakes and rivers. In 1987, just when the path of the main diversion canal had been chosen, the project was put on ice. The canal would have been on the order of 2,400 km long (the distance between Boston and New Orleans), 200 m wide, and 110 m deep.[67]

As if the scale of the canal were not enough to dampen the diverters' enthusiasm, IVP scientists raised the danger associated with the diversion of Siberian microbes, bacteria, insects, and toxic wastes to the South. Typhus and dysentery might spread to areas where arid conditions had precluded them. Encephalitis would migrate southward in ticks. Everyone involved in constructing the diversion system, from surveyors to engineers, from carpenters to concrete pourers, and their families, as well as the cities built to support them, were at risk. A special epidemiological service had to be established to prevent the spread of disease and pests with water.[68] Despite all this, the engineers of nature arrived at a shocking conclusion: "The negative consequences for the environment in the North, and environmental changes in transportation zones and distribution of flow, will not have a global character. Changes . . . will occur in a relatively small territory and basically can be anticipated, removed, or compensated for."[69]

Voropaev and his colleagues remained confident to the end. Their investigations "never raised any doubts about the utility of diverting waters into the Volga basin." Diversion in itself, however, was insufficient. Conservation was also required. Measures to ensure the efficient use of local water resources included water recovery systems, more effective water supply, and water-efficient technologies, especially irrigation systems; drainage of irrigated lands with repeated use of that water; better cleaning of wastewater; and so on.[70] But the benefits to the Caspian, Aral, and Azov Seas; to agriculture in Central

Asia, where meat, milk, egg, fruit, and fish consumption were significantly below the norm, where droughts are frequent and the demand for irrigation water far exceeds supply; for the health of the Volga and Dnepr river basins; for all these reasons Voropaev's colleagues recommended moving ahead, draining at least 25 or 30 km^3 annually for both projects, at a cost of untold billions of rubles. Later, under public scrutiny, they reduced their estimates to 5 or 6 km^3 annually.

High-level government and party support kept the project alive in the face of growing scientific opposition. The government placed jobs and economic growth in the South ahead of environmental concerns. The Brezhnev "southern strategy" ensured that the Ministry of Agricultural Industry, or Minagroprom, and Minvodkhoz ignored skyrocketing costs and scientific uncertainties. The Central Asian republics and Ukraine, not surprisingly, also hoped to see diversion come to fruition. Academy president Anatolii Aleksandrov, a fan of big technology projects, helped keep objections out of the limelight until after Gorbachev's rise to power.[71] He may have done this on the orders of the Central Committee after Minvodkhoz did extensive lobbying of the Academy presidium. One of Akademgorodok's founders, S. A. Khristianovich, defended diversion as he would defend two other pet projects: hydroelectric stations in general and a huge dam system to protect Leningrad from floods.[72] Perhaps it was merely institutional momentum, a central feature of Soviet science, that kept the project on target. It would be no easy task to wean 200 institutes (160, with sixty-eight thousand engineers, in Minvodkhoz alone) from Siberian water.

"PROJECTS OF THE CENTURY" FALL ON HARD TIMES

In the fights over diversion and Baikal, a small but influential group of Russian writers joined scientists in providing crucial criticism. The writers had come to see virtually any water project as a threat to nature and rejected outright Soviet-style industrialization, with its capital-intensive, urban-serving diversion, its hydroelectric power, chemicalization of agriculture, and nuclear energy projects.

The writers' criticism was quite unexpected in official circles. Since the founding of the Soviet state, writers have highlighted the signposts of socialism; in the genre of "socialist realism," they have emphasized the central qualities of the new Soviet man—tirelessness, disinterested pursuit of increased production, unmitigated hostility to enemies perceived and real, and devotion to the Communist Party and its programs. Economic and technological determinism, that is, the development of the economy and technology's central role in that process, have been central themes. In this genre even nature is an enemy. "In the popular literature and the press," historian Douglas Weiner wrote, "antipathy toward harsh nature frequently led authors to anthropomorphize nature. Nature was portrayed almost as a consciously antisocialist force which needed to be suppressed. . . . [N]ature had to be transformed and bent to human will—from the roots up."[73] The list of authors who wrote about "water management" in their literature is impressive: Maxim Gorky, with his authors' collective, produced *Belomor*, a work about the heroic construction of the White Sea-Baltic Canal; Bruno Iasenskii described the Vakhshskii Canal in *Man Changes His Skin*; and, in their short stories, Andrei Platonov, Evgenii Evtushenko, Valentin Kateav, Iurii Trifonov, and Konstantin Simonov also depicted man's glorious struggle with water.

Socialist realism gave way in the 1950s to slightly more subtle praise of the engineers of nature. Writers were still called on to glorify the achievements of the state. Yet bands of them were sent to Siberia in the 1950s and 1960s to chronicle the heroic mastery of nature's four powers: water, ore, coal, and forest.[74] The pen of economic determinism filled the press with vivid reports of huge Siberian construction projects—hydroelectric stations, industrial enterprises, Akademgorodok, and the communist future in general, even about the Selenginsk paper mill.[75]

When it came to Baikal, to Mother Russia, to her rivers and lakes, however, many writers rejected Bolshevism's technological determinism. Aleksandr Trifonovich Tvardovskii, editor of *Novyi mir* during the Khrushchev years, twice visited Irkutsk and Baikal. During his first trip he was a frequent guest at the home of Mikhail Kozhov, who was a bridge between the literary and scientific communities. In 1959 Tvardovskii wrote Kozhov, reminding him of a promise to write a piece for *Novyi mir*, one not too specialized, on the growing danger that economic development posed for Baikal.[76] Kozhov, a modest man, never complied. Mikhail Sholokhov, Nobel prize winning author of *And Quiet Flows the Don*, spoke in defense of Baikal at the twenty-third party congress: "Perhaps we will find in ourselves the courage to reject the cutting down of forests around Baikal, the construction there of cellulose enterprises, and instead of them build those that will not threaten to destroy the treasure of Russian nature, Baikal? [Applause.] In any case, it is necessary to take all steps to save Baikal. I am afraid that our descendants will not forgive us if we do not preserve the 'glorious, sacred Baikal.'"[77]

Literaturnaia gazeta, the weekly newspaper of the Union of Writers of the USSR and usually a source of allegiance to the state's economic goals, also sided with nature. In the late 1950s, in response to the debate provoked in the West by C. P. Snow's treatise on the myth of two cultures, the writers turned to an analysis of science's place in communist society. The resulting discussion between *fiziki* and *liriki* could not avoid considering environmental issues. The writers ran a series of articles critical of planners' enthusiasm, for example, a 1958 piece opposing the Grigorovich proposal to blast the outlet of Baikal wide open and the call by the Ukrainian writer Frants Taurin for Baikal to become a national park. "Baikal belongs not only to us but to our descendants, to people of the epoch of communism," Taurin wrote. "We are obligated to preserve Baikal."[78] Throughout the 1970s and 1980s the weekly served as a conduit for public discussion. In 1985–86, after the rise of glasnost, discussion of the unique lake expanded in the press with the assistance of *Pravda* and *Izvestiia*. Citizens' letters poured into the central papers, critical of diversion and of Baikal's ruin.

The RSFSR Writers' Union first addressed the question of water resources directly in December 1985, including in their program a plank on the need for "fixed attention to ecology." They were prepared to face down the government, airing their concerns in public if necessary. They enlisted such sympathetic scientists as geophysicist and Academician A. L. Ianshin to defend rich monuments of Russian architecture, culture, and history. Two writers who served as deputies in the Supreme Soviet, S. Mikhalkov and I. Bondarev, were instructed to introduce a resolution in the Supreme Soviet attacking diversion.[79] The politicization of writers drew momentum from the growing pace of perestroika. Indeed the entire January 1989 plenary session of the Union of Writers of the USSR was devoted to "Land, Ecology, and Perestroika," criticizing the extensive development projects and the government's and Academy of Sciences' role in those projects.[80]

The most important group of authors to develop antitechnological themes in their writings were representatives of the newly formed Union of Writers of the RSFSR. These men of letters, in a new genre called "village prose," embraced nature, orthodoxy, and the simple agricultural life of the village. Siberian Valentin Rasputin, one-time adviser to Gorbachev and modern-day Luddite, uses his stories to praise the slow-paced, even preindustrial life of the Russian village while criticizing the irreversible damage that modern technology inflicted on that life. In *Farewell to Matyora* Rasputin describes a town's final days before it is inundated by waters from a newly constructed hydroelectric power station. In his writings Baikal was a symbol of Russian purity that the communists had deflowered with industry. Rasputin took pleasure in ridiculing Minlesbumprom, not so much for having polluted Baikal, illegally dumped wastes, exceeded limits, changed limits to meet newer and higher targets, and failing to produce cord as planned but also for the ministry's absurd justifications for doing all those things. Backed up by Zhavoronkov's "scientific expertise," ministry officials asserted that the tons of untreated minerals discharged into Baikal made the water, which was low in iodine and other important salts, more healthful to drink! Rasputin scoffed at the claim that atmospheric discharges were excellent fertilizer. He concluded that the ministry had waged a secret war against its people and its country. But he also felt that the Russian people themselves were to blame: "When individuals who think that the earth is improperly constructed come forth among the masses, believing it is necessary to modify it, what is dangerous is not that they appear but that we permit ourselves to follow them like prophets."[81]

While Rasputin defended Baikal, Sergei Zalygin—current editor of *Novyi mir*, founder and chairman of the organization "Ecology and Peace," and Supreme Soviet deputy—assumed generalship in the battle against river diversion. He criticized the "global egoism" of Soviet natural resource development and called for the adoption of measures amounting to "war eco-communism" (a take-off on the "war communism" period of Soviet history, from 1918 to 1921).[82] A hydroengineer by training, Zalygin, born in 1913, brought special technical education to his polemics. In 1939 he graduated from Omsk Agricultural Institute and then worked as a hydrologist in the Siberian division of the Academy of Sciences. He then became a correspondent for *Izvestiia* in Siberia and wrote stories including several about Akademgorodok and its beautiful forests. According to Zalygin, two events convinced him to reject diversion projects out of hand. First, he had worked under P. Ia. Kochina on a study of the lower Ob river basin to calculate the impact of a new hydroelectric station destined to flood 300,000 acres. The project would have gone ahead had not oil people stepped forward to defend Arctic drilling sites. Later, Zalygin attended a public dissertation defense by an IVP diverter-in-training. The dissertation celebrated the merits of diversion, yet the candidate could not answer the simplest questions about the cost or effectiveness of his proposals. To Zalygin this typified the Soviet philosophy of technological determinism: "If it can be done, it will be done." Financing inevitably led to design which inevitably led to construction.[83]

In an emotional diatribe published in *Novyi mir* in January 1987, Zalygin painted all scientists and engineers associated with water melioration projects as incompetent, dishonest, and out of control. It was simply a case of "us" versus "them," of "society" versus "ministerial interests."[84] Like Rasputin, Zalygin viewed all "Army Corps of Engineers" projects with great disdain, referring to the geological engineers—the canal builders and river diverters, the Baikal paper combine operators, as "professionals of gigantomania."[85] He acknowledged that gigantic Stalinist construction projects served an immensely important ideological purpose in Soviet history:

> This uni-directedness, this path [of development] gave us Kuzbass [coal fields] and [Magnitogorsk], Turksib and DneproGES with its "furious tempos" of construction, the Cheliabinsk and Stalingrad tractor factories. The whole world was stunned by our achievements, and it really was an achievement of universal significance, it showed what man was capable of doing, what the people were capable of doing, when roused by the ideas of universal revolution.[86]

In the Brezhnev years this dynamism gave way to what Zalygin called the "new conservatism," what I have defined above as "technological momentum," the prolonging of projects long after their usefulness has expired. In spite of his awareness of the power of this phenomenon in the Soviet context, Zalygin resorted to personal attacks in an effort to affix blame for the diversion projects. He singled out Voropaev as being responsible for all erroneous calculations and actions.

Indeed, a prime factor in the extent of environmental degradation in Russia was the momentum that construction trusts acquired. Angarastroi, Sibakademstroi, Sredazgiprovodkhlopok, and Bratskgesstroi commenced operation with relatively narrow, specific plans and turned into bulldozers of "geo-engineering." Bratskgesstroi, for example, formed in 1954, had six thousand employees by 1955 and thirty-five thousand by 1961, and the town where most of its workers lived had grown to fifty-one thousand. The creation of Akademgorodok, the reader will recall, engendered the establishment of Sibakademstroi, whose construction efforts soon extended throughout Siberia and included ICBM silos. The construction of hydroelectric stations, paper mills, cement factories, nuclear reactors, particle accelerators, and the towns for the workers who build and operate them required thousands of employees and millions of tons of equipment. To avoid unemployment and significant investment, transportation, and other costs, Soviet planners naturally found new work for construction trusts already in place with nary a concern about environmental consequences.

Zalygin discovered unexpected opponents to his antitechnology campaign in the intelligentsia of Central Asia who viewed diversion as a matter of life or death. Representatives of Sredazgiprovodkhlopok (the Central Asia Scientific Research Institute on Irrigation and Melioration) and Uzbek writers rejected Zalygin's arguments in a series of articles with an unusual point of view published in 1987 and 1988 in the literary monthly of the Uzbek Writers' Union, *Star of the East* (Zvezda vostoka). They suggested that the dispute over diversion was a matter of the North (Russia) discriminating against the South (Central Asia). They argued that central government investment patterns promoted the development of cotton and other industries beneficial to their Russian brothers in the empire, while inadequate investment in food agriculture had left the Central Asians unable to produce enough food for a burgeoning population. (Fertility rates in Soviet Central Asia were three to five times national levels.) Without adequate water for irrigation, the consumption of meat, milk, and fruit, already below national norms, would decline further. They declared proudly that under Soviet power, the amount of land under cultivation had increased twofold and that Soviet irrigation systems "earned respect the world over."[87] The president of the Uzbek Academy of Sciences, P. K. Khabibullaev, had an intriguing solution to the Aral Sea problem, namely, to increase the availability of birth control devices in order to lower the birth rates and thereby decrease water use.[88]

Eventually a number of influential Academy scientists joined Zalygin in his polemics against diversion supporters. One staunch opponent was A. L. Ianshin, Trofimuk's deputy director in the early 1960s at the Institute of Geology and Geophysics, vice president of the Academy for earth sciences, and later president of the Moscow Society of Naturalists, a leading environmental group. At one time, Ianshin advocated diversion from the Ob and Enisei Rivers. By the late 1980s, however, he had a change of heart. He then singled out Voropaev for criticism and served on a temporary commission appointed by Gorbachev that ultimately led to an August 1986 resolution halting diversion. So clearly had the Ianshin commission demonstrated that diversion costs far exceeded benefits that Academy president Gurii Marchuk wrote, "To argue with the conclusions of the Ianshin commission—this would be the same as arguing over multiplication tables." But Ianshin's contention that the danger from diversion surpassed even that from Chernobyl strained credulity.[89]

Ianshin was not the only Siberian scientist to change sides. Abel Aganbegian had cautiously endorsed Siberian river diversion as "one of the greatest projects" of the third millennium, calling it the "organization of a 'green bridge' for Central Asia," although he admitted that "careful research" was required to ensure the full, scientific grounding of the program.[90] His ambivalence had disappeared by the early 1980s when the Institute of Economics issued a report rejecting the project on economic grounds—to the astonishment of Moscow and the Central Asian nations. Then several leading mathematicians—L. S. Pontriagin, G. I. Petrov, V. P. Maslov, and A. N. Tikhonov—criticized the methods used in forecasting water needs, water levels, and the salinity of the Caspian. Finally, in 1987, writing in *Nash sovremmenik*, a right-wing, nationalistic literary journal, Pontriagin and Petrov called for the removal of Voropaev as director of IVP.[91]

In spite of the intense, personal criticism directed at him, Voropaev maintained his faith in the science of water melioration, believing the criticism was unjust, based simply on a misreading of the intent of the project as well as on an inaccurate description. He was stunned that support for his research was so suddenly ended in 1986. Currently deputy director of the institute and chairman of the scientific council on the interdisciplinary study of the problems of the Caspian Sea, he continues to see his "rational techniques" of redistribution as indispensable to solving the water use problems currently facing Russia.[92]

Lacking access to the publications that Zalygin commanded, Voropaev generally had to respond to his attackers in classified reports with limited circulation that were written by his associates. The reports denounced as emotional the "groundless slandering of the achievements of water management science and practice." So troubled were they by the one-sided treatment they had received in the press, Voropaev, D. Ia. Ratkovich, M. G. Khublarian, and eleven other IVP scientists sent a letter to the Central Committee asking for redress. They assailed the personal attacks, repeated their concerns about the mismanagement of water resources symptomatic of the Brezhnev years, and called for further study of diversion. They defended Voropaev's scientific independence and cited their attempts to invite the writers—Zalygin, Rasputin, Bondarev, and Belov—to the Fifth All-Union Hydrological Congress in Leningrad in October 1986 to discuss truth and responsibility in the media. That Voropaev invited scientists who were known environmentalists to serve on the State Advisory Commission of Gosplan, which he chaired from 1983 to 1987, tended to support his claims.[93]

When these calls for discussion fell on deaf ears, the engineers of nature, in 1988, published a two-volume defense and criticism of their opponents as "falsifiers" of history. Taking a chapter out of Zalygin's writing, they likened criticism of their work to that used by Lysenko against his opponents. One scientist wrote that the rejection of water management science achieved a "scale of disinformation on the part of individual writers and scholars" that was "comparable to the sad memory of recidivists in the area of genetics, the rejection of cybernetics, and [modern] forestry." A geodesist from Tashkent questioned Zalygin's insistent call for public involvement in scientific disputes. After all the public, in its infinite wisdom, had rejected the Brest-Litovsk Treaty in 1918 but had supported Lysenko.[94] Zalygin's opponents were particularly offended by his having placed his slanderous article "Povorot" in *Novyi mir* next to Daniil Granin's "Zubr" (Bison) about Timofeeff-Ressovsky and Lysenkoism (see chapter 3) so as to put diverters in the same category as "pseudoscience."

Misrepresentation, exaggeration, diatribe, and the so-called quote and club method employed by Zalygin, Voropaev, and their respective allies have been standard fare throughout Russian history. Much could be said, of course, for the effectiveness of these techniques in raising public awareness as well as increased understanding of the economic and ecological uncertainties of such a project. But in a society without open discussion of the variants, costs, and benefits, technology assessment was nearly impossible.

BAIKAL, DIVERSION, AND SIBERIAN DEVELOPMENT
UNDER GORBACHEV AND YELTSIN

The Communist Party began to mirror public concern about Baikal and diversion shortly after Gorbachev became general secretary. In August 1985 Boris Yeltsin, then Central Committee member, interrupted a busy fact-finding mission on the Russian economy to drop in on Grigorii Galazii at the Limnological Institute. Yeltsin expressed particular concern over the state of Lake Baikal. He listened attentively to scientists as they described the impact of paper mills, other factories, and BAM on the health of the lake. He then visited the BTsBK and spoke with its director, E. G. Evtushenko, who preached to him of the civic and moral duty to modernize pollution control technology. Evtushenko described the factory's efforts in this regard but confessed a weakness for the production of rayon cord. Yeltsin was impressed only by "the poor work of the Institute of Environmental Toxicology which for ten years has primarily developed norms of discharges of various chemical substances and has not devoted any effort to the creation of real measures to defend Baikal from pollution." Those guilty of violations, Yeltsin believed, should be punished.[95]

Shortly thereafter, in February 1986, the party appointed a special government commission to investigate diversion under N. V. Talyzin, candidate Politburo member, a first deputy chairman of the Council of Ministers, and chairman of Gosplan. The commission's work resulted in an August 1986 Central Committee and Council of Ministers resolution to halt all diversion efforts. The resolution ordered the State Committee on Science and Technology, the Academy of Sciences, and VASKhNIL to continue their studies, however, and to explore economic stimuli to water conservation measures.[96] In Voropaev's words, all his preparatory work had been "extinguished." Yet since the resolution did not outlaw diversion once and for all, it triggered an outpouring of public recriminations pitting the IVP scientists against their Siberian colleagues and Russian writers. Growing evidence that the "project of the century" was outrageously expensive, environmentally unsound, and overly ambitious filled the Soviet press.

Gorbachev's government then followed a hallowed Soviet tradition in dealing with the mounting din. First, on December 26, 1986, the Central Committee devoted an entire meeting to the Baikal situation, attended by other interested parties: Gosplan and ministerial officials, including M. I. Busygin, the minister of forestry, and O. F. Vasiliev, a corresponding member of the Academy and minister of water melioration; writers, including Valentin Rasputin; the chairman of the State Committee on Hydrometeorology (Goskomgidromet), Iu. A. Izrael; and several scientists, among them vice presidents of the Academy, academicians A. L. Ianshin, B. N. Laskorin, and V. A. Koptiug, and biologist O. M. Kozhova. Talyzin gave the main report in which he described successful steps already implemented to lessen the impact of "economic activity" on the Baikal basin. He noted the failures to decrease industrial discharges into Baikal and to preserve surrounding forest, and he admitted, "It is now clear even to a fool that the cellulose factory on Baikal should never have been built." Egor Ligachev, one-time secretary of the district party committee in Akademgorodok, and much later Gorbachev's rival, concluded the meeting with a call to preserve Baikal yet somehow permit its rational utilization. Deceitfully trying to tar others with the brush of responsibility while ignoring the Central Committee's direct complicity, he extended his usual criticism of various ministries to the Academy of Sciences for its "passive attitude." The Central Committee reached the do-nothing decision to instruct Gosplan to prepare in short order still more recommendations.[97]

On April 13, 1987, the Central Committee passed another resolution to protect the Baikal basin while making its resources available for exploitation, and appointed yet another commission to determine how to do this. Valentin Afanaseevich Koptiug (b. 1931), a geophysicist, chairman of the Siberian division, and academician since 1980, was a logical choice for commission chairman. Koptiug, a long-time member of the scientific establishment and a technocrat, was unlikely to call for rapid change at this stage of perestroika. Like Zhavoronkov, infamous for his role in polluting Lake Baikal, Koptiug graduated from Mendeleev Moscow Chemical-Technological Institute. He became head of a laboratory of the Institute of Organic Chemistry in Akademgorodok in 1960 and joined the party in 1961.

Koptiug traditionally defended central economic planning and large-scale projects, even when he sat on a series of environmental commissions in the 1970s and 1980s that moved ever so cautiously to rectify decades of planners' plunder. At the height of Brezhnev's rule, Koptiug touted the "Siberia" development program for its "complex utilization of the rich natural resources and effective development of the productive forces" of Siberia, and Akademgorodok's leading role in these matters.[98] As Siberian division president, Koptiug assumed direction of "Siberia" and rode comfortably astride its engine of development. Altogether some forty-six institutes and organizations of the Siberian division contributed their scientific efforts to resource development through "Siberia," for example, the efforts of ministries of coal and ore extraction were joined with those of Akademgorodok research institutes in such programs as "Coal of the Kuzbass" and magnitohydrodynamics.[99] With the emphasis on economic programs, environmental concerns were put on the back burner.

The extent to which fundamental scientific research was directed toward economic tasks was clear from the organizational structure of the six sections of "Siberia": mineral resources; biological resources, with emphasis on forestry and agriculture; regional economic development; technologies of materials and transport; energy; and "programs of particular complexity and scale." This last section included projects noteworthy for their immense scale and potential environmental cost: BAM (the new trans-Siberian railroad; see chapter 6), diversion, and Baikal development. O. F. Vasiliev—in the words of former minister of the environment N. N. Vorontsov, a "real canal builder and big diverter"—served as chairman of two subprograms, which seemingly required contradictory approaches: the first was diversion; the second, ecology and nature conservation of Siberia. The predominance of physicists and chemists on the councils once again highlighted the obstacles ecologists and other life scientists faced in having their voices heard.

Not surprisingly, Koptiug's diversion commission returned ever so cautious recommendations. At a series of meetings in the fall of 1987 the commission reported to the Council of Ministers, the Academy of Sciences, VASKhNIL, and finally to the Politburo on December 24, 1987. The Politburo approved the commission's recommendations for measures to improve water resource utilization. The report covered familiar ground: data on river flows; data on the catastrophic pollution of Siberian rivers; criticism of reservoir and dam construction, and of profligate industrial and agricultural uses; and shocking indications that between 20 and 40 percent of all home water use was wasteful.[100] Any visitor to Russia will testify that many Russians never bother to turn off the water or repair broken faucets. To the public, and to an increasingly vocal Union of Writers of the RSFSR, the August 1986 resolution should have sounded the death knell on diversion research. Instead, some saw the recommendations as an attempt to prolong the life of the nature creators, while glossing over past errors.

Still there were grounds for hope. Perestroika and glasnost permitted decentralization of policy making and criticism of past practices on the local level. Soviet environmental policies that had stressed large-scale, capital intensive projects in the name of resource development and had given inadequate consideration to potential social and environmental costs came under attack. A July 1988 session of the Central Committee under Egor Ligachev on the failure to enforce the most recent resolution resulted only in a decision to use the media to generate public interest in enforcement.[101] But President Yeltsin put a stop to further development with a moratorium, and it seems the paper mill will be closed and a national park established.

Regional authorities were among the first to jump on the bandwagon in criticizing Soviet gigantomania. At a party *aktiv* (a high-level convocation of leading officials intended to promote a specific economic or political campaign) in Irkutsk attended by All-Union and Russian ministers and leading scientists and planners, A. G. Melnikov, head of the Central Committee Department of Science, heard Irkutsk officials attacking planning organizations for lying to the public for some twenty years. The officials deplored the "serious errors in the siting and development" of Baikal industry, the failure to preclude or limit, let alone monitor, atmospheric or water discharges. Many workers—even leading ministerial and party officials—had been fired, others disciplined, but nothing seemed to help. Those who worked in Baikalsk, no longer having to fear being punished for their honesty, offered personal testimony to violations of Soviet law.[102]

In the fall of 1988 the chief physician of BTsBK, A. V. Boldonov, finally acknowledged that children in the region suffered from allergies, upper respiratory tract infections, perhaps even scoliosis, possibly as a result of inadequate filtering equipment that permitted the discharge of chlorine dioxide, chlorine, sulfur dioxides, and heavy metals into the environment. Boldonov, still seeming to be on the defensive, angrily added that the level of these illnesses was no higher than anywhere else in the country.[103]

Toxicologist Beim was also angry. Refuting the claim that half the children of Baikalsk suffered from motor disorders, he said, "The comrades are mistaken!" Perhaps heavy metal industry might have that effect but the operation of a paper mill could not produce such results. He believed that the rejuvenated newspaper *Sobesednik*, what he called a rag of perestroika, besmirched his institute by publicizing the unsubstantiated claims of Galazii.[104] He asked the locals to defend Baikalsk and its children against the false charges.

Changes in policy under Gorbachev allowed even Valentin Koptiug to relax his prodevelopment stance and side with the public against Beim and other paper mill officials. Although hesitant at first, Koptiug finally recognized the need for significant change in environmental policy. Out of political necessity to appear to be a progressive in the Gorbachev years, Koptiug became a conservationist from a position of "sustainable growth." He now believes that the roots of Russia's environmental crisis are to be found in the structure and management of the Soviet economic and political system that left local organs powerless against the center and placed economic development ahead of social concerns.[105]

For Koptiug, the transition to these views required a metamorphosis seemingly impossible only a decade earlier. Although he has certainly not abandoned communist ideology, he now calls for "steady state development" corresponding to human values. His new ideology requires international cooperation in science, culture, industry, and the environment while also emphasizing national concerns: economic well-being for future generations and the defense of natural resources from squander. The major danger to the Russian environment today, he believes, is an open-door policy to western firms for the development of mineral and natural resources without adequate economic and ecological controls. Like Francis Bacon, and hundreds of other technocrats before him, Koptiug believes that the foundation of the new ideology is "scientific and technological progress" oriented toward "the interests of society" and the "ecologization of production." In a speech to a general assembly of the Academy in December 1988 he emphasized the importance of fundamental research on the biosphere, including the development of ecotoxicology as a subdiscipline, and argued that scientific experts must take the lead in solving ecological problems.[106]

According to Koptiug, independent organizations of scientists must play a major role in the "ecologization of production" using "technology assessment" based on cost-benefit analysis. As an example of what might be achieved, he singled out the Siberian division's thirty-year record in environmental research, including the study of forests and soil, the ecology of hydroelectric energy, ecological programs for the ferrous metallurgy industry (the Norilsk plant in Northern Siberia), for individual cities—Krasnoiarsk, Kemerovo, and Novosibirsk—and for all of Siberia. Now the division would contribute to the development of innovative, environmentally safe technologies, for example, the Institute of Chemical Kinetics and Combustion would develop an aerator for pesticides that would reduce their use tenfold; IIaF would develop a radioactive treatment of seeds to avoid the use of chemical pesticides and herbicides; and powerful electron accelerators, also manufactured by IIaF, would clean industrial wastes from the air and water.[107]

Central to Koptiug's program were so-called ecological passports that provided each enterprise with information about specific pollutants, including the levels of control achieved in the most advanced countries. The passports enable local and regional administrations to fight pollution rather than distant, disinterested ministries and committees. For example, at the request of Novosibirsk Oblispolkom, the Siberian division developed an ecological passport for a fossil fuel cogeneration plant, the TETs-3.[108]

Koptiug maintains his faith in progress and science and finds most contemporary discussions of environmental issues alarmist in advocating the closing of industrial enterprises. Although he admits that growing populations and economic development make some environmental degradation inevitable, he draws the line when it comes to diversion, recalling how the Siberian division, under his chairmanship, was alone in its battle against diversion, despite intense pressure from several Central Asian republics.[109]

We have yet to see the practical results of Koptiug's ecological passports on Baikal or on diversion. His Baikal commission published standards for a new environmental regime in keeping with his gradual approach: there would be no new sources of pollution, old factories would be closed, heavy fines would help prevent discharges, central heating plants would be converted from coal and oil to natural gas, and a national park would be established in the Baikal basin.

Throughout 1988, however, the commission was riddled by haphazard enforcement. Some members asserted that Galazii continued to exaggerate the dangers of pollution at a time when progress was being made in fighting to restore Baikal. Galazii, for his part, asserted that "Minlesprom SSSR and its minister, M. I. Busygin, have begun to comply with the resolution with their usual deceit." Even Trofimuk has come to embrace Galazii's absolutist position: the only way to save Baikal is to reject development.[110]

The final chapter on Baikal and diversion has yet to be written. To this day scientists in the Limnological Institute hesitate to claim victory. They are concerned that new appointees in the paper industry will unquestionably accept the protestations of BTsBK directors that their plant is ecologically pure. The mill easily meets established norms but lags well behind world standards. Ministry officials continue to exaggerate social problems of economic dislocation were the mill to be closed. They seem to have convinced policy makers that a rayon cord shortage is impending.[111]

Yet the Baikal and diversion disputes marked a turning point in Soviet history. For the first time since the 1930s a major political dispute was played out in public. Various interest groups pressed the Communist Party Central Committee to reverse its decisions, yet none suffered the fate of dissidents under Stalin. Still today they fight to remove the offenders. Scientists who previously dismissed environmentalists as being "antiprogress" now recognize that the technological momentum of Soviet economic development had to be abandoned. Such scientists as Andrei Trofimuk from Akademgorodok and Grigorii Galazii from Irkutsk, products of the Soviet system, rejected their heritage for the environment and were able to draw on the media for support.[112]

Other forces now plague Siberia. The empire's edges are frayed. Russia can no longer rely on the former USSR republics either as sources of finished goods or markets for raw materials. Big science has lost its technological momentum, although more for reasons of economic downturn than direct political action. The rickety infrastructure of BAM, hydropower stations, oil derricks, and paper and steel mills is aging none too gracefully. Reliance on prefabricated construction materials produced and assembled by Soviet workers even as long as thirty years ago has left industry in need of capital investment. The cost of environmental cleanup is steep and may take a backseat to investment in industry. Meanwhile, closed nuclear and chemical cities have discarded wastes without regard for future generations, and forestry and construction has cut through the fragile taiga. Some argue that Russia must attract foreign partners by relaxing environmental protection laws; others caution against the offers of western firms to develop Siberia's resources at breakneck speed. Trofimuk and Galazii have retired, and Koptiug, as Akademgorodok's chief administrator, grows increasingly conservative.

A decade from now, when writers describe Siberia, will they speak of Lavrentev's Golden Valley or the taming of the Ob River for Central Asian agriculture? Perhaps both. But whatever the future holds there is reason for hope. Western investors, having discovered Siberia's mineral wealth, will now be required by local and national officials to follow accepted western practices in their extraction of Siberia's natural resources. Private citizens no longer tolerate gigantomania and its costs. To ensure the rational use of resources, Akademgorodok and the Siberian division have called for the establishment of international research centers to attract foreign specialists in a wide variety of fields. And in the meantime Siberian scientists have managed to derail diversion and to put on hold any further development of Lake Baikal.

The Siberian Algorithm

IN THE EARLY 1960s Tatiana Zaslavskaia, a sociologist with a background in economics and physics, wanted to determine the quality of life in the Siberian countryside, the amount of personal property owned, the extent of the farmers' loyalty to the regime, and why outmigration from the village to the city was increasing. Interviewing a poor Siberian villager, she asked, "How many pigs do you own?" "How many do you want me to have?" her suspicious respondent answered, unused to the openness required in survey research. Such was the modest inception of social science east of the Urals.

In Akademgorodok economists and sociologists moved on from these humble beginnings to apply quantitative research methods and modern survey techniques to the study of the Soviet system. They ventured to Siberia to escape the stultification of social science by stodgy party scholars in Moscow and Leningrad. Social science, more than any other branch of knowledge in Stalin's time, was forced to toe the Marxist line. Whole fields of research, methodologies, and western literature were taboo. Research that might cast a negative cloud on the regime or have the slightest taint of bourgeois social science theory was considered heresy. No one wished to hear evidence of the maldistribution of goods and services, backward economic performance, the concentration of political power in the hands of a privileged elite, or the excessive costs of collectivization or industrialization in the 1930s. From philosophy and epistemology to political theory and economics, no discipline was spared the party line.

After Stalin's death conditions were more propitious for the development of modern social science, and Akademgorodok provided a hospitable environment for its resurrection. Although Khrushchev assisted in revitalizing a number of fields, quantitative approaches such as those used in linear programming, in computer applications, and in survey research still remained beyond the pale. New approaches in social science were explored within the walls of the Institute of Economics and the Organization of Industrial Production (Institut ekonomiki i organizatsii promyshlennogo proizvodstva), or IEiOPP, by economist Abel Gezevich Aganbegian and sociologist Tatiana Ivanovna Zaslavskaia. Aganbegian applied multivariate analysis and linear programming to create optimization models of economic growth. Other economists elaborated mathematical models on national, territorial, and branch industrial levels. The rebirth of mathematical economics and of management science promoted the development of new theories of planning while questioning traditional Soviet planning and management techniques.

Zaslavskaia embarked on a study comparing the quality of life in urban and rural Siberia using survey research. Although she and her colleagues lacked the rigorous requirements of question design and sampling techniques used in western studies, their sociological investigations were impressive, revealing the negative effects of modernization, industrialization, and outmigration on the countryside; questioning the regime loyalty of the increasingly well-educated worker whose expectations for rewards under the terms of the Soviet social contract had changed; and suggesting that the intelligentsia had also lost faith in the ability of the Soviet system to meet its needs.

For Akademgorodok sociologists and economists alike it was difficult to break away from the hyperrationalization of the Soviet tradition that sought to use allegedly objective, empirical standards to manage all processes. For decades Soviet economic planners had viewed citizens as interchangeable with other "inputs" in the economy, like machinery, equipment, or mineral ore. Economics and sociology were intended to achieve state goals for heavy industry. The question was whether Aganbegian, Zaslavskaia, and their associates could break away from this tradition and redefine such economic concepts as efficiency, rationality, and utility. The problem was that the evaluation of these concepts was not strictly an empirical task. Given different social goals, political programs, and resource constrictions, the determination of the criteria becomes a normative consideration. Researchers at Aganbegian's institute never fully broke from the Soviet tradition. They maintained the desire to manage all economic and social processes on the basis of results produced by multivariate modeling. At one point they even conducted engineering studies almost "Taylorist" in their embrace of what they considered scientifically determined objective standards.

Still significant methodological innovations took hold in Siberian economics and sociology. For Akademgorodok social scientists, the new approaches required an accumulation of data to test hypotheses and to verify the power of new mathematical models. The data revealed just how poorly the Soviet economy operated and how inequitable life was in the countryside. Algorithms introduced in a series of models to assist planners in optimizing industrial production instead showed fissures in the Soviet system. Indeed the work of Aganbegian and Zaslavskaia, what I refer to metaphorically as the "Siberian algorithm," suggested the need for fundamental reform in the Soviet economic system. Surely no one could have anticipated the revolutionary changes brought about in the USSR under Gorbachev, including a complete repudiation of Soviet Marxism if not socialism itself. Yet years earlier, within IEiOPP, economic and sociological investigations had already questioned the basic tenets of the system. And for a short time these researchers had found Gorbachev's ear.

ECONOMIC SCIENCE FROM STALIN'S MOSCOW TO AKADEMGORODOK

The heyday of Soviet economics was the 1920s. The development of mathematical economics paralleled the rapid development of statistical approaches in mathematics and physics. Many economists had been trained in the prerevolutionary years when mathematics was prominent in several university departments. They knew mathematics well. Their field seemed to have a bright future. Economists accumulated significant data and results based on their analysis of the small-scale market system that was allowed to flourish during the New Economic Policy (1924–29). For example, S. G. Strumilin, considered one of the founders of the five-year plan, studied time budgets in the 1920s to advance a series of recommendations for planning labor inputs. Much of the western postwar economics research on input-output, growth models, and the economic problems of developing countries grew out of the 1920s.[1]

Changes in the face of the discipline were inevitable with Stalin's "Great Break" of the late 1920s and 1930s—a revolution he initiated from above to change in short order Soviet politics, economics, society, and culture. The Stalinist economic program centered on the development of heavy industry, extraction of investment capital from the agricultural sector, reliance on planners' preferences, a centrally planned economy rather than market mechanisms, and the use of more labor than was needed to provide goods and services as a means of avoiding unemployment. Policy makers believed that speed was essential to create industrial self-sufficiency and a working class and to avoid "hostile capitalist encirclement." To achieve these goals, five-year plans were introduced to raise output per input of capital. The emphasis on heavy industry distorted the economy, leaving investment for consumer goods, housing, and health care far behind. Furthermore, Stalin and his allies determined that the centerpiece of the endeavor should be Marx's labor theory of value, which posits that the value of a given commodity or service should be based on the quantity of labor required to produce it. This undervalues "free" resources such as air, water, and land, and allows little leeway for prices to reflect scarcity values. It should be noted that Marx borrowed from Smith and Ricardo, intending his work to be a criticism of capitalism, not a model of socialism. The debate among scholars over the implications of Marxian notions for socialism had yet to be decided when Stalinist policies put an end to discussion.

Tentative five-year and annual plan targets were established by Gosplan but could be changed by enterprises through negotiation—or deceit. Enterprise managers were central in formulating and implementing the plan by calculating how much labor they needed, average and total wages, and requests for materials. Managers were likely to hide labor or capital reserves and to avoid overfulfillment of plans because they feared even higher targets (the so-called rachet effect). They also had little influence on prices as these were set by the government, as were wages. Targets were usually defined in terms of production quotas, leaving little incentive to produce quality goods. Gosplan, having to review hundreds of thousands of prices annually, a physical impossibility even with modern computers, soon became overburdened and had to resort to spot checking. Thus each year prices came less and less to reflect actual costs. In addition, the tendency to rely heavily on undervalued capital and natural resources resulted in their excessive use and hampered technological progress. Bribes and connections played a major role in the acquisition of goods, the black market assumed central importance in distributing scarce goods, and people stood long hours in lines to receive them. To deal with these problems, the Communist Party created dozens of bureaucracies to oversee the economy from the top down by directly controlling prices and the flow of goods and services.

This economic reality was reflected in the social science research program. Stalin sharply criticized economists whose work challenged his model of development or, worse still, those who supported his opponents. At a conference of Marxist agrarian specialists in December 1929, he attacked Nikolai Bukharin's theories of balanced economic growth (as opposed to hyperindustrialization), at the same time ridiculing planning up to then as "playing with numbers." (Not only did Bukharin have the audacity to offer an alternative economic vision, he was also Stalin's chief political opponent after Leon Trotsky's exile and was ultimately executed for treason in the Stalinist purges in 1938.) At that same conference Ernst Kolman, a competent mathematician but an ideologue as well, criticized stochastic methods in quantum mechanics. By the mid-1930s any effort toward the rational use of resources by means of price criteria and mathematical economics itself were seen as bourgeois science, if not outright falsification. Optimization techniques necessary for resource allocation were now held back both for scientific reasons (the absence of general algorithms, computational technology, and data) and political ones. During the purges dozens of talented economists were accused of "wrecking" and were arrested, many of them shot. Restrictions on the number of Jews admitted to mathematics and economics departments were enforced and adhered to through the 1980s, depriving those fields of a number of promising researchers.

Status quo Stalinist economists, who valued the descriptions of processes that included citations of Marx, Engels, Lenin, and the great Stalin himself, preferred sitting in their offices imagining what the socialist economy ought to be like rather than learning the preferences, motivations, or complaints of workers and farm laborers firsthand.[2] Of course planning the Soviet economy was not possible without mathematical economics. One Soviet economic historian wrote: "The objective need for the rational use of resources existed and, one way or another, the problems of how best to distribute these resources, particularly capital investment, were solved. Frequently, however, the solutions were administrative and not economic in nature."[3] All this stalled the development of mathematical economics until the late 1950s when it was again taken up at IEiOPP, NIITrud (the State Scientific Institute of Labor), Leningrad, Gorky, and Ural Universities, and the Central Statistical Administration.

Some research in mathematical economics managed to continue under L. V. Kantorovich, V. V. Novozhilov, and V. S. Nemchinov. In 1939 Kantorovich, a talented young economist who later moved to Akademgorodok, introduced linear programming to solve the problem of distributing different kinds of raw materials among various machine tools with the goal of maximizing output for a given assortment. In the summer of 1940 Kantorovich met Novozhilov, and the two jointly taught a seminar at Leningrad Polytechnical Institute. Building on Kantorovich's ideas, Novozhilov studied problems of the comparative advantage of capital investment. In 1942 he escaped the Leningrad blockade for Yaroslavl where he completed a manuscript on economic calculations of the best use of resources, but because of Stalinist academic strictures, neither Gosplan nor Moscow's Institute of Economics allowed the manuscript to be published until 1958. He did manage to publish several articles in the late 1940s, however.

V. S. Nemchinov, later director designate of the new economics institute in Akademgorodok, showed great courage at this time in his defense of mathematical economics and, at the 1948 Lysenkoist conference, also defended modern genetics for which he was removed from all administrative posts. In 1948 ideologues in the Academy's Institute of Economics asserted that a tie existed between Nemchinov's defense of "bourgeois ideology of Weissmanism-Mendelism" and "mistakes of a formal mathematical nature on the basis of questions of statistical methodology." But Kantorovich's surprising award in 1949 of a Stalin prize enabled mathematical economists to continue publishing articles on occasion.[4] Why he received this award is unclear. His connections with the Department of Political Economy of the Central Committee's Academy of Social Sciences (1947–57), his membership in the presidium of the USSR Academy (1953–62), and his chairmanship of SOPS (1949–64) may have given him leeway to popularize input-output analysis.

As in the field of genetics, Khrushchev played a central role in the history of economics, specifically in the rejuvenation of management science. He and his followers argued that scientifically determined production norms and quality control would increase industrial efficiency. Seeking to strengthen party control over managers, he sought to establish "scientific" standards for managers' behavior. He also wished to decentralize the top-heavy, Stalinist economic system by dismantling the central ministries and reorganizing them under territorial economic councils, or *sovnarkhozy*. Khrushchev believed that mechanization, standardization, and automation from simple items to prefabricated apartment buildings and even nuclear reactors were the keys to modernization. Of course many of his ideas reflected his fascination with large-scale technologies rather than a real understanding of how to promote efficiency.

The twentieth party congress in 1956 triggered rapid changes in the economic sciences as it had in genetics, philosophy, and cybernetics. The Institute of Economics, founded in 1930, was now joined by the Institute of the World Economy and International Relations (IMEMO) to study "bourgeois" economic science and capitalist economies. IMEMO studied, translated, and published western management materials that had been classified as "secret" since the early 1930s. Nemchinov then founded the Academy's laboratory of economical mathematics in 1958, which later became the Central Economic Mathematical Institute (TsEMI). In 1959 he persuaded the Central Statistical Administration to compile an input-output table for the USSR. By that time forty institutes were engaged in mathematical modeling, computer-based methodologies, and other quantitative research. Cyberneticians, led by Aksel Berg, who had presided over the rebirth of cybernetics with Aleksei Liapunov (see chapter 4), saw quantification as proof that "management was already becoming an independent branch of scientific activity."[5]

At this time, in an edited volume of essays, Nemchinov urged the Academy's social science division to sponsor a national conference on mathematical economics, linear programming, statistics, and computer applications, which it did in April 1960.[6] Fifty-nine talks were duplicated and distributed to participants in advance of the conference. Discussions between the participants revealed a deep-seated mistrust between those who saw the taint of bourgeois ideology in mathematical economics and those who believed in the promise of new theoretical approaches to solve Soviet economic problems. The former were concerned on several fronts. First, they believed that the "mathematical" theory of prices, that is, the use of prices as simply numbers to assist decision makers in arriving at optimal solutions, was contrary to the labor theory of value. Second, they argued correctly that planning was not a mathematical problem but a political, social, and technological one that required conscious decisions by planners. They believed that focusing attention on the rational organization of production, which involved the application of mathematical techniques, would lead to the neglect of attention to production relations. And they also resented implied criticism of traditional planning methods.[7]

Supporters of mathematical economics contended that the increasing mathematization of economics was proof in itself that economics had become an exact science. They pointed out that Gosplan's Computer Center was at work at solving optimization models. Leningrad and Moscow Universities (soon to be joined by Novosibirsk University), the Leningrad Financial Economic Institute, and the Moscow Institute of Economics were training young "economic cyberneticians." Mathematical economics was part and parcel of the process of recognizing the laws of nature and society and putting them at the service of social progress. Since the increasing productivity of labor was based on modern technology, mass production, and automation, why not use these as the basis of the planning process itself? The economic cyberneticians, as they came to be called, pointed to mathematical economics in the West, not as a sign of its ideological decrepitude but of its universal, objective application, although a critical approach was always called for with respect to the work of bourgeois scholars. They argued that only scientific management could handle the problems associated with a modern economy. In the words of one specialist, "[The socialist system] demands a more perfect system and technology of management."[8] Still the conservative critics of mathematical economics were correct on one count: they recognized, as the proponents did not, that optimal planning was not merely a technical problem to be solved with new methods but one that required social and political choices about goals, priorities, and strategies.

The proponents gained the upper hand over the next thirty years. Unfortunately the insistence on administrative measures to solve economic problems limited the impact of any reforms. Each measure that was intended to increase the enterprise manager's autonomy, encourage initiative and risk taking, accelerate innovation, or lessen reliance on planners' preferences in setting prices was countered by bureaucratic intransigence, ministerial interference, and half-hearted attempts at implementation during the Brezhnev years. Paradoxically, those who favored new techniques had a hand in the continued reliance on central control mechanisms. They saw new computer technologies as a panacea for the USSR's economic planning problems, believing that a series of mainframes located in Gosplan could facilitate instantaneous input-output analysis for the entire economy. The mainframes, linked with every enterprise, store, and farm, could select, store, transmit, and analyze information drawn from all over the country, leading to a radical improvement in management. Ultimately computer management systems were favored by conservative economic planners who recognized the great control they acquired through them. They could identify bottlenecks and nonperformance, as well as punish the guilty, all without moving from Moscow. Thus an atmosphere of tension between reformers and traditionalists provided the environment for the development of Siberian economics and Abel Aganbegian's new beginnings east of the Urals.

AGANBEGIAN ARRIVES AT AKADEMGORODOK

In 1957 economics lecturer Abel Gezevich Aganbegian, speaking to a group of mathematics students at Moscow State University, was asked, "To what extent is economics an exact science?"[9] Of course a simple answer would not suffice and was also not Aganbegian's style. The resultant discussion prompted him to sit in on mathematics courses with university freshmen, which then led him to study number theory and linear programming and to apply these to his work. Over the next three years he lectured on political economy at the university,[10] an experience that prepared him well for his future immersion in the natural sciences at Akademgorodok.

A captivating speaker with looks that won him a reputation as a lady's man in his youth, Aganbegian was born in Tbilisi in 1932, graduated from Moscow University in 1955, and joined the Communist Party in 1956. Until 1961 he worked on wage and labor problems for the State Committee of the Council of Ministers of the USSR. With V. F. Mayer he published *Salaries in the USSR* (Zarabotnaia plata v SSSR; 1959) and edited *The Application of Mathematics and Computers in Planning* (Primenenie matematiki i elektronnoi tekhniki v planirovanii; 1961). Feeling hemmed in by the tediousness of his work at the State Committee, largely directed toward shortening the work day, increasing salaries, and publicizing the supposed advantages of the Soviet system such as the absence of unemployment, free day care, and universal health care, he was drawn to the novelty of mathematical economics.[11]

Aganbegian was long aware of the poverty of the Soviet existence. He recalled that when he married in 1953 he temporarily worked in a textile factory in the small town of Sobinka in Vladimir region. He called on his wife's relatives in the village of Zhokhovov in the same region. The village was 80 kilometers from the railway, and the dirt road to it was impassable mud in autumn and spring and covered with snow in the winter. Aganbegian walked almost 20 kilometers on foot. Though 150 kilometers from Moscow, Zhokhovov did not have electricity. Its one shop was open only twice a week and carried little beyond sugar and salt. Local workers earned little and depended mainly on their family plots for food.[12] It is not surprising that Aganbegian was attracted to work intended to improve the lot of the Soviet citizen or that he was deeply committed to overcoming the inequities of life in the countryside.

Reading of Akademgorodok in the newspapers in 1958, he headed for Siberia in November 1961, when his daughter was three, and remained there for twenty-five years. His son was born there. At first he lived in Novosibirsk and traveled on weekends to Akademgorodok, then just one muddy road and a mass of construction. But spiritually he could feel the thaw and the energy of the people working there. Youth organizations imbued with the spirit of de-Stalinization dominated life in the city of science. Through the Komsomol, the Council of Young Scientists, the physics-mathematics school, the social clubs, and the institutes, the young workers explored ideas, emotions, art, literature, and science in a manner impossible just five years earlier. Akademgorodok was homogenous, overwhelmingly intelligentsia, so that many of the class-based confrontations that are standard in any setting were a rarity.

Akademgorodok's intellectual climate invigorated Aganbegian as much as its social life. He saw interdisciplinary research—biophysics, mathematical economics—blossom before his eyes. Unlike in the West, where interdisciplinary work was common and research was tied to teaching in the universities, in Russia, from the start, academia and universities developed under different bureaucracies: the Academy, founded in 1725, separate from Moscow University, established in 1755. Science was parcelled out in central research institutes divided according to narrow branches of knowledge. At Akademgorodok Lavrentev had integrated science and education from above both methodologically and geographically through mathematics.

Several aspects of mathematical economics were of concern to Aganbegian. He pointed out, for example, a rather obvious dilemma: if final production targets or assortment were improperly established, no amount of optimal planning would lead to plan fulfillment. It was necessary both to introduce mathematical methods on a broad front, working toward a system of mathematical models for the national economy, and to define final targets correctly, based on sophisticated understanding of consumer demand. In his work on wage and labor problems for the State Committee in the late 1950s Aganbegian had wrestled with this very problem. In association with several other research institutes and computer centers, he studied data on more than twenty-four thousand Soviet families in a survey conducted by the Central Statistical Administration. Using the survey to correlate consumer demand with income level, he hoped to fix accurate production targets and only then develop optimization models. Just such a dynamic component was needed in the work of Kantorovich and Novozhilov and in linear programming in general.[13] And it was toward this end that Aganbegian and his associates worked at IEiOPP.

ECONOMICS EAST OF THE URALS

Most of the prewar economic research in Siberia involved the accumulation of geological and geographical data of the area's vast natural resources which was conducted under the jurisdiction of SOPS. By the mid-1950s fewer than twenty specialists worked in the Department of Economic Research of the West Siberian branch of the Academy of Sciences, and only two were candidates of science.[14] The continuing tension between the more traditional Soviet approaches to planning and modern mathematical economics is reflected in the history of the two major postwar actors: IEiOPP, the first economic research institute to be founded east of the Urals, and Aganbegian's Laboratory of Economic Mathematical Research (Laboratoriia ekonomiko-matematicheskikh issledovanii), or LEMI.

On June 7, 1957, the presidium of the Academy of Sciences approved the organization of an Institute of Economics and Statistics in Akademgorodok. Nemchinov was to head the new center, but illness prevented him from doing so. In May 1958 the name was changed to the Institute of Economics and Organization of Industrial Production, and G. A. Prudenskii, corresponding member of the Academy, took on the task of its organization. Prudenskii, a leading specialist in industrial economics, studied how to raise labor productivity within existing labor and technological reserves. His work on the free time of workers contributed to the rebirth of sociology in the USSR.[15] The institute's main functions were to analyze the organization, planning, and distribution of production, including the analysis of labor reserves and pertinent statistics. It ultimately established economic laboratories in the largest organizations in the region.

In 1961 LEMI, under Aganbegian's direction, became affiliated with the institute. LEMI's activities were closely tied to those of the Computer Center and the Institute of Mathematics. Aganbegian and Kantorovich, who had helped the Computer Center get started in Moscow but was happy to move to Siberia since he continued to have trouble publishing in the new field, joined Aganbegian in organizing a seminar to train specialists in mathematical economics, management science, and computer applications. Surrounded by powerful mathematicians, the two began to train twenty-nine potential economists.

Prudenskii's institute was located in Novosibirsk. Through close contacts with the Novosibirsk party committee, Prudenskii was able to provide his staff with attractive apartments in the city. In fact, when the institute finally opened in Akademgorodok, many on the staff did not want to move, preferring ministerial calm to academic uncertainty. Aganbegian's LEMI, on the other hand, having established an independent existence in Akademgorodok, grew rapidly in conjunction with the expansion of economics facilities at the university and quickly attracted more than a hundred employees. Prudenskii's followers "became jealous of us," Aganbegian recalled, and a schism developed pitting Prudenskii and the old school against Aganbegian, the Institute of Mathematics, and the Computer Center.

The tension was played out in meetings of the institute's methodological seminar, the body responsible in part for ensuring that the scientists adhered to the ideology of Soviet Marxism. At one such meeting a party member called for the censure of Aganbegian for his one-sided criticism of the country's economic insufficiencies. While insisting on more criticism of bourgeois economic theories, he expressed concern about the poor attendance at the seminars.[16] Prudenskii himself spread rumors that LEMI was "hostile to Marxism," and ultimately the obkom began to take notice. In 1964–65 LEMI was removed from the institute, whereupon Lavrentev, in the presidium of the Siberian division, announced that LEMI was to become an independent laboratory at Akademgorodok with Aganbegian as director. The new laboratory was intended to develop mathematical methods in planning, development, and distribution of the production process, to undertake research on the tempo and proportions of Siberian economic growth, and to apply statistics and mathematical methods to the analysis of social processes. In protest of Lavrentev's actions, Prudenskii's allies in LEMI quit.

In April 1965 the disagreements between Prudenskii and Aganbegian became public, reflecting a national debate over the purposes economics was to serve. Prudenskii, an accomplished if pedestrian scholar, preferred politically safe, concrete, narrow research topics concerning regional (Siberian) economic development. To assist in these studies he called on such central economics organizations as SOPS, Gosplan RSFSR, the Institute of Economics, NIITruda, and Gosplan itself. He recognized that new mathematical research tools were needed but did not see their utility in his institute nor did he believe that the new institute could focus both on mathematical economics and Siberian development.[17] Moreover, he had a "suspicious attitude toward computers" and, in the words of a contemporary, made administrative decisions "by rumor." Compounding the problem, he surrounded himself with similarly inclined individuals, for example, L. V. Starodubskii, who earlier had spoken against cybernetics.[18]

The effort to analyze the economic problems of the Stalinist legacy now came to a head. The lack of coherence in research direction resulting from the dispute between Aganbegian and Prudenskii was aired at a series of meetings of the institute's party apparatus. Prudenskii and his followers expressed concern about the proliferation of institute themes and sectors reflecting increasing specialization in modern science. The information sector, for example, had created a sociological group. Was this overlap really necessary? V. I. Pandakov observed that junior staff had a particularly hard time doing their work when "themes changed constantly." According to G. A. Zakharov, because of the organizational problems, not one senior staffer had written "a solid work." Finally, inadequate facilities, limited access to computer time, and poor housing combined to slow the nascent research program and created difficulties in staffing the institute. Prudenskii admitted there were problems in the focus of the research. As of December 24 the director had yet to receive annual reports from any sector, let alone plans for the next year's activities. Skirting responsibility for the disorder in the research program, he made the incongruous suggestion that researchers spend more working hours in the institute rather than providing them free days for research and writing at home or at the library.[19] Only when Aganbegian assumed leadership did research move firmly in the direction of modeling national, regional, and branch industry.

Aganbegian's allies, half the economists at IEiOPP, resigned to go with him to LEMI. Prudenskii, now critically ill, had no choice but to leave his depleted institute along with several older researchers, and returned to Ukraine. Aganbegian, the only remaining corresponding member of the Academy of Sciences, was the logical choice for director of the new institute formed by the melding of IEiOPP and LEMI. After Prudenskii's death, out of sympathy for his family, staffers at IEiOPP collected mock prize money among themselves as a small gift for his widow.[20]

Organizational turbulence now gave way to problems with facilities and housing.[21] IEiOPP was intentionally one of the last institutes to be moved from Novosibirsk to Akademgorodok. The building to house IEiOPP, as well as the presidium, raikom, raispolkom, and a library, was completed only in mid-1963.[22] Lavrentev's "commandment" that all research was created equal, that it was to be unified through mathematics, and that all scientists were to have equal access to facilities was not obeyed. The social scientists were treated like second-class citizens, even refused entry at the Computer Center except during their assigned period—3:00 A.M. to 5:00 A.M.

Yet despite internal intrigue and housing problems the institute grew steadily. From only 116 researchers in all of the Siberian division's economic organizations in 1959, including just one corresponding member of the Academy and one doctor of science, by 1984 the economic institute had grown to 510 staffers of whom 210 were scientists, with 140 doctors and candidates of science among them.[23] Reflecting the party's success historically in attracting members from the social sciences, if not the exact sciences, on average one-third of the staffers belonged to the party and one-sixth belonged to the Komsomol.[24]

IEiOPP's scholarly productivity grew hand in hand with the national development of economics as a discipline. Of the nine economics institutes, ten economics scientific councils, and eight economics journals in the USSR Acad-

TABLE 6.1
Annual Number of Copies of Soviet Economic Journals, 1980–1990
(in thousands)

		Journal		
Year	EKO	Voprosy ekonomiki	Planovoe khoziaistvo	Izvestiia AN SSSR, Seriia ekonomicheskaia
1980	76	50.0	36	3.5
1981	86	53.0	38	3.4
1982	100	44.0	32	3.4
1983	115	44.0	32	3.5
1984	136	44.0	32	3.4
1985	141	43.0	33	3.3
1986	158	43.0	33	3.1
1987	155	44.0	34	3.2
1988	147	45.0	36	3.1
1989	160	60.0	38	3.6
1990	171	73.5	39	4.5

Source: The Special Fond of the IEiOPP.

emy of Sciences in 1975, IEiOPP was fourth in terms of staff size, fifth in the number of doctors of science, seventh in the number of candidates, eighth in the number of dissertation defenses, and first in the number of books published per staff member. Within the institute's first eight years, its researchers published 100 works, including 22 monographs, 17 collected volumes, and 17 brochures, most often in Moscow since academic publishing in Novosibirsk was not fully established until the 1970s. During the ninth five-year plan (1971–75) IEiOPP researchers published 184 articles, including 75 in *Izvestiia Akademii Nauk SSSR, Seriia ekonomicheskaia* and 60 in *EKO*. Like their colleagues at the other institutes, researchers promoted awareness of their new science through public lectures and television and radio appearances.[25]

Most significant in publicizing the new science was the creation of a new economic journal, *EKO*, soon one of the most influential economic journals in the USSR. Planned for 1968, *EKO* first appeared in March 1970. Its initial run of 8,000 copies grew to 171,000 by 1990 (see Table 6.1). The editors, wishing to create a journal that would become standard reading for economists and managers, chose a lively format and nonacademic style, with emphasis on management science and how to think "economically" (po-khoziaistvenno) by applying modern, rational, long-range mathematical and computer planning techniques. Articles, interviews, conversations, round-table discussions, photographs, and unique graphics filled each issue. An early article was intended to offer a graphic on the "increase in production of oil and gas," but instead showed "fur and gas." One editor commented, "Well, at least production grew."[26]

With the growth of IEiOPP, a series of affiliates were established in Siberian city centers that focused on issues of regional development, avoiding topics that might question national investment and pricing policies. They included the Krasnoiarsk economic laboratory; the Department of Regional Economics and Distribution of Productive Forces of East Siberia in Irkutsk (1968); the laboratory of economic forecasting in Kemerovo (1968); the economic laboratory in Tiumen (1968), reorganized in 1972 as the Department of Economic Research; the laboratory of the economics of industry (1972) in Barnaul; and in Kyzyl, the Tuvin economic laboratory (1974). Economic departments were later established in the Iakutsk and Buriat branches of the Siberian division of the Academy of Sciences. Other economic centers have been organized in connection with problems of resource use: the coal-rich Kuznetsk basin, hydroelectric power, Baikal, and BAM. Ultimately an entire network of economic bureaus was established throughout Siberia, under the direct or indirect supervision of IEiOPP, and often with offices or laboratories in major industrial enterprises, higher educational institutions, and ministerial research institutes.[27]

Unlike its association with the physics, chemistry, and mathematics institutes, Novosibirsk University only belatedly became a source of personnel for the new economics institute. A humanities department was created in 1962 with a Department of Economic Cybernetics. A university laboratory of economic mathematical research followed. In concert with the IEiOPP and the Institute of Mathematics, the laboratory earned hundreds of thousands of rubles through contracts with dozens of enterprises, design bureaus, and branch research institutes through the development of computer-based models and optimal variants for regional automobile transportation, coal mining, and the construction of cement factories and apartments. An independent economics department was formed in 1967 with four specializations: economic cybernetics, economics and statistics, planning, and political economy. Lavrentev proposed opening special departments in all major universities to train mathematical economists. The hope was that in Moscow, Leningrad, and Novosibirsk at least 250 specialists could be trained by 1970. There was some basis for optimism. A survey of eighteen thousand teenagers from the Novosibirsk region who intended to enter the university showed that mathematical economics occupied a close third place behind physics and chemistry in terms of interest. In fact, between 1962 and 1974 more than five hundred individuals graduated from the Novosibirsk University mathematical economics program. The program stressed "broad-profile training," a kind of cooperative educational approach where games, case studies, semester-long projects, and research from the fourth year on for Akademgorodok institutes, all involving computers, were central to training. At the suggestion of Aganbegian and V. N. Lisitsin, the deputy chairman of Gosplan RSFSR, and with the welcome endorsement of university rector Spartak Beliaev and the head of the Siberian division, Gurii Marchuk, a center for retraining mid-career professionals in economic mathematics and computer techniques, the so-called *spetsfak* (special department), was also created. About twenty-five individuals from various branches of government and industry attended three-month courses based on the case study method, an approach similar to that used at Harvard Business school.[28]

SIBERIAN ECONOMICS RESEARCH UNDER AGANBEGIAN

Unlike life sciences and exact sciences, the social sciences had no streetlights to illuminate research owing to thirty years of Marxist stultification, isolation from international trends, and the avoidance of "bourgeois" methodologies. Certain approaches simply had to be avoided. Microeconomic analyses that revealed poor performance at the enterprise level were preferable to macroeconomic ones that tended to find fault with industrial branches, if not the entire Soviet system. Similarly, survey research, mathematical models, and the application of computers had to be used with great care on two counts. First, as social scientists quickly learned, new sources of data were bound to demonstrate discontinuity between the public vision of communist plenty and the private reality of standing in line for basic necessities. Second, economics and sociology had long been controlled by individuals whose careers depended on traditional approaches.

Economic science at IEiOPP was given the thankless task of analyzing the declining economic performance of the Soviet system and suggesting ways to improve it. The latter task was further constrained by the unwillingness of the party and its planners to embrace necessary far-reaching reforms. In their study of the socioeconomic system, Zaslavskaia, Aganbegian, and their coworkers could not avoid the conclusion that the Stalinist system, with its top-heavy bureaucracy, planners' preferences, and emphasis on heavy industry, which remained largely intact, was the source of most of the problems. Fortunately much of the institute's work was carried out for Gosplan RSFSR and Gosplan SSSR, both of which had a vested interest in macroeconomics and were more open to research in that area.

Work at the institute took two main directions in its first five years. The first concerned the distribution of industrial production and perspectives for the complex development of Siberia and the Far East. By 1968 scientists working in this area had succeeded in completing research that outlined the basic directions of Siberian economic development to 1980. They made branch by branch and territorial recommendations, focusing on increasing the tempo of the growth of capital investment in construction, manufacturing, ferrous and nonferrous metallurgy, electrical energy, and chemical industries. In addition, they studied the manufacture of technologies appropriate for harsh, northern climates.[29]

The second area of work at the institute was the "perfection" of the organization of labor, production, and the rational use of labor time. This work was intended to make good on scarce capital and labor and to find "reserves" of machinery, equipment, and labor for existing industry. In a speech in Tselinograd in March 1961, Khrushchev drew attention to the problem of the rational use of labor resources and called for the strengthening of economic science in Siberia to study the question. This was a problem of particular importance in Siberia because of the region's climate and its labor force, which was not nearly as well educated as that in the European USSR. Moreover, attracting and keeping workers in the region was problematic because of inadequate social services. Thus institute studies focused on how to increase workers' free time without shortening the working day. In addition to official statistics, the researchers used such novel data, by Soviet standards, as photographs, interviews, and surveys. The studies reflected the institute's proclivity to strive to manage economic and social problems through scientific intervention.[30]

TAYLORISM REVISITED

The Scientific Organization of Labor (Nauchnaia organizatsiia truda), or NOT, the Soviet form of Taylorism, found its rebirth in Siberia under the roof of IEiOPP in search of labor reserves. Within two years a major study, *Problems of the Scientific Organization of Labor in Industrial Enterprises*, had been published, and the institute's Taylorist laboratory received international recognition and was included in a UNESCO study. In its early years NOT zoomed to popularity on the coattails of fascination with western management science. Dozens of enterprises in the Novosibirsk region alone established special NOT "laboratories," or offices from which Taylorist specialists attempted to bring their science to the shop floor.

NOT was originally a neoclassical doctrine of management founded in the 1920s and 1930s under A. K. Gastev, inspired by the time-motion studies developed by Frederick Winslow Taylor, an American engineer, to determine the most efficient way for workers to accomplish their jobs. Standing in the factory next to the worker, Taylor and his white-smocked followers examined all aspects of the production process—the bending, lifting, and stacking motions, the interaction between worker and tool, and the location of material to be worked (wood, fabric, steel, and so on). They believed they could determine objectively how best to organize production, claiming to be friend to both worker and manager. Siding with neither party, strikes and slowdowns would become things of the past.

Workers had good reason to believe that the goal of Taylorism was to control the entire job situation through the adoption of uniform standards and the "de-skilling" of labor. They saw the time-motion studies as one more effort to increase owners' profits by extracting more labor in less time. Managers supported Taylorism since it placed more knowledge, and thus more power, in their hands. Yet the time-motion experts actually knew little about the work they were studying. They arbitrarily decided the "best" motions and "shortest" length of time that workers needed to complete a given task. Case after case, where Taylorists were enlisted to introduce their techniques—at the Watertown Arsenal and in the Boston Naval Yards—workers rejected the new management science out of hand. Rather than tolerate the engineer's stopwatch, workers went on strike. In the Soviet Union in the 1920s and 1930s, and again in the 1960s, however, they had little choice in the matter as party leaders endorsed the creation of NOT organizations.[31]

Taylorism found an ardent follower in Aleksei Gastev. After the revolution, industrial production in Russia fell to pre-1913 levels. When workers were advanced rapidly into positions of management, production fell even further. Gastev, a writer of proletarian poetry and an engineer, saw scientific management as a way to bring a new culture of labor to Russia, at the same time raising industrial production scientifically. He proposed to create the Central Labor Institute, under the jurisdiction of trade unions, to transform the Russian industrial culture through ideas developed by Taylor, Frank Gilbreth, and Henry Ford. Lenin, initially ambivalent to Taylorism, came to see it as useful for resurrecting industry, and he supported the creation of the Central Labor Institute. In that institute, Gastev and his followers developed a typology of workers from the most to the least qualified, based on time-motion studies with stopwatches and cameras, so that they would "become increasingly mechanized and standardized, like cogs in a vast machine." Proletarians would become numbered and classified, as depicted in Aldous Huxley's *Brave New World* and captured even more brilliantly in Evgenii Zamyatin's *We* (1920). The Workers' Opposition, led by Aleksandra Kollantai and Vladimir Shliapnikov, resented the diminution of workers' control inherent in Taylorism. Others to the Right in the party feared the technocratic prominence of engineers and what they believed were the utopian expectations of the Taylorists. Still, as late as January 1936, Sergo Ordzhonikidze, Commissar of Heavy Industry, placed Gastev in charge of preparing cadres for the Stakhanovite movement, and nearly a million industrial workers may have been trained in his institute. Gastev, however, rapidly fell from favor. He was purged in 1938 and died shortly thereafter.[32]

Given Khrushchev's fascination with science, it is not surprising that Taylorism was reborn after Stalin's death. NOT found its first home in the newly established Scientific Research Institute of Labor under the State Committee on Problems of Labor and Wages in 1955. Aganbegian joined the staff of the State Committee shortly thereafter, worked in a different sector, but was aware of the Taylorists' work. Aleksandr Gastev was also rehabilitated at this time, and his reputation was restored in 1962 when Aksel Berg, the leading cybernetician and close associate of Liapunov, wrote a laudatory article on NOT in *Pravda*. Over the next few years a series of NOT laboratories was formed to study western management techniques and to explore quantitative research approaches. The rebirth of NOT, and management science in general, did not occur without attacks from such conservatives as P. N. Fedoseev, head of the Institute of Philosophy, for its "narrow empiricism" and for displaying "elements of managerialism and technocratic theory." But the economists had the upper hand and by the early 1960s succeeded in generating support for "Ekonogorodok"—a city of economics modeled on the Siberian division's economic sciences.[33]

Aleksei Kosygin, prime minister of the USSR after Khrushchev's ouster, considered himself a specialist on industrial management and defended management science. He soon lost favor with the inner circle as being too liberal for Brezhnev's rule. Before his departure he managed to promote a series of economic reforms, including management training programs, which ultimately proved to be ineffective. In December 1965 the presidium of the Academy of Sciences recognized management science as a legitimate field of scientific inquiry. In June 1966, for the first time since the second NOT conference in 1924, more than a thousand specialists from over five hundred institutes convened in Moscow to discuss the administrative science. NOT grew rapidly over the next five years, with a center created in Sverdlovsk and with laboratories established in higher educational, ministerial, and academic institutes.

No sooner had IEiOPP opened its doors than its NOT researchers—Evgenii Maslov, Luidmila Zudina, and others—produced results of national significance. Economists analyzed more than two thousand photographs taken at thirty-five machine-building and repair factories which revealed that technological and labor productivity was low. They concluded that more rational utilization of inputs already in place—including improvements in the organization of labor—would increase productivity 20–25 percent in two to three years and save hundreds of thousands of rubles. In Krasnoiarsk, in September 1963, IEiOPP scientists announced their intention to create NOT laboratories in industrial enterprises throughout Siberia and the Urals.[34]

For several reasons, however, the NOT sector in IEiOPP was disbanded after seven years. First, the IEiOPP specialists in NOT, including its director, P. F. Petrochenko, left for other research centers owing to the institute's macroeconomic focus. Others had already left with Prudenskii. Second, it was NOT specialists who were fired first when enterprise managers needed to cut costs. Managers resented the "scientific" specialists peering over their shoulders. NOT reforms rarely went further than the factory gate. Third, the research of sociologist Tatiana Zaslavskaia revealed a far more compelling reason for low productivity of labor, namely, worker dissatisfaction, a problem NOT could not hope to solve. Most important, economic specialists allied with the Brezhnev political leadership, critical of the alleged narrow empiricism of management science, and fearful of "bourgeois" influences in the USSR eventually brought the NOT experiment to an end. *EKO*, however, had no qualms about publishing articles on NOT through the mid-1980s.[35]

KANTOROVICH AND LINEAR PROGRAMMING IN SIBERIA

Whereas Taylorist NOT lived a short existence in Akademgorodok, the linear programming techniques of mathematical economics found fertile ground, largely under the tutelage of Leonid Vasilievich Kantorovich. Upset with the pace of reform of economic science in the USSR—even this future Nobel laureate had trouble publishing certain results—Kantorovich came to Akademgorodok in 1960 with a number of his graduate students to establish the laboratory of mathematical economics in the Institute of Mathematics. With the support of Lavrentev and Sobolev, he joined scientists in the Computer Center, the mathematics institute, and IEiOPP to elaborate on the techniques he had developed in the late 1930s.

A talented mathematician with interests in biology, economics, and computers, Kantorovich (1912–1986) finished his first publication at age sixteen. He specialized in approximate methods of analysis, functional analysis, mathematical economics, and the problems of planning and organization of production and was one of the founders of the theory of semi-ordered spaces. When only twenty-two he became a professor at Leningrad University. In 1935 he was given his doctorate without defense. (In that year scholars had gained the right to receive degrees for the first time since most academic degrees were outlawed as elitist and bourgeois in the revolutionary enthusiasm of 1918.) His work on functional analysis in proximate calculations earned a Stalin prize in 1949, and in 1958 he was elected corresponding member of the Academy of Sciences. By the mid-1960s Kantorovich had published more than 150 articles and 12 monographs, many of them appearing abroad. He and Aganbegian joined forces to form an economics department at Novosibirsk University in 1968, creating a "school" of economic mathematical modeling with a national and international reputation.[36]

Kantorovich's pioneering work was in the use of linear programming for economic planning, his first works dating to 1938. At that time a laboratory of a plywood trust turned to Leningrad University with what seemed at first to be a simple problem: At eight machine tools it is necessary to work five kinds of material. The productivity of each machine tool is known for each kind of material. The goal is to standardize the work so that different kinds of material are worked in the given proportions and in maximum quantities. Solving this problem in classical mathematics requires endless calculations. Kantorovich succeeded in deriving an effective new method to solve the problem, later given the name "linear programming." (He himself has actually credited Novozhilov with being the first economist to recognize the significance of linear programming.[37]) In 1939, when Kantorovich presented his results at a seminar, Marxists, convinced they were warding off dangers on all fronts, accused him of using "mathematical methods" that were comparable to "capitalist apologetics."

Kantorovich's *Mathematical Methods of the Organization and Planning of Production* (Matematicheskie metody organizatsii i planirovaniia proizvodstva) (1939) summarized his optimization work in linear programming. His 1959 monograph, *The Economic Calculation of the Best Use of Resources* (Ekonomicheskii raschet nailushchego ispol'zovaniia resursov), for which he received a Lenin prize in 1964, was his most famous work on mathematical methods of optimal planning. A reviewer of this work commented, "It is almost impossible to determine if the author is a mathematician, completely familiar with economic problems, or an economist, who has made a significant contribution to mathematics."[38]

In the Khrushchev and early Brezhnev years, young researchers at IEiOPP (and at TsEMI, the Central Economic Mathematical Institute) widely believed that linear programming would "help identify reserves and more fully utilize the advantages of a planned socialist economy over capitalism."[39] They saw linear programming as a panacea for planning problems and welcomed mathematicians to join in on their discussions. The scale, concentration, and growth of industrial production created challenges for planners at all economic levels, from Gosplan to individual enterprise directors. The question was how to make optimal use of labor and capital inputs, and the answer was mathematical methods. The research program of the Institute of Mathematics focused on introducing mathematical models in planning. One of the early efforts to apply these models was carried out jointly with the Siberian Institute of Mathematical Economics (SibIME) and involved the development of a methodology to optimize the structure of machine-tractor parks throughout Siberia and the Far East. The researchers then produced models with millions of operations and thousands of equations to optimize rolling and pipe mill production for branch industries. Under Kantorovich's direction, the mathematicians created one of the first automated management systems in the USSR, ASU-Metall, for the state procurement organization (Gossnab). Kantorovich's laboratory next applied new algorithms to optimize parameters for a wide range of products in concert with the main research institute of the State Bureau of Standards (Gosstandart). The mathematicians published two volumes based on this work.[40]

A goal of this research was how to avoid a problem that was endemic in Soviet enterprises, namely, "storming" (*shturmovshchina*) in order to meet output targets. Because of a lax work ethic, alcoholism, supply bottlenecks, and so on, the production in Soviet enterprises was very uneven. As a result managers had to exhort workers to reach targets with threats, appeals to their socialist ethics, promises of bonuses, and access to scarce goods in factory stores. The result was *shturmovshchina*, intense periods of production in order to meet plans. With the assistance of a group of researchers under the direction of Kantorovich, the Novosibirsk Instrument Factory set out precisely to determine a more rational use of labor, capital inputs, and equipment to avoid "storming." But without financial incentives or access to good housing, it is difficult to imagine how workers could be convinced to work harder, to avoid bottlenecks (and the bottle), or managers to avoid threats. And perhaps the problem was not inefficiency of labor at all, as both planners and economists assumed, but rather overambitious targets. Further, Kantorovich's mathematicians had no time to go into the field and troubleshoot in order to see whether Siberian enterprises actually embraced their work. Finally, the complex models were most likely incomprehensible to enterprise managers. *Shturmovshchina* undoubtedly remained a part of life of the Novosibirsk Instrument Factory, as well as for the economy as a whole.[41]

Finding a warm environment for his interests, Kantorovich joined Aganbegian in October 1962 to organize a conference at Akademgorodok on the application of mathematical modeling techniques and computers in economic planning. Reflecting the increasing intellectual excitement over mathematical economics during the Khrushchev thaw, the conference was attended by 150 mathematicians and 350 economists, as well as several engineers, all somehow finding the time to discuss 120 papers (a third of them by Akademgorodok scientists) and 200 other presentations. Akademgorodok mathematician Sergei Sobolev opened the conference. Prudenskii followed offering hope for the future of mathematical methods in planning but cautiously calling for the accumulation of appropriate data before its application. Kantorovich himself delivered a talk entitled "Mathematical Problems of Optimal Planning," analyzing the far-reaching development of mathematical methods. In the paper Aganbegian presented, "Optimal Planning of Distribution and Specialization of Production," he drew attention to the work at LEMI on this problem and also discussed mathematical and computer methods in planning. He expressed satisfaction in the creation of new computer-based research organizations in Akademgorodok, Kiev, Minsk, and Tartu.[42]

The conference participants agreed that there were four barriers to the effective application of modern mathematical economic methods: insufficient theoretical and methodological elaboration of modeling methods; a limited number of finished algorithms and programs at this early date; shortfalls of machine time and computer technology with limited memory; and few trained specialists. Nevertheless, specialists in mathematical economics managed to accumulate significant data from their studies of the machine building, metallurgy, and chemical industries. Using an M-20 Computer, they developed programs to optimize serial production in machine-building enterprises which, like most such factories, produced a limited run of equipment. Early research was promising; the economists calculated how to increase output at the Novosibirsk Turbogenerator Factory by 39 percent. (The factory had built the VEP-1 accelerator for the Institute of Nuclear Physics.) Using traditional methods this kind of research could have involved millions of calculations, thousands of years, and repeated trial and error. With computers, and using linear dynamic programming, game theory, and algorithmic calculations, optimal solutions were found in short order. Results were applied in mills, the cement industry, coal mines, agriculture, forestry, and the fishing industry, often even without getting specific parameters from the industries.[43]

Another project was completed for Sibriproshakht, the State Siberian mining engineering organization, to forecast optimal five-year targets for coal mining in the coal-rich Kuznetsk basin south of Novosibirsk. Economists proudly proclaimed that their theoretical studies had produced substantial savings in coal production and increased profitability without supplementary capital investment—through the more efficient organization of production. For the Barnaul Fiber Chemical Combine and a Barnaul design bureau, linear programming methods were applied to optimize the manufacture of rayon. Work was also undertaken for a variety of military enterprises, although the precise nature the work is unclear.[44]

Several factors militated against the success of the new discipline in Akademgorodok. First, Aganbegian had little regard for the value of theoretical economics. As he put it, the profile of the institute was not so much about theoretical research as about practical attempts to construct working models. Therefore the economists focused more attention on gathering data to crunch in their new models on computers rather than on creating new theoretical approaches. Relying on "our trusted helper—the Computer Center—which carries out 2–3 billion operations per day,"[45] they repeatedly studied a hundred Siberian enterprises and surveyed a thousand directors to develop dynamic models. They emphasized methodology, for example, optimization techniques, once again seeing themselves as technicians who had to come up with the "right answer." Still, theoreticians were pleased that they had the opportunity to undertake research unique in the USSR and to work from a rich database.

Eventually researchers advocated placing terminals in enterprises to gather up-to-the-minute information on orders, supplies, and so on. The goal presumably was to develop a nationwide system of mainframes and minis to link Gosplan's input-output tables for all branches of the economy down to the level of the shop floor. The creators of this system clearly intended to have computers solve millions of equations simultaneously to rationalize planning. On two counts, however, the plans were doomed to failure. First, computers were incapable of simultaneously solving so many equations in less than a few years. This also assumes that all the data can be input at superhuman speed. Second, managers rejected the establishment of centrally controlled computer stations in their enterprises. They feared that the computer systems would tie their hands, eliminating any flexibility they might have to acquire scarce goods through bribes and connections. Boris Orlov, a leading Siberian theoretician at IEiOPP, told me that "mathematical economics is a dead end since it is impossible to optimize society. Everything is muddled today because we have no theoreticians and no one is needed for applied research. Too much emphasis has been placed on applied research related to Siberian economic development, with no real results. Practice could not adapt to scientific concepts. Science understands that practice is its falsifier. It is sad that I spent thirty years on unnecessary things."[46]

Applied Economic Research for Siberian Development

At the time Aganbegian assumed directorship of IEiOPP in 1966 its research program focused on five major areas, most of it geared toward Siberian development. The first involved the creation of a dynamic, interbranch model to determine the optimum tempo and proportion of economic development throughout the country. This work revealed just how distorted the economy was with 50 percent of investment going to heavy industry and the military.

The second area of work concerned the distribution and development of economic resources. The third was sociology. The fourth involved the application of mathematical models and computers to research. And the fifth fell under the "Siberia" program, which involved the study of the productive forces and labor resources of Siberia and the Far East.

Together with specialists at TsEMI and SOPS, institute economists worked on a national project called "Basic Positions for Optimal Planning of the Development and Distribution of the Productive Forces." K. K. Valtukh completed research on optimal tempos and balances of industrial development. N. F. Shatilov created a dynamic model of interregional balances. A. G. Granberg, later director of IEiOPP and then economic adviser to Gorbachev and Yeltsin, together with his colleagues created an interdistrict model of the distribution of production that combined linear programming with a model of an interbranch balance. M. K. Bandman, in the optimal territorial planning department, developed district models for territorial-production complexes. Work was also done on the structural changes in the U.S. economy with an eye toward analogous changes in the Soviet economy. Dampening the success of all this work, the researchers acknowledged, was the "predominance of empirical methods of research when the internal structure of the model was something like a 'black box.'"[47] In other words, reliable projections were unlikely.

In the 1970s the economists continued to elaborate an interbranch dynamic model using computer-mathematical techniques. The model had grown to include 180 branches and was intended to offer various scenarios for economic growth. The institute strived to tie its work to practical problems, for example, finding solutions for Kuzbas coal development, the Siberian cement industry, and east Siberian forests. In all, researchers overfulfilled their own plan, analyzing some seventy practical problems and publishing dozens of monographs and volumes of essays. Research had moved from the development of independent optimal models to their application in solving Soviet macroeconomic problems. This meant the expansion of existing models as well as their combination with other models. The creation of a dynamic economic model based on thirty branches was the capstone of achievement, and the economists worked on a six hundred branch model with economists at TsEMI and SOPS.[48] Unfortunately, by all accounts, something nebulous cropped up in the data employed to create the models, a phenomenon western specialists now refer to as "garbage in, garbage out."

Economic research on regional, territorial, and branch production remained the mainstay of the institute through the early 1980s. Researcher K. K. Valtukh produced a series of papers to clarify the connection between the rate of growth of productivity and changes in its structure under the influence of so-called scientific-technological progress. Z. R. Tsimdina and L. P. Bufetova used interbranch models of metallurgical production and macromodels to calculate ways to increase the quality of steel. A. G. Granberg supervised studies on the role of territorial factors in the socioeconomic development of the Soviet Union. M. K. Bandman analyzed production complexes. For Gosplan, the institute produced a specialized optimization interregional interbranch model with dis-aggregate data for the fuel-energy, machine-building, chemical, forestry, build-ing, informational, and agricultural sectors.[49]

Research on the development and distribution of the production forces of Siberia and the Far East complemented the work of Zaslavskaia's sector on the formation, migration, and utilization of labor resources. Much of this work involved extensive survey research in the machine-building and construction industries of western Siberia, among agricultural workers of Novosibirsk re-gion (conducted together with the Central Statistical Administration RSFSR), and among workers in the new Tomsk and Tiumen oil fields to understand factors surrounding their hiring and retention.[50] The major issue here was the need to deal with labor shortages in any way possible.

Certainly Siberia's harsh climate led to rapid labor turnover. But standard Soviet economic practices were also a cause. The authorities had to pay salaries two to three times higher than normal to attract workers. Inadequate machin-ery and equipment made their work particularly demanding, requiring the hiring of more service personnel. From interviews of twenty-eight hundred workers in Tomsk in 1968 and 1969, institute surveys revealed that workers came to Siberia out of feelings of obligation to society, the hope to find interest-ing work, the desire to live a romantic existence, and to earn higher salaries. But these levers on the whole were ineffective in retaining them. More atten-tion had to be paid to social factors, that is, to better living and cultural condi-tions. Initial increased outlays for social services would ultimately result in lower costs as more and more seasonal workers decided to stay on. The costs of training and attracting workers would drop. The question was how to encourage enterprises and the oil and gas ministry to invest money at the start to lower costs in the future. Because of limited funds for capital investment and the high cost of labor, enterprises were unwilling to develop local infra-structure. Indeed, Zaslavskaia's research showed that higher wages led to higher turnover rates in northern Siberia. People came to make a lot of money fast to buy a car or a cooperative apartment and then returned to better living conditions.[51]

One evening in Akademgorodok I dined in my hotel restaurant with an oil rig worker and his wife. They had flown to Novosibirsk from Tiumen, part of the way by helicopter, to spend two weeks at what they described as "the only good restaurant within two thousand kilometers." Both were long-term workers, intent on saving money to buy an apartment in St. Petersburg. They had been in Tiumen nearly two years and were tired of the dreary, concrete town they shared with five thousand others. The stores were empty, the apart-ments bare.

Like Tiumen itself, the sector of the institute that focused on Siberian eco-nomic problems was not a real hot spot. Researchers complained that no one surveyed their interests, that the choice of research was imposed arbitrarily from above. The sector had to focus on all branches of industry, even the service sector, and requests for new research came in without warning. The organizations even imposed differing methodologies on the researchers which, Starodubskii declared, "drags the work of the sector backward."[52]

The solution to branch and territorial economic problems was tied closely to the Siberian question. A special Siberian sector was organized under R. I. Shniper, a leading economic specialist, to coordinate the Siberian themes—labor resources, migration, and the development of productive forces. Research often entailed special month-long summer expeditions to the North, to Magadan, to the Far East to Vladivostok and Kamchatka, to the Enisei and West Siberia. "This was the period of the flowering of Siberia when we traveled every summer," Aganbegian recalled. Leading scholars who represented all specializations participated in these expeditions, as did researchers from Irkutsk, Krasnoiarsk, Kemerovo, and Tiumen, and economists in the Buriat, Iakut, and Chita regions.[53]

Ultimately, like the other scientific institutes in Akademgorodok, the institute's fundamental research took a backseat to applied research in order to foster Siberian economic development. The institute's party committee was instructed to keep an eye on this desideratum. The institute was given planning responsibilities for the "EniseiLes" (Enisei Forest) combine and other enterprises to accelerate technological progress in the forestry industry. Researchers recommended the creation of floating sawmills for use in reservoirs that backed up recently built hydropower stations. Like the floating atomic power stations designed by the nuclear establishment, these floating sawmills symbolized the extreme faith of Soviet economists, planners, and engineers in the rational management of nature itself. Floating sawmills would accelerate the assimilation of forest tracks that were under assault by insects and fire hazards, not to mention the reservoirs built by the engineers of nature themselves.[54]

A reason for the increased emphasis on applied research may have been the failure of researchers to complete fundamental projects as scheduled. The institute's party committee was deeply troubled that two-fifths of the projects were not completed on time. The problem was endemic, from the top down, and reflected, according to archival documents, "poor leadership." According to the sociologists, researchers were remiss in fulfilling socialist obligations and in meeting publication deadlines. It was not that they "do not work hard," one party member declared, "but they do not do what is set forth in the plan."[55]

A more important reason for the growing interest in applied research concerned the normal inclination of Akademgorodok institutes to turn toward applications under pressure from central political and economic organizations, especially after the promulgation of the "Siberia" national economic development program. The "Siberia" program (see chapters 3 and 5) attracted a majority of the IEiOPP scientists to undertake a series of state-ordered technology assessment reports, environmental impact statements, planning and management documents, and surveys. The outlines for "Siberia" seem to have originated in a document prepared by IEiOPP. Then, in 1979, a scientific council to study the assimilation of natural resources and the development of industry and agriculture in Siberia was established by a central party directive with the geophysicist A. A. Trofimuk as chairman. Serving as his deputies were Aganbegian, the biologist D. K. Beliaev, the chemists G. K. Boreskov and S. S. Kutateladze, and others.

Aganbegian was chairman of two subsections of this council. The first, the section of regional economic programs, had ten subdivisions that focused on the Arctic, the West Siberia oil-gas complex, the Novosibirsk region itself, agribusiness in the Altai region, sociological research on the quality of life of Siberians, and several others on regional territorial production complexes. The second subsection focused on large-scale complex programs such as river diversion, the environment, Baikal, and BAM. IEiOPP personnel directed each study.

"Siberia" had a central place in the grand Brezhnevite vision of Soviet modernization and military might to the year 2000. Like other programs, it acquired significant bureaucratic and technological momentum. Directives rained down on the Siberian division and its institutes from the Central Committee to tackle the vexing problems of the development of Siberian natural resources, the formation of "territorial-productive complexes," and the speeding up of scientific-technological progress. Yet the researchers complained that those sitting in Moscow learned little from their studies. For instance, they warned that labor shortages would hamper "Siberia" at every step. Aganbegian criticized the "Siberia" program for overemphasizing extractive resources at the expense of local industry and social services. At a party meeting in January 1979 he complained that the central economic, scientific, and party apparatus had ignored the recommendations of Siberian division scientists on a wide range of issues. His institute had prepared a sixteen hundred-page report on Siberian development, but this was dropped from the final "Siberia" plan. For example, the chemical and machine-building industries were not included. While ore exploitation was supported, the ferrous metallurgy industry was left out. The center kept on expanding "Siberia" to fifteen, twenty-four, then thirty subprograms but continued to ignore the "social aspect"—labor resources, demographic questions, quality of life. The central government was determined to overlook scientific study that suggested measured development in favor of the usual emphasis on heavy industry: ore, oil, coal, and gas.

Just as critical, no institute was designated with responsibility for environmental issues. These were left to a voluntary Siberian division commission when serious, professional interdisciplinary research and advice were required. Aganbegian proposed the creation of a special institute to study water diversion; the small group in Trofimuk's Institute of Geology and Geophysics was inadequate to the task. Furthermore, IEiOPP's own laboratories and departments in Tiumen, Krasnoiarsk, and the Altai were understaffed and poorly equipped. The work at IEiOPP itself was deflected from considering environmental issues by the burden of the West Siberia oil and gas complex and the Altai agricultural complex. Siberian scientists wanted to call a conference on these problems but could not do this without the approval of the presidium of the Academy of Sciences in Moscow which "for all sorts of bureaucratic reasons was being held up."[56] One reason was BAM, the new trans-Siberian railroad that contributed both to breakneck Siberian development and environmental degradation.

BAM—MAGISTRAL OF THE CENTURY

Like Stalin and Khrushchev before him, Brezhnev signed off on grandiose projects. Stalinist hydropower stations like Dneprostroi, factory towns like Magnitogorsk, huge, ornate neoclassical pieces of architecture like the metro and Moscow skyscrapers gave way to Khrushchevean monuments to science: rockets, reactors, particle accelerators, and cities of science. Brezhnev required similar testimony to the glory of his rule. He chose the taming of Siberian resources as a symbol of "developed socialism." In January 1976, at the twenty-fifth party congress, he basked in his achievements as first party secretary. At his beckoning, the congress participants designated BAM (the Baikal-Amur Mainline), a new second trans-Siberian railroad, as the "project of the century" to be finished in short order. The taming of Siberia required the linking of "Siberia" and "BAM." These two undertakings clearly exemplify how technological, bureaucratic, and institutional momentum came to characterize the gigantic Soviet development projects.

BAM was the key to unlocking the natural resources in Siberia and the Far East. Stretching 2,800 miles, at two points reaching one mile above sea level, BAM was intended to facilitate the assimilation of more than one-sixth of Soviet territory. In its first decade, Glavbamstroi, an organization of 30,000 workers and 2,600 high school and university students in "shock" brigades, moved more than 400 million cubic yards of dirt, built 2,400 miles of auxiliary roads, 2,237 bridges, 1,525 drains, and 2,200 miles of railroad. The Kuznetsk Metallurgical Combine produced new, specially treated rails.[57] Like "Siberia," BAM required the input of skilled economists. IEiOPP was designated the "head institute" for BAM, calling upon roughly half the institute's research effort. Economists turned away from modeling research which had secured the institute's international reputation, and toward applied research which was the trademark of Brezhnevite science policy. The party leadership expected "science" to show how to build each kilometer of "the road into the twenty-first century." Science was needed since "everything along this road is difficult, how to lay the rails and coexist with nature, how to assimilate copper mountains and sow wheat fields, how to build factories and preserve the health of the new inhabitants, how to search for minerals and manage the organizations of the planned economy."[58]

As with electrification and hydroelectric power, discussion of the new "magistral" began in the 1920s. In April 1932, as part of a Siberian development package, minor construction commenced that follows the present route fairly closely. But investment in heavy industry and the military sector west of the Urals limited investment in BAM. Only a hundred tractors, twenty-four cars, thirty units of heavy equipment, fifteen locomotives, and assorted machine tools were provided. Surveyors succeeded in finishing aerial photographs of 39,000 square miles (smaller than Pennsylvania). Construction virtually ceased during World War II. Then, obsessed by the sting of the Nazi invasion and determined to place Soviet industry far from the hands of any future marauder, Stalin ordered equipment back to work on July 20, 1945. The cost was exorbitant, however, and construction ceased after Stalin's death.[59] Not until the Brezhnev years were "BAM" and "Siberia" put back on the front burner.

The Siberian division was expected to solve all the problems associated with BAM and resource development; at a series of meetings of party, planning, engineering, and scientific personnel in the 1970s, the Siberian division's responsibilities in managing the railroad were increased. IEiOPP prepared all major planning documents for the railroad. Initially, twenty-six Siberian division institutes, joined by some forty other branch, design, and ministerial research institutes, participated in BAM. The number grew rapidly to more than three hundred organizations, all under IEiOPP's supervision, with the economists preparing increasingly extensive reports, for example, "The Problems of the Economic Assimilation of the Baikal-Amur Magistral Zone" and "The Scientific Foundations of the Complex Program of Economic Assimilation of the BAM Region," two reports prepared under Aganbegian. Akademgorodok research institutes were critical in solving engineering problems, developing a region-by-region plan, suggesting how to use local ore, forest, and water resources, and studying labor problems. Economists developed software called "BAM-control" to keep track of construction. The Institute of the Earth's Crust studied seismic and geological conditions and where to site tunnels, bridges, villages, stations, and rights-of-way.[60]

IEiOPP research on BAM dovetailed neatly with efforts to develop models for interbranch, interregional, and territorial planning with regard to natural resources. R. I. Shniper, head of the Department of Optimal Economic Territorial Planning of Siberia and the Far East, was responsible for directing much of the research. His group studied the availability of oil, coal, forest, asbestos, and hydropower resources. The research served the interests of nearly sixty different ministries and organizations. Shniper complained, however, that these organizations had narrow interests and rarely took a systematic approach to assimilation. They failed to recognize the need to cooperate with other branches of industry, preferred investment in areas with capital and labor already in place, and ignored research recommendations. To avoid the typical short-sighted approach of the relevant players to BAM, Shniper called for a long-range, thirty-year plan to ensure attention not only to natural resources, capital, and labor needs but also to the social and environmental consequences of BAM.[61]

The task of assimilating resources proved an endless source of trouble. Not the least of the problems was the failure of construction to keep pace. Typical of such Soviet large-scale projects, little was done to provide workers and their families with the amenities of life in harsh Siberian conditions. The standardized II-49D-BAM and 94-BAM housing, which was intended to give settlements the "atmosphere of a capital," did not do the trick. The first villages and towns along the way had few comforts, let alone streetlights. In spite of "complex study" and long-range planning, severe construction problems mounted. Bridges failed to meet weight and strength requirements, foundations were poorly built in the permafrost, tunneling a total of twenty miles moved at a snail's pace. Efforts to standardize construction practices, equipment, and materials to deal with labor and equipment problems were premature. Workers were exhorted to push harder in campaigns asking "Who will be first?" Like the Stakhanovites of the 1930s, who set superhuman norms as examples to the industrial proletariat, I. N. Varshavskii's brigade laid nearly ninety miles of track in one year and A. V. Bondar's laid 3.3 miles in twenty-four hours. Ninety percent of the workers participated in "socialist competitions."[62] Yet soil erosion, industrial pollution, the accidental and purposeful discharge of oil, fuel, and PCBs, and destruction of the permafrost accompanied BAMs every spike. The Soviet system of centralized control of goods and services and administrative fiat was unable to overcome problems of poor infrastructure, overburdened administration, and bottlenecks of resources, inventory, and finances.

For Brezhnev and the Communist Party, the display value of BAM proved to be more valuable than its reality. Though unfinished, poorly constructed, socially and environmentally unsound, BAM was declared to be completed in 1984. To this day it is unclear when BAM will function as planned. Owing to the ongoing economic crisis in Russia, BAM workers struggle with inadequate resource services. In 1991 more than seven thousand workers quit Bamtransstroi. Only in January 1992 did workers reach the 3,110 kilometer marker, almost to the Pacific Ocean. According to a Russian journalist, BAM, Brezhnev's "project of the century," exists only in "fanfares and hearty songs. It is a forgotten and unnecessary, indeed empty outlay of resources and money."[63] Yet current leaders have not lost the feel for construction projects of great physical presence, if not social utility. In February 1992 Russian President Boris Yeltsin signed a document outlining measures to finish BAM and to begin work on AIaM (the Amur-Iakutsk Mainline) with public and private funding and companies.

Why was it that despite all adversity BAM had to be finished? In addition to the institutional momentum that Soviet large-scale projects acquired owing to the nature of the economic and political system, the very nature of Soviet economics itself lent its inertia to Siberian development. This was its hyperrational self-image, an image built on applied research topics and on the belief that any problem, including the identification of capital and labor reserves or the mitigation of environmental devastation, could be managed scientifically. This notion, so central to economics in IEiOPP, was a prominent view in sociology.

SOCIOLOGY REBORN: TATIANA ZASLAVSKAIA
IN AKADEMGORODOK

Tatiana Ivanovna Zaslavskaia came to Akademgorodok in 1962. Within five years she had established the field of "economic sociology," almost single-handedly resurrecting sociology in the post-Stalinist USSR. She compiled extensive data that documented an imbalance in resources and the quality of life between the city and the countryside. She intended to identify the reasons for the significantly higher outmigration from state and collective farms than planners and policy makers desired. She conducted a series of formal and informal surveys, some of which were anecdotal, that revealed high job dissatisfaction, especially among the young. These studies led her to believe that the Soviet social and economic systems may have been appropriate for the 1930s and 1940s but no longer met modern-day needs.

Zaslavskaia shared her views with Mikhail Gorbachev in April 1982, just after he had become Central Committee secretary for agriculture, at a meeting of specialists he had called to discuss the country's "food program." Gorbachev, probably already formulating the basic tenets of perestroika, confirmed his interest in reform through Zaslavskaia's findings. They impressed each other, and Gorbachev consulted with Zaslavskaia periodically. She introduced him to Aganbegian who became Gorbachev's leading economic adviser and headed a series of councils investigating economic reform. Zaslavskaia believes that she and Gorbachev had common ideas not because they borrowed notions from each other "but because we were both students at the same university at almost the same time, and both of us studied agricultural economy and sought ways to raise its efficiency, particularly with regard to the human factor, that is, by overcoming workers' alienation from their labor."[64]

A charming, talented, and unassuming scholar, Zaslavskaia's eyes are alert, showing no sign of the heart attack she suffered in 1988. She now lives in Moscow, having left Akademgorodok to become director of a national survey organization. She combines breadth of knowledge with an understated yet effective approach to complex problems. She was appointed chairperson of a national committee for the development of Soviet villages. She is the president of the eight-thousand-member Soviet sociological society. (In all the USSR there were roughly fifteen thousand to twenty thousand sociologists.) In 1968 she became a corresponding member of the Academy and in 1981 an academician, one of only five women among 220 Soviet academicians. Like Aganbegian, she founded a special interdisciplinary journal, *Sociological Research* (Sotsiologicheskie issledovaniia), which became the major forum for the rebirth of sociology.

Who would have predicted in the 1920s that sociology would suffer the fate of mathematical economics, theoretical physics, genetics, and quantum chemistry? Spurred on by the revolution, Marxist scholars, many of whom were social scientists, turned their attention to the study of social structure, classes, and social relations. Similarly, followers of Weber, Durkheim, and others no longer felt constrained by reactionary tsarist educational policies in studying modern sociological works. Within the universities, the Institute of Red Professoriat, and the Communist Academy of Sciences, social scientists debated the nature of social groups, interests, morals, and values.

The Communist Party officially encouraged studies of the relationships of different groups of peasants toward the party; of workers' attitudes in such cities as Moscow, Petrograd, Ivanov-Voznesensk, Kostrom, and Tambov, and of the impact of industrialization generally; of Gastev's time-motion studies; and so on. Indeed, the party debates over industrialization and collectivization required input from sociologists who represented a wide range of political views. Three major issues were of interest at this time: the study of the transition from capitalism to socialism, the withering away of the state and of classes, and the role of economics in socialist society.[65] That the Communist Party clamped down on the social sciences is all the more surprising since it was well represented among sociologists and economists. Of 25,286 "scientific workers" in the USSR surveyed in 1930, only 2,007, or 7.9 percent, belonged to the party, with but a handful in the exact sciences, whereas two-thirds of the scientific workers who claimed party membership were social scientists.[66]

With the Great Break, however, sociology became a minefield of disputes between various shades of Marxists. Sociologists were joined by psychologists and economists in discussing "nature versus nurture," that is, the extent to which human behavior is determined by genetic circumstances or by environmental issues such as upbringing, education, and so on. If man's behavior was largely biologically conditioned, then what hope was there to produce a new Soviet man? If Lamarckian notions of the inheritance of acquired characteristics were added to the equation, then it would be possible over a few generations to introduce new behaviors or modify old ones to create a new Soviet man with all the attributes needed to create socialism. Finally, if environment were the central factor in determining human behavior, if personality had plasticity, then changing the economic basis and the superstructure of social, political, and cultural institutions would necessarily lead to the formation of a new psyché, perhaps one that was collectivist and hard-working in all its manifestations.

After 1929 the problems connected with rapid economic and social change in a backward peasant society came to a head. The resolution of these problems had an impact on law, education, literature, art, and sociology. The issue is usually phrased as one of "spontaneity versus consciousness," of "genetics versus teleology." What this meant in practice in Soviet society was whether to place emphasis on the tendencies imminent in the situation (a deterministic attitude) or whether policy choices should be directed toward the goals that had been fixed for the masses. In the Soviet context, "reliance on spontaneity" meant sitting passively by and letting automatic laws do the work or divert the masses from the task at hand. "Consciousness" meant that leaders, armed with historical materialist study of human culture and the individual (sociology, history, and psychology) would intervene to coordinate the role of institutions and individuals toward the overall goals of the state. The dominant conception of man became that of an increasingly purposeful being who was more and more the master of his own fate and less and less the creature of his environment. Sociologists were not needed to study man in the case where man was the master of his own fate. Nor were they needed to study class structure, since Stalin decreed in the mid-1930s that there were no more classes. In such a classless society there should have been no repression, but the presence of "enemies of the people" at home and abroad required vigilance—and a brutal response. Given this political, philosophical, and academic environment, sociologists had to treat issues of personality, career, parental influence, upbringing, and education in terms of their effect on individuals and society with the utmost care.[67] The sole purpose of sociology was to help the state in its social engineering.

Prohibitions against certain kinds of social science research abated in the 1950s. Such economists with a "social inclination" as G. A. Prudenskii, V. D. Patrushev, and others sponsored economic sociology in their institutes and programs. E. G. Antosenkov studied fluctuations in regional labor forces. V. N. Shubkin conducted research on career choice and job attraction. And Zaslavskaia was responsible for the rebirth of sociology per se, as well as its interdisciplinary confluence with economics and survey research in Akademgorodok. Still, Soviet sociology grew in fits and starts. An Institute for Social Research was established in the late 1960s but fell prey to stricter ideological controls and safer research topics under Brezhnev, and its leading researchers were dispersed to institutes throughout the nation to study alone. This left the IEiOPP—under Zaslavskaia—and the Center for the Study, Forecasting, and Forming of Public Opinion—under the Central Committee of the Georgian Communist Party—as the two most progressive survey organizations in the country.

Zaslavskaia's path to sociology and Siberia was filled with personal tragedy, political turmoil, and academic change of direction. Born in Kiev in 1927, she moved to Moscow where her mother was killed in the first air raid on the city during World War II. Like many other Muscovites she was evacuated and then returned to finish secondary school. She then entered the physics department at Moscow State University, which at the time was under the control of conservative, anti-Semitic scholars who saw idealism lurking in quantum mechanics and relativity theory. She was in her third year of study when she switched to economics. Active in the Komsomol, she inadvertently got caught up in intrigues. Upon her graduation in 1950, however, she managed to gain a position in the agricultural sector of the Academy's Institute of Economics to begin graduate work. In 1954 she became a party member. Remaining in the Institute of Economics for twelve years, Zaslavskaia became a senior scientist with a reputation for first-rate work. She completed several studies on how economic conditions, for example, family budgets, affected collective farm workers. Although satisfied with her employment, she grew bored with the repetitiveness of her research. A number of her friends moved to Akademgorodok in the late 1950s and encouraged her to follow. The ecologist Pavel Oldak urged her to join Lavrentev's endeavor and she became smitten with intellectual curiosity. Then a trip to Sweden in 1957 showed her a lifestyle quite different from her own and shattered any idea she might have had that working people in the West were suffering.[68]

In 1961 the political economist Ia. R. Kronrod invited Zaslavskaia to join his sector in the Institute of Economics (in Moscow) where she remained for two years before being ordered to return to the agricultural sector. She could not "face that swamp" again. Oldak and Aganbegian dropped in to see Zaslavskaia on her "name day," and persuaded her to take a look at Novosibirsk. Aganbegian promised her a three-bedroom apartment to replace the cramped Moscow flat that housed four people. Walking through the forests of Akademgorodok, Zaslavskaia was enchanted by the bright sun, the snow, and the cold, not to mention the apartments, but especially by the spirit and youth of the city. The institute, although under construction, was far from the Central Committee, both geographically and intellectually. Here standard economic thought could be revised with modern mathematical techniques, something Zaslavskaia easily embraced. Only a few days after arriving, she agreed to move to Akademgorodok.[69]

From the start she was captured by the intellectual spirit and academic freedom of Akademgorodok. She recalled her first trip to the post office. There were six people in line. The two in front of her, physicists, were talking about Budker's new particle accelerator. The two behind her, biologists, discussed DNA. Life had an "intensely collective form and substance," she told me. "In Moscow we sat at home, exhausted, at the end of each day. Here people meet in clubs and cafes, in each other's apartments every evening. We all gladly welcomed each and every visitor."[70] Of the intellectual environment, Zaslavskaia wrote, "The brief Khrushchevean 'thaw' was a clumsy and inconsistent attempt at returning the country to the road of socialist construction. Historically speaking, it was ill prepared and had little chance of success, although it can be said that without it there would have been no restructuring today."[71]

Until the spring of 1968 the spirit of Akademgorodok remained unchallenged. The obkom was thirty-five kilometers away, although it seemed like light-years. Then the sociologists, doing their part to inflame the party, wrought havoc. They invited Pavel Makonin, director of the Institute of Marxism-Leninism at Charles University in Prague, to visit the Scholars' Club to discuss his work, in light of the "Prague Spring"—Czechoslovakia's short-lived experiment with economic, political, and cultural liberalization that was put down in August by Soviet and Warsaw Pact troops. In anticipation of Makonin's comments, the club was packed; people "hung from the chandeliers," Zaslavskaia said. In his talk he discussed a recently published book on the sociology of Czechoslovakia that focused on "groups" established by empirical research, not by standard Marxist categories of class. He wished to extend his work to the USSR for a comparative aspect. Then he animatedly answered the many questions flung at him about the Prague Spring. Zaslavskaia recalled, "We saw glasnost. We couldn't believe it would ever happen here." The Novosibirsk obkom was miffed by the lengthy talks about liberalization. Initially it did nothing to punish the participants in Makonin's seminar, but the seeds of the crackdown on Akademgorodok had been planted. "The mother of the entire clampdown," Zaslavskaia told me, "was Czechoslovakia."[72]

The festival of bards tipped the balance against Akademgorodok. In May, in all auditoriums, in the university, before the young, the students and the old, the professoriat, in apartments like Zaslavskaia's, and finally in the eight-hundred-seat "Moscow" movie theater, such folk singers as Vladimir Vysotsky, Bulat Okudzhava, Iulii Kim, and Aleksandr Galich sang ballads. Most were neutral, praising such values as friendship, love, and compassion, but all of them engendered spiritual freedom and a few cast "daggers in the heart of communism," drawing loud applause and triggering even more critical ballads. Standing ovations were common, perhaps inspiring even Lavrentev to stand. The party could not forgive this expression of political solidarity. A storm was surely brewing, although at first the party apparatus liquidated only the council of the Scholars' Club. Then when the Voice of America and the *New York Times* spread the word that Akademgorodok scientists had signed a letter protesting the persecution of Soviet dissidents Galanskov and Ginzburg on trumped-up charges, the so-called *podpisanty* (signatories) affair, the KGB descended on Akademgorodok. For years afterward the party used this affair as a pretext to maintain vigilant scrutiny of Zaslavskaia's sector, since one of the signatories had worked there. "The obkom's patience had run out," Zaslavskaia said. "All opposition centers were to be destroyed."[73]

As institute director, Aganbegian naturally worried about the crackdown. For one thing, Lavrentev was getting older and although he resented the party's growing interference in Akademgorodok, he could not fight it effectively and relations with the obkom worsened. Those with the central committee were no better. Aleksei Kosygin, although a supporter of economic reform, disliked Lavrentev because of his defense of Baikal. Brezhnev's disdain for the city of science was obvious: he never visited the city even once. And Academy of Sciences president Mstislav Keldysh, worried about challenges to his authority because of Akademgorodok's growing prestige, failed to come to the city's aid.

In August 1968, as Zaslavskaia and her associates journeyed nearly four thousand miles through Siberia doing survey research, they learned that Soviet tanks had invaded Czechoslovakia. This tragedy had immediate repercussions for IEiOPP. Makonin sent a telegram saying he would be unable to attend an institute seminar because of "technical difficulties." Zaslavskaia later went to Czechoslovakia but was not permitted to see him.

The obkom's reach spread rapidly. A group of American sociologists presented papers at an IEiOPP seminar that fall on the American youth movement against the war. But they failed to include in their talk any mention of "the working class in America." The seminar ended at 8:00 P.M. At 9:00 the next morning, a member of the science department of the Novosibirsk obkom called to inquire about the oversight.[74] This chilled the souls of those at Akademgorodok. Clearly the KGB had put spies in their midst. From then on all aspects of their work were subject to heightened censorship. The state censorship agency, Glavlit, examined every publication. To ensure central control, all social science institutes were forbidden to use local printing facilities for their publications. Data on birth, mortality, alcoholism, migration, pregnancy, and marriage rates became state secrets. The most striking example of this state control is that while the 1959 census was published in dozens of volumes, the next census, published in 1970, appeared in only five volumes, and the 1979 census in only one.

Now Zaslavskaia found encouragement for her work only in Akademgorodok, receiving little political support and virtually no public interest. Her research was expected to paint a glowing picture of the plodding Brezhnev regime, not the need for socioeconomic transformation of the Union. As a result, she later wrote, "Social science was transformed into one of the stagnant zones of Soviet science." Rather than studying areas of growing discontent, group conflicts, or potential solutions, sociologists were relegated to questions of secondary importance: migration and fluctuation of labor, attitudes toward work, choice of a profession, and quality of life, especially in the countryside. Paradoxically, even these studies generated a wealth of data indicating discontent and social inequities in the Soviet social order. Zaslavskaia felt she had to avoid studying "the root causes of negative tendencies" that were to be found in "the social mechanisms of power, property, and distributive and managerial relations."[75]

Fortunately for Zaslavskaia, her sector found the protection of R. O. Simuzh from the Central Committee's agricultural department, a man she later described as her "angel of mercy." He enabled the researchers to maintain a modicum of freedom, permitting them to mimeograph up to five hundred copies of books and reports without turning to Glavlit. Still, as for all studies throughout Brezhnev's empire, four copies of each had to be sent to the Central Committee. The sociologists worked in fear that someone in the central party apparatus would interpret some unintended misstep as being "slanderous" to agriculture and censure them, but Simuzh assured Zaslavskaia that she could rest easy. Why he protected her is unclear. Perhaps he appreciated her objective appraisals of village and urban lifestyles based only on empirical research without the use of horror stories. "We had the advantage of being away from Moscow and Leningrad, away from provincial centers where the obkom controlled the money, the books, and the paper. We had the Siberian division's support," Zaslavskaia told me. On the other hand, a book she wrote with M. I. Sidorova on comparative labor productivity in the United States and the USSR was never published owing to its pessimistic conclusions. "Two years of highly intensive work had been wiped out of my life," she said.[76]

In this tense intellectual environment, Zaslavskaia and her associates began pioneering work on job satisfaction in the Siberian countryside, assisted by the fact that virtually no one paid attention to the concerns of village residents. Party officials usually only met with the director of a kolkhoz and perhaps one or two other local bureaucrats. "We were the first to deal with the 'little people,'" Zaslavskaia said, "and we interviewed thousands of families. It was hard to get honest answers." She and her associates encountered naive and incredulous village inhabitants who were accustomed to directives from above and not to inquiries about their daily lives. They were asked about their lifestyle, their families, living conditions, career patterns, and aspirations and usually needed to be coached in their answers. "They believed that if they answered that they lived poorly, we would somehow make their lives better," Zaslavskaia said. Some would respond, "Write down just what you need to."

Some villagers wanted terribly to be interviewed and could not fathom why they were not selected. "Come sit, I'll tell you everything," one would hear. "Why are you talking with Volkov? He's a drunkard. He doesn't know anything!" Problems also arose out of misunderstandings. When villagers were asked questions concerning any "difficulties in the home" or about "home economics," they assumed they were being asked about their sex lives. This elicited chuckles, guffaws, or the statement, uttered proudly, "No problems whatsoever!"[77]

When the sociologists first descended with their questions, the villagers brought out all their finery. They shared their favorite delicacies, their sour cream, pickles, vodka, and home brew. At their second or third visit, however, when locals realized the sociologists were powerless to change their lives, they refused to talk. "Siberia is a big place, so we rarely met with the same respondents," Zaslavskaia said. Unfortunately this often denied them a dynamic component to their research. Zaslavskaia only participated in initial interviews, feeling that her age and a hearing impairment would interfere with difficult journeys into the nineteenth-century Siberian countryside.

Iurii Voronov, a short, mustachioed, and attractive man and an energetic, talented organizer, began his career at the IEiOPP working with Tatiana Zaslavskaia. He completed extensive research on vocational choice. One of the first research projects in which Voronov participated was undertaken for "Lespromkhoz." This forestry concern was interested in tackling the problem of rapid turnover and learning why individuals changed their place of work, residence, and occupations, and what role, if any, orders from above played in those decisions. Zaslavskaia's crew was given eight dirty dump trucks with lousy shock absorbers. The young researchers spent thirty days combing the countryside in decrepit old vehicles, bouncing around on hot, dusty roads, living in spartan conditions, and doing interviews. They were even arrested on one occasion by local police in Vengerovskii district who had no sense of what the researchers were doing and were not about to permit it in their district without precise instructions from the authorities.[78]

DISCONTENT IN THE SOVIET COUNTRYSIDE

The development of Siberia required more than mere cataloguing of the region's great natural resources. While Trofimuk's Institute of Geology and Geophysics studied Siberia's oil, gas, and mineral resources and Beliaev's Institute of Cytology and Genetics surveyed the region's natural flora and fauna, Zaslavskaia's sector tried to understand how to attract and hold workers to develop those resources in an area marked by harsh climate, economic hardship, and cultural deprivation. Such an analysis required an interplay between personal, microeconomic, and macroeconomic decisions of individuals, families, enterprise managers, farm directors, and the government. Little did Zaslavskaia and her colleagues realize how quickly they would uncover a growing dissatisfaction with Soviet life, particularly among those with a higher education and among all agricultural laborers. The Siberian sociologists deserve deep respect for their dogged study of society, but they failed to avoid an engineering mentality of managing social processes, standard in Soviet scholarship.

In her early work in Moscow in agricultural economics, Zaslavskaia was already aware of the significant lag in the quality of life in the Russian countryside. She observed firsthand the grueling hours people worked to rebuild Russia after World War II. She calculated that collective farmers received an average of one kopek for a day's work. She saw farmers hiding their animals in mountain pastures to avoid tax authorities, risking severe punishment in doing so in order to hold on to some hope—and property. She discovered that "the advantages of socialist distribution of income" meant that a social class comprising about 40 percent of the population was paid practically nothing for its work. She then studied the impact of changes in the administration and economic management of the kolkhoz and sovkhoz initiated under Khrushchev, such as the elimination of the Machine Tractor Stations. She had no doubt that the Stalinist system, with its emphasis on heavy industry, was responsible for the poorly developed economy and social services in the villages.[79]

In 1963, as newly appointed head of the IEiOPP social problems department, Zaslavskaia foreshadowed the conclusions of a major study on ways to increase Soviet agricultural productivity in an article published in the leading Soviet economics journal, *Problems of Economics*. She argued that salaries pegged to such "economic" measures as qualifications, experience, educational level, and quantity and difficulty of work did little to encourage effort. She rejected the leveling of salary differentials that had occurred over time. There was nothing wrong with offering higher salaries in line with qualifications and experience, so long as there was a "scientific" basis for determining pay scales. But Zaslavskaia concluded that it was more effective to estimate a task's "social value" and to offer remuneration, consumer goods, and better services accordingly. In particular, she advocated greater attention to workers' material well-being, arguing that animal husbandry and agriculture were important to society and merited not only higher wages but greater access to goods and services.

"The optimal wage system," she wrote, "is one that guarantees the greatest material interests of the workers and in the final analysis the most rapid and harmonious development of production. But material interest is a social phenomenon that depends on many socioeconomic factors." This was not yet a precise, mathematically determined magnitude but an interval "that defines the wages that guarantee the most successful combination of [workers'] personal interests."[80]

In 1966 in *The Distribution of Labor in Collective Farms*, Zaslavskaia definitively identified social factors, and not economic or administrative ones, as the source of low agricultural productivity. She concluded that a new system of material stimuli was required to encourage higher productivity. The current practice of extracting funds from kolkhoz income for general social needs was "economically unfounded." Other economic and administrative measures created similar disincentives. She concluded that workers' income should depend on the general income of the kolkhoz. She further noted that harvest targets failed to consider the cyclical nature of agriculture or the influence of climate, for example, draught, which led to a sharp divergence in kolkhoz performance across regions.[81]

The development of methodology went hand in hand with the accumulation of data. Aganbegian and Shubkin had published an article on sociological research and quantitative methods in 1961. A handbook on methodology, first published in Novosibirsk in 1964 in a limited run, was revised and republished in 1966 and included chapters by such leading sociologists as Aron Vinokur, Vladimir Shliapentokh, and Inna Ryvkina. In November 1966 Zaslavskaia and her sector held a seminar on the application of quantitative methods in sociology. The seminar participants outlined differences between so-called objective and subjective methods (documents and statistics versus interviews and surveys), quantitative and qualitative methods, and other ways to gather information (observation, experimentation, surveys, and questionnaires).[82] Zaslavskaia grew fascinated with sociological methods to study labor force mobility in the countryside, reading the works of and meeting with Shliapentokh and Shubkin. Relying on her training as a physicist and economist, she introduced quantitative research methodologies in all future studies. Mathematics was the "logical filter" for this methodology. The material of sociological analysis that passed through it "was placed in the foundation of science." The sociologists applied regression analysis and factor analysis to the example of the migration of Siberian rural inhabitants.[83] By western standards this was a modest beginning but at least it was a start.

Zaslavskaia's subsequent research revealed a series of significant disincentives to increased labor productivity in the Soviet system. The state wished to regulate migration and slow job turnover. Requirements of job permits and internal passports notwithstanding, Zaslavskaia showed that Soviet workers had acquired great freedom in selecting their place of work, residence, and profession. The state provided universal free health care, day care, and education. Tenure of service and performance, once central in securing housing and higher salaries, was now less important, available to productive and nonproductive workers alike. Workers moved where they could, knowing their benefits would be the same anywhere. The leveling of salaries encouraged workers to take on easy jobs. Highly qualified workers from eastern and northern regions of the country were moving toward central, southern, and western regions for less responsible positions. Every year 1.5 million inhabitants left the countryside for the city, and 1 million moved from one city to another. In Siberia, 17 percent changed residence every year and up to 40 percent did so in many other cities. Only 25 percent of Siberian rural residents lived where they were born.

For three years in the mid-1970s Zaslavskaia, Inna Ryvkina, and their colleagues carefully studied the system of disincentives to higher productivity of labor in the Siberian countryside. They analyzed six major subsystems of social life: production, lifestyle and consumer needs, personal supplementary income, demography (sex, age, etc.), spiritual needs and leisure time, and education. Their research identified the need for a wholesale reorganization of village life. They described the impact of industrialization, electrification, and mechanization on agriculture. In general they alluded to great achievements, but the details were troubling: there were great lags with respect to all major capitalist countries in terms of productivity of labor, capital intensity of work, and quality of food processing and delivery. In 1972, for example, 95 percent of the inhabitants of Novosibirsk oblast had a "trade center" or store but only 52 percent had a dining hall; 50 percent had public baths; 44 percent had access to clothing repair and tailors; 34 percent had a beauty parlor; and 23 percent had access to dry cleaning. In most cases these facilities were miles from work, their business hours conflicted with working hours, and one could only get to them by overcrowded, filthy buses that rarely ran on time.

Zaslavskaia assembled a catalogue of poor living conditions: in rural Siberia only 6.9 percent of families had running water; 4.6 percent, waste plumbing; 6.5 percent, central heating; 3.8 percent, a full bathroom; and 1.1 percent, hot water—all this in comparison with 50–60 percent in all categories for urban inhabitants. Housing was "dysfunctional": 29 percent of the population, those with six or more family members, occupied only 18 percent of available space. Paradoxically the villagers' diets, which should have been augmented by nearby farm products, were much less complete than in cities in terms of fats and protein, whereas sugars were consumed in excess. Salaries lagged. A source of these problems was the overbureaucratization of agricultural life. Zaslavskaia identified seven different ministries, several other state committees, and still other organizations whose efforts to mediate these problems were poorly coordinated. They made decisions without adequate information and never collaborated. They promoted environmentally unsound practices, such as the intensive use of chemicals in agriculture that affected the health of villagers.[84]

Zaslavskaia rejected traditional administrative and ideological means of regulating migration and job turnover. Stalinist coercion no longer did the trick.

The work and residence permits that every individual was required to carry had limited impact. Ideological exhortations to work for the common good, to engage in socialist competitions, and to earn the honorary title of "communist laborer" had lost their effectiveness. Such administrative measures were ineffective "during a period of scientific technological progress." Workers were better educated, less inclined to tolerate heavy-handed managerial pressure. A "perestroika" of labor management was needed. Zaslavskaia's research indicated that whereas most individuals who "migrated" in the 1950s did so for economic reasons, in the 1970s only one-sixth cited economic factors as a primary motive in their decision to move. During the Stalin years, a period of "hostile capitalist encirclement," the emphasis of state programs was logically directed toward the rejuvenation of backward industry. In the modern Soviet Union, however, the human factor took precedence. Society needed to mobilize the workers' spiritual and physical forces to serve society. In Marxian terms, the "social factor" required greater attention than "the development of the productive forces."[85]

The further the sociologists examined Siberian labor mobility, the more complex the issue became. First, less manpower was available than in any other region of the country. How could "reserves" be identified? How could productivity of labor be raised? Second, by the early 1970s, outmigration of able-bodied young people assumed mass proportions as village residents could no longer tolerate the harsh living conditions. The government thought that increased investment of any kind would be accompanied by growth in labor productivity. Zaslavskaia determined, however, that labor productivity was largely dependent on social factors: better services, not higher salaries; workers' increased responsibilities and initiatives, not more managerial reforms.

The village itself was an eyesore. Few roads were paved, there were no parks, and dirt and garbage littered the streets. In the summer, streets were saturated with heavy machinery, trucks, and motorcycles that choked pedestrians with dust. Building schools was inconsequential if the schools were inadequately equipped and the teachers poorly trained. Higher salaries were meaningless since store shelves were bare, stocked with goods of the worst quality, especially foodstuffs and other necessities. The poverty of Tsentrosoiuz, the bureaucracy charged with getting goods to the village, was another sore point. I myself traveled on local routes to some of the villages. After standing in line for thirty minutes on a very cold day, a bus would finally arrive. Any semblance of a line gave way to anarchy. We all broke for the door. More than a hundred of us managed to board the bus and were packed so tightly that I could not reach into my pocket for change to pass up front for my fare. Nor could the man next to me shift his weight to unwedge his briefcase from between my legs as the bus teetered through the countryside. The main department store of Iskitim, thirty miles south of Akademgorodok, was palatial and well stocked by Siberian standards; empty by mine. Like most stores it was understaffed, and virtually all necessary consumer goods such as vacuum cleaners were unavailable.

Other goods—pianos, movie projectors, and the like—could be had but at great cost, and no one wanted them. Fishing lures abounded but no rods or reels. Rural residents naturally viewed the city as the promised land, a style of life to be emulated. By the end of the Soviet period, only a quarter of rural residents had no tie to the city through their jobs or relatives. Zaslavskaia had concluded early on that only improvement in the lifestyle of the village could slow or reverse outmigration. The Marxist urban-centered worldview of Soviet policy makers and planners was unlikely to favor changed investment patterns to help the Siberian peasant.[86]

Throughout the late 1970s and early 1980s Zaslavskaia continued her research on this subject, identifying sources of social and economic instability in the Soviet Union. Her research was listed in official institute research planning documents under the official rubric of "theoretical problems of the creation of the material-technological basis of communism, the perfection of production relations of developed socialism, and the strengthening of the Soviet way of life" (also known in Brezhnevite descriptions of research as "theme 4.2.1.1"). No longer content to accumulate data, Zaslavskaia now tried to set the agenda for national investment programs in such works as "Overcoming Social Differences between City and Countryside." She elaborated the concept of "the human factor of production" in a series of reports based on survey research. In essence, this research, which measured the quality of life of Siberian inhabitants and was prepared for the Russian and state labor committees, Goskomtrud RSFSR and Goskomtrud SSSR, the Novosibirsk city party committee, and Sibakademstroi, was concerned with social engineering, that is, with the scientific management of a social problem.[87] Zaslavskaia was now prepared to pull twenty years of research together in what turned out to be an indictment of the Soviet system that rocked the establishment.

THE NOVOSIBIRSK REPORT

In August 1983 a paper Zaslavskaia wrote for a closed IEiOPP seminar gained world attention. It was so critical of Soviet economic management practices that westerners assumed it had been written for a seminar organized by the economics department of the Central Committee. Zaslavskaia believes this rumor started for two reasons. First, two extra copies of the paper were circulated in Moscow and Leningrad by scholars unhappy with its limited distribution, without a cover page providing any information on the author or origin. Second, the macroeconomic approach, with its suggestion of systemic inequities, differed sharply from safer, microeconomic methodologies that focus on local social issues and therefore had been the methodology of choice among most academics. A paper in Moscow with a macroeconomic bent must have originated, so the thinking went, at the order of the Central Committee. In the end the paper caused such a scandal that it indeed came to the attention of the Politburo where Gorbachev read it. Zaslavskaia feels certain that it greatly influenced his thinking, especially since she had met with him to discuss the Brezhnev food program.

The preceding fall Zaslavskaia and her closest associate, Inna Vladimirovna Ryvkina, had completed a research proposal on "Social Mechanisms of Economic Development." They hoped to complete an empirical and theoretical study on the subject by 1987 and planned to include a number of sectors of the economy in their analysis. Copies of the proposal were sent to ten different institutes in Moscow, Leningrad, Perm, and Sverdlovsk to generate support, and a seminar was held in Novosibirsk in March 1983 to discuss the project. The forty to fifty invitees all wished to have greater details so Zaslavskaia set out to write a discussion paper, which she worked on in January and February 1983 while in a hospital recovering from an illness. When her sister read the paper, she said, "I think it's good. But this is not a paper, it's a manifesto."

In order to distribute the essay to the seminar participants Zaslavskaia and Ryvkina had to get permission from the state censor, Glavlit. The local Glavlit in Novosibirsk took its time in responding and at the last moment refused permission. To get around this problem, Aganbegian suggested mimeographing the essay, marking each copy "classified" under his signature, numbering each copy to keep track of its recipients, and holding it in the institute's safe. There would be no extras, and all copies would be collected after the seminar. Somehow at least two of the numbered copies disappeared from the chancellory. Several sociologists from Tashkent also copied Zaslavskaia's paper by hand overnight. When it came time for discussion, the atmosphere of the seminar fell within the Akademgorodok tradition. Speakers were given seven minutes, although many spoke for five times that long. Discussion went beyond the bounds permitted in Moscow; the true fabric of Soviet social relations was given microscopic scrutiny. The sociological "cream of Soviet science" furiously debated the spirit, concepts, and ideas of Zaslavskaia's project, arguing "far into the night—in hotel rooms and foyers and colleagues' flats."[88]

In May representatives of the KGB arrived in Akademgorodok, went straight to Aganbegian's office to examine the list of proposal recipients, and wanted to know where all the copies were. Zaslavskaia had to write a letter explaining how such a secret document could have gotten beyond the institute's walls— indeed she could not explain it—but it was already too late. Zaslavskaia's paper was reported by Dusko Doder in the *Washington Post* on August 3, 1983. Extracts were printed in the *New York Times* on August 5, 1983, and in *Die Welt* on August 18–20, 1983. BBC issued reports. Radio Liberty's *Arkhiv samizdata* published most of the Russian text on August 26. (In 1984 a copy was published in English in the journal *Survey*, with an introduction by Philip Hanson, an economist at the University of Birmingham, England.) Western specialists were stunned by the open criticism of the Soviet social system; Siberian officials were horrified. Valentin Koptiug, then head of the Siberian division, warned Zaslavskaia that there would be hell to pay. She did not have to face the fire, however, as she came down with pneumonia and was hospitalized for two months. By then the storm had passed.

In the early days of perestroika Zaslavskaia reread the "western version" of the paper and discovered indeed a duplicate of what she had written. Why had her "manifesto" caused such an upheaval? Simply put, a party member argued in the report that the Soviet economic system had changed little since the 1930s, was no longer appropriate to modern conditions, and was the main cause of declining Soviet economic performance. The system had "repeatedly been readjusted, renewed, and improved, but not once has it undergone a qualitative restructuring." Zaslavskaia also criticized the high level of centralization in economic decisions and the weak development of market relations so that "the prices of goods in demand and the means of production bear no relation to their social value."[89]

She implied that a shift to market rather than administrative allocation of resources was desirable and called for an analysis of the social and political forces that resisted change in the system. She stressed the social as much as the economic unsuitability of the present system for its damaging effects on people's moral attitudes, initiative, and readiness to work. As economist Philip Hanson argued, Zaslavskaia's support for a decentralized system was the least striking feature of the paper, since the recommendation that enterprises should cease to be given targets from above and should select their own suppliers and customers (a key element in Hungary's New Economic Mechanism, or NEM) was quite commonly put forward. More striking was her analysis of why the system worked to retard growth and her suggestions of how to employ economic sociology to analyze problems and to pave the way for change. In turgid Marxian terms, she had identified the problem as the "lagging of the system of productive relations behind the development of the productive forces." The solution was to reject socialist heterodoxy and allow for market relations and the existence of antagonistic social groups—in a society in which "classes" were supposed to have disappeared in the 1930s.

The Stalinist social system may have been appropriate to train the efforts of poorly educated and often illiterate workers and peasants toward one goal: industrialization. However, the economy had become more complex, and management from the center could not keep pace with its responsibilities. Furthermore, Zaslavskaia argued, the command system worked for an obedient, passive, poorly educated labor force but no longer fit a better-educated and materially more secure labor force whose individual rights and sense of justice had increased. Technological advances and universal education had created a class of workers, more developed and economically free, who were now "unable to make sufficiently effective use of their labour potential and intellectual resources, unable to ensure a high level of labour, production and plan discipline, high quality work, effective use of technology, or to assure positive modes of conduct in the managers, accountants, and supply technicians."[90]

Zaslavskaia contended that the social consequences of the existing economic system were damaging to performance. Excessive reliance on impracticable control from above generated laziness, neglect of quality, low moral standards, social passivity, and a propensity to "departmentalism." *Shturmovshchina*, low labor productivity, theft, alcoholism, and low morale are familiar phenomena to students of Soviet society. However, Zaslavskaia's call for radical reorganization of economic management, her study of the interests of individuals and social groups, in a word, her insistence on taking into consideration "the social aspects of the process of improving production relations under socialism," were fundamentally new criticisms of the Soviet social order.[91]

In her essay Zaslavskaia had clairvoyantly identified the opposition to reform of the Soviet economic system that was to be seen during Gorbachev's efforts to promote perestroika. This opposition came primarily from entrenched bureaucrats in middle-level economic administration, especially branch ministry officials, those who were somewhere between the central planners and the enterprises themselves. These were individuals who "occupy numerous 'cozy niches' with ill-defined responsibilities but thoroughly agreeable salaries," while both central planners and managers have lost power.[92]

Zaslavskaia concluded her essay with a plea for government support to develop "economic sociology," a discipline whose research would be central in repairing the social mechanisms of Soviet society. The new field would study the economic structure of society; investigate the self-awareness of various economic groups, their values, needs, interests, and motivations; analyze the patterns of their behavior; study the economic interactions between various government and social bodies, groups, and individuals; and establish methods to ensure more effective use of labor in production. Economic sociologists would cooperate with political economists, legal specialists, psychologists, and cyberneticists. Zaslavskaia was especially concerned that, except for IEiOPP, the connection between sociological and economic research in the USSR had been poorly explored.[93]

During the Gorbachev years Zaslavskaia repeated the central ideas of her 1983 manifesto in a number of forums. She criticized the "economic methods of management advanced in the 1960s" and embraced Gorbachev's call for thinking and working in a new way (*novoe myshlenie*). For the citizen, she believed, this meant recognizing each individual not as a productive force like ore, energy, or technology, that is, as something to be managed mechanically, but as the subject of economic, political, social, and cultural activity who merited equal access to all goods and services of society. The old system of privilege, bribes, black market, and connections had to be destroyed.

The discussion of Soviet economic, political, and social ills under Gorbachev initially found a place in such literary journals and weeklies as *Novyi mir*, *Oktiabr'*, *Neva*, *Znamia*, *Ogonëk*, and such "leftist" newspapers as *Moscow News*, *Komsomol'skaia pravda*, and *Moskovskii komsomolets*. By 1988, even such party organs as *Pravda* published critical analyses of Soviet social problems, including an article by Zaslavskaia entitled "Perestroika and Sociology" in which she reiterated the need for social science research to assume a vanguard position in the struggle to restructure social relations. Social science had lagged, she wrote, "confining itself largely to reiterating, explaining, and approving party resolutions that have already been adopted." As a rule, few sociologists had been consulted in technology assessment for such major projects as river diversion or BAM. They were needed to provide "full, accurate, and truthful information" about the real state of affairs to facilitate correct decisions in accelerating economic development: which administrative measures are superfluous; which types of individual labor merit support and which should be stopped; appropriate levels of remuneration; determination of individual, group, and public rights, interests, and duties.[94]

Zaslavskaia argued that more surveys were needed to overcome the "extremely low level of reliable social statistics." No data were available on crime, suicide, alcohol and drug abuse, and ecology to help identify the "human factor" in reform. Zaslavskaia acknowledged that increasingly systematic data were being developed, but the absence of properly trained sociologists hindered systematic investigation. Whereas in the USSR the first hundred professional sociologists would graduate with higher education in 1989, in the United States alone there were some 260 sociology departments, with six thousand graduates annually. Still, more than four hundred articles on the subject of "economics and society" were published between the 1960s and the early 1980s, with the number of articles on social problems of economic mechanisms increasing 7 times, on deviant behavior 7.5 times, and on consumer demands and lifestyle 6.5 times. In addition, a new journal of economics and applied sociology was launched by the Siberian division in 1984.[95]

At the start of the 1990s Zaslavskaia and her colleagues breathed new life into sociology. Taking glasnost and perestroika at face value, she pushed to make empirical "economic sociology" the leading sociological science. The first steps of economic sociology called for the creation of a theoretical foundation, a scientific language, and the establishment of university curriculum. Zaslavskaia hoped to build on foundations established at Novosibirsk University in 1982–83 when economic sociology was first introduced in the economics department. She published syllabi and lecture notes so that other higher educational institutions might follow suit. At first, entrenched interests fought Zaslavskaia's efforts to spread the gospel of economic sociology. At Leningrad University, for example, B. Ia. Elmeev considered such sociological categories as "the position of social groups," "behavior," and "interests" as "subjectivism that is not needed in political economy." Zaslavskaia deflected such closed-minded comments like those of Elmeev by pointing out that criticism was valuable to her endeavor, for "it is not coincidental that Marxism asserts that only critical science is constructive, whereas apologetic science is not only useless but harmful."[96]

In a book outlining the major aspects of the new discipline, *The Sociology of Economic Life* (Sotsiologiia ekonomicheskoi zhizni) (1991) Zaslavskaia and Ryvkina attempted to provide the theoretical foundations of the discipline. In their words, economic sociology is the study of "the laws of the functioning of social groups in a system of economic relations, and correspondingly the development of the economy as a social process. Neither political economy nor sociology studies this." Neither area "contains a special theoretical, let alone methodological apparatus for research on the structure of the subject of economic development, the basic forms of economic activity of that subject, tendencies observed in that activity, the mechanisms of that activity, and the influence of economic development on sociology." Zaslavskaia and Ryvkina therefore called for the study of group formation, behavior, and structure for all social layers in Soviet society in order to evaluate the relationship between position, wealth, values, norms, and other variables, and to understand economic behavior of various groups: scientific-technical intelligentsia, managers, and workers.[97]

Soon after Gorbachev came to power Zaslavskaia and Aganbegian both left Novosibirsk for Moscow. For Zaslavskaia, Abel Gezevich's departure had changed the face of the institute. Work had become dry and gray; the institute had lost its vibrancy. Having spent twenty-five years in one place, taking on institutional, obkom, and university responsibilities, she had had no time to do her own work. She starting coming to the institute only on weekends, fearfully avoiding the administrators. She felt "mired in the same place, following the same path, working with the same empirical basis." When she was invited to head the new center for public survey research in Moscow, she jumped at the opportunity. Her position as academician gave her clout and funding.

Zaslavskaia became director of the Russian Center for Public Opinion Research in 1988. Owing to the propiska system (state-provided residence permits), she could not take many staff members from IEiOPP with her to Moscow. Her new staff of two hundred employees includes sociologists, economists, political scientists, psychologists, and computer specialists, a hundred of whom work in thirty-six departments in the former republican capitals and the largest industrial centers of Russia and Ukraine (including Voronezh, Stavropol, Gorky, Toliati, Perm, Novosibirsk, Kemerovo, Magadan, Norilsk, Dnepropetrovsk, Simferopol). The center does contract research for the government and media organizations and has close contacts with research, media, and government organizations in the United States, western Europe, South Korea, and Japan.

Zaslavskaia's pollsters conduct monthly, quarterly, and annual surveys of four thousand individuals at ninety-two points throughout the nation to keep tabs on the pulse of Russian public sentiment. Representative samples of workers, leaders of state enterprises, farmers, entrepreneurs, and experts on socioeconomic policy are also drawn. The center publishes a monthly bulletin, *Economic and Social Change: Monitoring Public Opinion*, based on the results of the surveys. The results indicate that popular support for economic reform has waned as the strains of inflation and unemployment make it difficult for many families to live within their means, requiring them to spend their meager savings. Many citizens fear crime. The surveys have charted the rise in Russian nationalism and the growing concern, especially among the less educated people in society, about the "threat of selling off Russia's national wealth" and the alleged "excessive influence of non-Russians in Russia's life." The surveys also reveal that most citizens have lost interest in politics and feel that ongoing economic and political changes have passed them by. Zaslavskaia, for her part, remains convinced that scientifically based sociological monitoring of economic and social change in Russia during its transition from post-totalitarianism to democracy will aid in the development of "socioeconomic policy appropriate to the sociocultural specificities of the country and mass consciousness of the people."

Aganbegian, too, has had the opportunity to create something new and exciting: He dreams of building an Academy of Economic Science in Moscow, an "Ekonogorodok," ensconced in a twenty-floor, million-square-foot, modern, multimillion-dollar complex. The trials of putting together this second Gorodok and his growing distance from the pulse of the nation have told upon his health. Meanwhile, with Zaslavskaia's and Aganbegian's departure for Moscow, what does the future hold for the Institute of Economics and the Organization of Industrial Production?

MARKET RELATIONS AND POLITICAL TURMOIL

The Communist Party built an impressive administrative apparatus to manage the economy. Its goal was to allocate resources efficiently in support of state economic development programs that emphasized heavy industry and the military sector. The party, and the bureaucracy it created, were unwilling to relax their control over the economy. Any reforms that would allow decision making to devolve to the enterprise level, such as in Hungary's New Economic Mechanism, were seen as a turn to market mechanisms and an anathema to socialism. Declining economic growth, decreasing capital and labor productivity, and repeated failures to implement reform by administrative fiat left the economy in shambles on the eve of Gorbachev's rule.

At the twenty-seventh party congress in February 1986 Gorbachev highlighted the reconstruction and modernization of existing enterprises as the main thrust of investment policy for the twelfth five-year plan, and indeed until the end of the century. He was most concerned with accelerating the introduction of modern technology—computers, microelectronics, and advanced materials—into the economy, a goal referred to as "uskorenie" (acceleration). The theme of uskorenie was based in part on Aganbegian's prescient analyses of the path perestroika should take to change the Soviet economy, its management system, and patterns of development radically. In agriculture, too, Gorbachev called for reform, for restoration of "the economic balance between town and village." Hearkening back to Zaslavskaia's work, he proposed the "social development" of rural areas—housing, medical care, a rural electrification and road building program, and increased investment toward improving transport, storage, and processing of agricultural produce.

But serious obstacles stood in the way of economic reconstruction. First, the existing capital of Soviet industry was so aged that modernization was required in nearly every sector and would gobble up investment funds. Second, the investment sphere of the economy was totally unprepared for the renovation policy: the necessary scientific, design, planning, and machine-building base did not exist. Such sunrise industries as robotics, artificial intelligence, and fiber optics had been neglected too long. The machine-building industry proved incapable of producing the needed replacement machinery and equipment. All the work of Aganbegian's institute in support of *uskorenie* went for naught as production of turbines, locomotives, gas and oil drilling equipment, diesel and electrical motors, metal-cutting machines, and transport and construction equipment fell precipitously. At the same time the need to redirect investment away from the military and toward the consumer sector, housing, and the environment became more pronounced. When it turned out that *uskorenie* was insufficient, Aganbegian realized that a change in the entire social environment was required, Zaslavskaia's message precisely.[98]

IEiOPP nonetheless seems to have a bright future. Siberia, with its vast natural resources, beckons to western companies and governments. They long for accurate information about Siberia in a machine-readable form, something the institute can provide. Its economists and sociologists have studied "the organization of production" and the "distribution of resources" in Siberia and the Far East for thirty years. The distinguished research personnel among its 300 specialists include 1 academician and 1 corresponding member of the Russian Academy, 21 doctors of science, and 148 candidates of science. Siberia's millions of square miles are home to twelve departments of the institute, four of which are in the industrial centers of Kemerovo, Irkutsk, Krasnoiarsk, and Barnaul, with forty-five sectors and groups in other major cities.

Yet the economic and political turmoil in Russia threatens to force staff and program cuts. The budget in 1991 was 2.5 million rubles indexed for inflation, primarily from the Siberian division, with another 500,000 from contract research. Unfortunately inflation outstripped indexing. Then the failed August 1991 coup led to the destruction of all former ministries and the Soviet Academy of Sciences and in their place the creation of new Russian ministries and a Russian Academy of Sciences. These organizations are still trying to resolve responsibilities, priorities, and funding schemes in the face of great uncertainty that has left institutes throughout the country short of funding. The assistant director of IEiOPP, Viacheslav Evgenievich Seliverstov, told me, "Our reputation remains high and we're doing okay, but some sectors have grown more poor than others, and there are tensions between groups." Moreover, many capable scientists have left for new cooperative endeavors and others have gone abroad. As colleague Cynthia Buckley, a sociologist at the University of Texas, told me, "At a time when ideology allows freedom in intellectual endeavors, the 'market' takes it away, as now all researchers have to scramble for sponsors." The illustrious tradition of the sociology department is maintained by capable, if young scholars who lack Zaslavskaia's reputation and productivity. Zaslavskaia and Aganbegian essentially left the institute high and dry; they rarely visit but are remembered fondly, if distantly. All this tells upon the institute's work.

The Gorbachev years brought about a change in the orientation of IEiOPP that occurred almost as rapidly as did those in the political and economic spheres. Reform could nòt occur without special expertise. New tax laws, greater reliance on market mechanisms, and the growing autonomy of republican, regional, and local economic units required research to bring about new planning, organizational, structural, and management tactics. The institute turned away from interregional, input-output models for contract research on short-term problems. R. I. Shniper and his colleagues examined new mechanisms for economic management in autonomous republics, regions, and districts. A. G. Granberg explored the economic relationship between self-financing enterprises in light of new taxes, laws, and other reforms. The research was intended to assist in deciding questions of socioeconomic policy between central, regional, and local organs of power. Most studies were sent to such central economic organs as Gosplan.[99]

When Granberg became one of Yeltsin's chief economic advisers, the institute staff undertook research for him on the reform of the Russian economy and the establishment of market relations. But this research was often outdated before it was completed. The Russian economy changes its face daily as it lunges from one problem—and one solution—to another. Small subdivisions of the institute have been liquidated and subsumed elsewhere to preserve their personnel, while others, whose "themes have lost significance," have simply been dispersed. In preparation for the transition to a market economy, the modeling of economics after capitalist countries (in terms of tempo and proportions of industrial production) was refocused into a group to study technological progress in capitalist countries.

In connection with the growing attention to the environment, a laboratory was created in 1988 to study the ecological and legal problems of the Baikal basin. No longer limited by Brezhnevite political controls on technology assessment reports, institute economists moved aggressively into the area of environmental impact statements. Under the Academy of Sciences program "Biosphere and Ecological Research to the Year 2015," institute economists explored methodological aspects of determining norms and their introduction to value and pay for land, air, and water resources and their pollution. Institute personnel completed technology assessments of the Katunskaia Hydropower Station, which led to the tabling of the project, coal mining in the Kuzbas region, river diversion, water and land resource use in the Omsk region, and environmental costs of oil production in West Siberia.[100]

Iurii Petrovich Voronov, a deputy editor of *EKO*, took a leave of absence from IEiOPP to capture the energy of the burgeoning market in contemporary Russia. He is now the editor of the *Siberian Commodity Exchange Newspaper* (Sibirskaia birzhevaia gazeta)—a vital source of information for Russian businessmen, compiler of the ten-thousand-word *English-Russian Learning Dictionary of Accounting and Commercial Terms*, and dean of a newly established Siberian business school. His newspaper provides information on how "business" is done in the West; risk and investment strategies; what contracts are and how to break them without paying fines; and the impact of recent laws and resolutions on economic activity. He also publishes prices of various goods in markets throughout the region, translations from European and U.S. economic journals and newspapers, and advertisements from banks, enterprises, and commodity exchanges. Voronov, a free market enthusiast, criticizes the speed of reform. He likens the current period to the "time of troubles" that followed Ivan the Terrible's death in the late sixteenth century. All government decrees seem determined to hold onto the major levers of the economy. The government sets "norms" and offers "prognoses" instead of "plans."[101]

Zaslavskaia has followed the unfolding of perestroika from her Moscow apartment. She, too, was troubled that the first stages of perestroika did not go far enough. Rapid economic development required "changes in the mechanisms for managing the economy," she wrote. Restructuring of social relations was as essential as radical economic reform. Zaslavskaia came to believe that nothing short of a second socialist revolution was required to overcome entrenched party bureaucrats and all-powerful state monopolies that had erected obstacles to change. They slowed delivery of mail, interfered with the circulation of money, and prevented the distribution of goods. Sociological research has demonstrated the untruth in the contention "deeply etched in our consciousness—that our society has completely abolished all forms of economic exploitation of some groups by others." It has revealed the tendency of a "socially degenerate stratum of officials" to turn into a ruling class. It has led, in Zaslavskaia's words, to the emergence of mafioso-type groups. One economy exists for the politically blessed, another for everyone else.

But the second revolution may never come. Zaslavskaia's public opinion studies indicate that the people continue to live "in fear and obedience, unquestioningly carrying out orders 'from above.'" They have become "alienated from social values, [have] immersed themselves in their personal lives, [and have become] engrossed in the accumulation of property." They have adopted a wait-and-see attitude to reform, unwilling to be receptive and unaccustomed to risk. A "low level of political consciousness and activity, the lack of democratic traditions, the immaturity of public consciousness which is full of prejudices and dogmas, the inadequacy of public opinion on many questions, and the low level of social morality" all waylay reform.[102] And far away from Moscow, economic uncertainty, social displacement, and profligate use of natural resources require the genesis of a new Siberian algorithm in the hands of the independent worker, farmer, manager, sociologist, and economist.

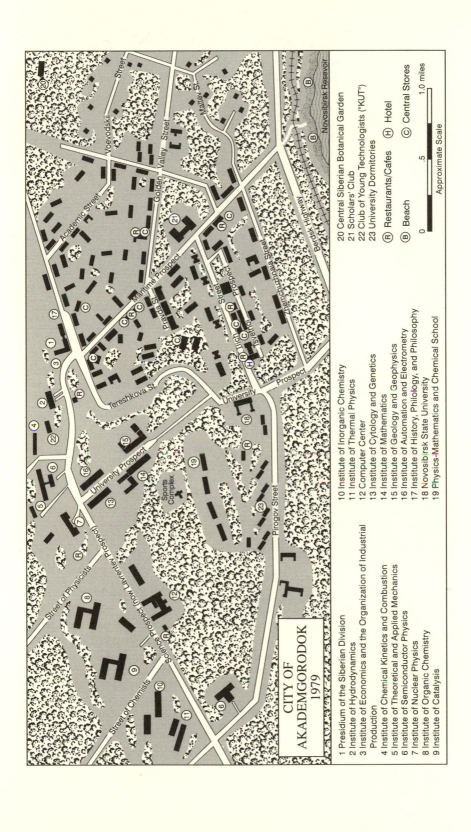

CITY OF
AKADEMGORODOK
1979

1 Presidium of the Siberian Division
2 Institute of Hydrodynamics
3 Institute of Economics and the Organization of Industrial
  Production
4 Institute of Chemical Kinetics and Combustion
5 Institute of Theoretical and Applied Mechanics
6 Institute of Semiconductor Physics
7 Institute of Nuclear Physics
8 Institute of Organic Chemistry
9 Institute of Catalysis

10 Institute of Inorganic Chemistry
11 Institute of Thermal Physics
12 Computer Center
13 Institute of Cytology and Genetics
14 Institute of Mathematics
15 Institute of Geology and Geophysics
16 Institute of Automation and Electrometry
17 Institute of History, Philology, and Philosophy
18 Novosibirsk State University
19 Physics-Mathematics and Chemical School

20 Central Siberian Botanical Garden
21 Scholars' Club
22 Club of Young Technologists ("KUT")
23 University Dormitories

(R) Restaurants/Cafes   (H) Hotel
(B) Beach   (C) Central Stores

0   .5   1.0 miles

Approximate Scale

Aleksandr Skrinskii, here teaching at Novosibirsk University, succeeded Gersh Budker as director of the Institute of Nuclear Physics. He presided over both the challenging effort to build bigger, more expensive particle accelerators and fusion devices, and the institute's difficult readjustments required by the breakup of the Soviet Union (top photo courtesy of Rashid Akhmerov). In June 1991, before the current economic and political crisis in the Soviet Union, Boris Yeltsin visited Veniamin Sidorov, deputy director of the Institute of Nuclear Physics, to voice his support for fundamental research in Russia (photo courtesy of the Institute of Nuclear Physics).

A number of distinguished scientists were trained in Akademgorodok. Petr Kapitsa (*left*) joined Gersh Budker (sans beard) at the Institute of Nuclear Physics at a banquet, most likely celebrating the successful dissertation defense of Roald Sagdeev (now a professor at the University of Maryland) (top photo courtesy of Rashid Akhmerov). The nuclear physics institute has always benefited from close ties with the nation's scientific elite. Shown here, in 1982, are (*from the left*) Sidorov, Siberian Division chairman Gurii Marchuk, and institute director Aleksandr Skrinskii welcoming Anatolii Aleksandrov, then president of the Academy of Sciences, to the institute's "round table" (photo courtesy of Rashid Akhmerov).

Dmitrii Beliaev saved the Institute of Cytology and Genetics from interference from followers of Trofim Lysenko, a pseudoscientist who rejected modern genetics. Beliaev, who lost an older brother to the Stalinist purges, never joined the Communist Party, but he became a member of the scientific establishment, harnessing his institute's research programs to the Brezhnev food program (top photo, Beliaev [right] with Academy President Anatolii Aleksandrov, and below, Beliaev with his beloved foxes, both courtesy of Rashid Akhmerov).

In spite of their distance from Moscow, Akademgorodok's scientists always felt pressured to conduct applied research at the expense of basic science. Specialists in hybridization, Petr Shkvarnikov (*left*) and Ivan Chernyi, had success at the Institute of Cytology and Genetics in producing new strains of wheat (top photo courtesy of the Institute of Cytology and Genetics). The institute's core research collective in its early years included Aleksandr Kerkis (*row 1, 1st from left*), Dmitrii Beliaev (*row 1, 3rd from left*), and Rudolf Salganik (*row 2, 5th from left*) (photo courtesy of the Institute of Cytology and Genetics).

The current director of the Institute of Cytology and Genetics, Vladimir Shumnyi, has lost a number of talented scientists to newly established genetics engineering firms, but he remains committed to the past style of administering science from above rather than encouraging more democratic means in order to lead the institute through uncertain financial conditions in the mid-1990s (top photo courtesy of the Institute of Cytology and Genetics). The institute's deputy director, Rudolf Salganik, headed the laboratory of nucleic acids and worked on nuclease antivirus preparations for the treatment of certain serious viruses in man, bees, and silkworms. He gave underground lectures on genetics in Moscow physics institutes before Lysenko's fall (photo courtesy of the Institute of Cytology and Genetics).

Biologist Zoia Nikora (top photo courtesy of the Institute of Cytology and Genetics) was one of the few researchers at the Institute of Cytology and Genetics to defend such scientific outcasts as Nikolai Timofeeff-Ressovsky and Raissa Berg when the institute took a conservative turn under Beliaev. Another politically conservative individual, geologist and Party member Andrei Trofimuk, opposed official policies—and Academy President Anatolii Aleksandrov—to destroy pristine Lake Baikal. But in other respects Trofimuk embraced science, particularly his own field of oil and gas engineering, as being crucial to breakneck Siberian development (photo courtesy of Rashid Akhmerov).

Economist Abel Aganbegian (top photo courtesy of Rashid Akhmerov) oversaw the development of economics in Akademgorodok with a distinctly Western flavor (as Tatiana Zaslavskaia had done in sociology), later becoming an adviser to reform-minded Mikhail Gorbachev. As was the norm under Brezhnev, Aganbegian and his institute were tied into the big science of resource development through the "Siberia" program. Here he is shown with Mikhail Lavrentev (*far right*) and Gurii Marchuk (*center*), who was Lavrentev's successor and later president of the Academy of Sciences (photo courtesy of Rashid Akhmerov).

Abel Aganbegian (*center, pointing*) also supervised scientific aspects of the costly and often delayed construction of the new trans-Siberian railroad known as BAM, which Brezhnev intended to be a monument to his enlightened rule (top photo courtesy of Vladimir Novikov). Current chairman of the Siberian Division of the Academy of Sciences and of Akademgorodok, chemist Valentin Koptiug (*left*), a conservative administrator and cautious scientist, often consulted on the "Siberia" program with Academy President Anatolii Aleksandrov (*right*) (photo courtesy of Rashid Akhmerov). To this day he strives to maintain the perquisites of the Siberian Division leadership, including the disposition of apartment buildings, stores, and land under their jurisdiction, at the expense of research institutes and the inhabitants of Akademgorodok.

CHAPTER SEVEN

Crackdown: The Communist Party and Academic Freedom in Akademgorodok

AFTER A DECADE of academic freedom rarely seen in the USSR since the 1920s, the Communist Party put an end to the golden years in the Golden Valley. The central apparatus in Moscow and its agents in Novosibirsk had grudgingly tolerated the vibrant, critical atmosphere in Akademgorodok. The de-Stalinization thaw had contributed to the environment of free thinking, not only in scientific research but in politics, literature, and the arts. To be sure, the party had freed scientists from the strict controls of the Stalin years. For instance, no longer were contacts with western scientists—in person, through the mail, or in the exchange of reprints—in and of themselves evidence of "idealism" or perhaps even "anti-Soviet activity." In fact, extensive contacts were seen as a sign of vigor. The party ceased drawing distinctions between "bourgeois science" and "proletarian science." No longer did it scrutinize every aspect of science policy. The scientific establishment was to be trusted.

Mikhail Lavrentev's close personal relationship with Khrushchev was critical in this regard. It ensured that Siberian party organs expedited orders, freed up personnel, and overlooked procedural liberties with less than the usual obfuscation. Khrushchev had known Lavrentev since the early 1940s in Ukraine, trusted him, and had ordered that he be issued an open pass to the Kremlin. The Siberian division even had its own Communist Party committee (partkom), which was dominated by scientists. The existence of the partkom meant that the Novosibirsk party organization kept its distance. Indeed the national, regional, and local party and economic organizations were more interested at first in facilitating the construction of Akademgorodok than in establishing political control.

All this changed in the mid-1960s. Khrushchev's ouster in the fall of 1964 was a bad omen. He had shaken the party apparatus with de-Stalinization and had sought to give more decision-making power to managerial experts. His sovnarkhoz program, which set up regional economic units intended to bridge ministerial barriers to the supply of goods and services, undermined the authority of central party cadres in making economic policy. But with Khrushchev gone, the new general party secretary, Leonid Brezhnev, intended to reestablish the supremacy of the central party apparatus. He called for "trust in cadres," that is, "party cadres." This meant reliance on the party apparatus, its mechanisms and personnel, and its "management techniques" imposed from above to ensure homogeneity in decision making. Brezhnev called for "perfection" (usovershenstvovanie) of existing mechanisms rather than reform.

Rocking the boat was dangerous; muddling through was better than reform or decentralization.

In terms of foreign policy, Brezhnev competed with the United States for influence in the Third World through adventurous military forays. He sought strategic parity that heightened the ideological tension between the USSR and the West. The party reformulated ideological precepts in most areas of Soviet life, growing more vigilant toward perceived internal threats and external enemies. The apparatus redoubled its efforts to achieve the goals of increased industrial and agricultural production at the expense of health care, consumer goods, housing, and fundamental science. Science and technology came to be seen merely as instruments to achieve higher industrial production and military might. The "bulldozer" science of river diversion, Baikal development, and BAM was considered more important than basic research. The measure of scientific success was its contribution to economic growth. All these changes told upon the life of academic science in general and Akademgorodok in particular.

For the Siberian city of science, the implications of these policy shifts were staggering. Akademgorodok had enjoyed a healthy balance between fundamental and applied research, in part because of the institutes' unhappy physical state: the absence of instruments, material, and equipment and the poorly constructed buildings all contributed to a theoretical profile. Akademgorodok's interdisciplinary organizational philosophy based on a common mathematical language reinforced this tendency. Now the value of fundamental science was seen in its ability to raise the productivity of labor, increase agricultural and industrial production, tame the seemingly unlimited natural resources of the vast empire, and wed the advantages of the Soviet system to the so-called scientific-technological revolution.

Second, as an adjunct to Brezhnevite "trust in cadres," the party apparatus exerted more control over Akademgorodok life, both through sheer growth in numbers and the incorporation of the Siberian division partkom into the local party apparatus. Thus the Siberian division lost its unique scientist-dominated party organization. Instructions now came from the center and they would indeed be heard. The Communist Party extended its control through party organizations in every institute. These organizations had important powers regarding the conduct of research, the hiring and firing of individuals, researchers' personal lives, and the political and philosophical discussions that took place in the institutes. One of the most critical changes with regard to "cadres" involved the effort of the national Communist Party and the Communist Youth League to weaken the influence of local youth organizations. These organizations, like the Council of Young Scientists and the Akademgorodok Komsomol, were leading sponsors of the economic, political, and cultural experiments in Akademgorodok. Together with the natural aging of Akademgorodok's population, this resulted in older, more conservative scientific administrators replacing those who had given vitality to the Siberian city of science.

Third, a direct effort was made to end Akademgorodok's special status as a symbol of the Khrushchev years, of the de-Stalinization thaw, of academic freedom, of free thought on a wide range of subjects. Akademgorodok had shown scientific independence from the start in physics, in genetics, in computers, economics, and sociology. The doors to the institutes were virtually left open; few paid attention to the "*propusks*"—ID cards—that were required currency for gaining entry to institutes in every other Soviet city. Leading personnel made themselves accessible; getting to see Lavrentev or Budker was a simple matter. Their administrative styles were democratic. Most colleagues normally were on a first-name basis, dropping the formal patronymic form of address. Close relationships even existed between students and academicians. Finally, foreigners were a vital component of the political, cultural, and social give-and-take of Akademgorodok. All this gave way to the more formal styles of Marchuk and Koptiug, and to stricter controls on international contacts.

Akademgorodok, well known for its informality, was also famous for its open satire of Soviet life. The Scholars' Club frequently held exhibitions of the works of Soviet painters that were normally locked in basement storage at other museums. At the social clubs—"Under the Integral" and "Grenada"—the scientists played cards, drank, and sang into the wee hours. For party officials whose formative years were the Stalin period, Akademgorodok was too young in terms of its inhabitants' age, worldview, and desire to accelerate rather than abandon the de-Stalinization thaw. In a word, Akademgorodok represented generational, ideological, cultural, and academic freedoms, all anathemas to the conservative leadership of the Brezhnev regime. These freedoms were curtailed starting in 1968 in response to three events: the open protest over the show trial of the writers Sinyavsky and Daniel, the appearance of singing bards whose lyrics challenged Communist Party dogma, and the Soviet invasion of Czechoslovakia. The face of science in Akademgorodok would continue to be distinguished from the Soviet norm, but its colors were drawn increasingly in gray and brown hues. Taken together, the Communist Party's policies toward Akademgorodok after the mid-1960s were responsible for the fact that the city of science never lived up to the promise of its first decade.

WHAT IF THEY GAVE A PARTY AND THE SCIENTISTS CAME?

The center dominates in a one-party system, with most decisions originating from above. Yet the manner, style, and earnestness of implementing these decisions often depend on local organs. Perhaps because the Communist Party had learned the hard way the costs of heavy-handed interference under Stalin, it trod less heavily in matters of research policy in the postwar years, reflecting policies that revealed the tenor of leadership and the personalities of Khrushchev and Brezhnev. The central apparatus created party organizations in all Akademgorodok institutes, but these organizations rarely embraced central party dictates with the fanaticism of the Stalin years.

I was one of three Americans who first received permission to work in the archive of the Communist Party of the Novosibirsk region, and the first to use materials for Akademgorodok. I was allowed to look at bound indexes of holdings, an honor unheard of only months earlier. Before perestroika, researchers had to rely on the kindness of archivists to find crucial materials, but because of instructions from above or out of laziness the archivists were usually obfuscatory and often pawned off documents that had already been published. I saw minutes of party meetings from the local to the regional level, annual statistical accounts, and personnel folders. Aside from usurious rates charged for the right to see files and even higher tariffs to make copies, which no Soviet citizen could afford, I have no complaints about my work in the party archive—except one. During the week I finished my research in the archive, the temperature ranged from a high of −26° F to a low of −37° F. I grew exhausted from the one mile walk to the train station and the hour-long trip to and from the archive in unheated railway cars. Fortunately, at the end of the week, a friend with his own sauna invited me for a personal de-Stalinization thaw.

Perusal of party archive materials revealed the central role of party personnel in all aspects of daily life. The party apparatus usurped many of the functions taken up by private citizens or other bureaucracies in civil society. It investigated drunken disorders, petty crimes, robberies, and murder; the occasional breakdown in the supply of hot and cold water; car and industrial accidents; the theft of "socialist property"; socialist obligations, competitions, awards, and honors; complaints about the low number of subscriptions to the party's periodical literature that allegedly reflected inadequate vigilance and political consciousness among the scientific workers in Akademgorodok; even many cases of food poisoning. The party used its authority to grant higher pensions to retired communists and veterans since the average Soviet retirement benefit never afforded comfort.

In Akademgorodok, local party organizations considered all aspects of the activities of the scientific institutes under its jurisdiction, from research and development to philosophical seminars, from the economics of innovation and funding to contemporary geopolitical issues (as interpreted in Moscow), to celebrations of such events as the upcoming party congress or a recent past one, the 100th birthday of Lenin, the 150th anniversary of the death of Engels, and so on. The raikom (the district party committee) was particularly concerned with the formal aspects of rule: regular meetings, record keeping, attendance, and getting the right reports and forms filled out. Endless meetings examined in endless detail the inner workings of subordinate party organizations. Judging from stenographic accounts in the party archive, the predominate attitude was science be damned. Attendance at meetings was more important than ongoing experiments. In some periods, one particular theme dominated committee meetings, usually on instructions from the central party apparatus, for example, labor discipline, absenteeism, or violations of the "socialist order."[1]

The shuffling of party secretaries at all levels of the apparatus, the creation of new party groups, the sucking up and spitting out of these groups by one another, transfers, and liquidations of party units even at the level of kinder-gartens drew much attention. The party also addressed far more sacred concerns including the *nomenklatura*—lists of individuals approved for promotion, with the implication that undesirables could be excluded from employment and others could be fired. It considered party membership at dozens of bureau meetings where comrades were either accepted into the party or approved as candidates for membership, while others were rebuked, disciplined, or ousted. Autumn turned out to be a good season for this activity, perhaps because it was also the season for national, regional, and local party gatherings. In sum, the Communist Party established control over all Soviet institutions through top-down administration. It used its powers of the purse and personnel to push economic performance and ideological homogeneity in any way it could: by exhortation, threat, and reward.

The party apparatus resembled a pyramid, with most major policies hashed out at the apex and sent by directives and resolutions to subordinate levels. The Politburo, at the apex, was made up of leading communists promoted from the next level, the Central Committee. The Central Committee consisted of more than two hundred officials who were elected from regional and local party organizations but whose candidacy was approved by the Politburo. Patronage was a key to successful promotion and was often based on personal ties dating back to service together in some regional organization. The Central Committee had a series of departments responsible for various aspects of the economy, policy, and culture (industry, agriculture, military and foreign affairs, ideology, education, science, etc.) whose personnel gathered information, conducted analyses, and wrote reports. A Politburo member usually served as secretary for each area. Below the Central Committee was the provincial or regional (oblast) level (the obkom), which is something like a county. City and district (raion) party organizations, or raikom, fell under regional jurisdiction. The districts had jurisdiction over primary party organizations (PPOs), some of which had the power to admit members, hire and fire personnel, and make policy.

Within each level of the party apparatus there were further divisions of authority. At the top sat party secretaries who were assisted by second secretaries and department heads. Each secretary or office had a staff to assist in reading reports, compiling statistics, writing speeches, verifying compliance with orders, and passing on information to higher and lower levels. Those who aspired to become first party secretaries usually worked for a first party secretary as the head of some department, for example, economics, agriculture, higher education, and so on. The first party secretaries often had candidate of science degrees, frequently in a technological or engineering specialty but often in party history or Marxist philosophy. Department heads performed analyses, wrote speeches, did the scut work for conferences and plenums, and read citizens' complaints. For Akademgorodok, the important regional and local bodies were the Novosibirsk regional party committee (obkom), the Siberian division's party committee (partkom) which was subsumed into the Sovet Region party committee (raikom) in 1965—that being the region where Akademgorodok is located—and the primary party organizations of the institutes.

Most Akademgorodok PPOs grew rapidly. By June 1962, for example, 40 percent of the staff of the Institute of Nuclear Physics were communists or komsomoltsy (Communist Youth League members). In 1962 alone the institute's PPO grew from 69 members to 147, 83 percent of whom were younger than thirty-five. Eighteen months later the organization had grown to 203 communists and 273 komsomoltsy out of eleven hundred total employees. In a practice considered to be an honor, the bureau of the raikom then asked the Novosibirsk obkom to allow the institute to appoint one of its employees as secretary of the institute's party organization, with the great privilege of paying him from institute, not party coffers. The Experimental Factory followed suit when it had grown to 152 communists and 139 komsomoltsy out of one thousand employees.[2] In some cases growth in numbers occurred because institute directors tried to pack the PPO with sympathizers who valued science over party directives. In other cases this occurred because the director, a true communist, saw growth in numbers as a sign of qualitative growth. But in most cases the institutes' party organizations grew because the institutes themselves grew, as did Akademgorodok, and because of central directives that fixed national targets for membership growth, temporary cutbacks, and more growth.

The relationship between secretaries of the various party organizations and Akademgorodok scientists was determined by their personalities and priorities. The secretary was less likely to interfere with scientific activities when other concerns predominated. The respective secretaries of the obkom, raikom, and gorkom (city party committee) used formal and informal means, persuasion, browbeating, connections, and even bribes to meet targets, expedite orders, and get around bottlenecks. If the secretary was an enlightened individual, he was inclined to give the scientists great leeway. Some secretaries met their responsibilities with culture and grace. They worked well with the scientists. Others were narrow-minded technocrats more interested in their careers than in Akademgorodok. Training in an engineering or agricultural specialty left them inclined to see the development of the productive forces, not scientific institutes, as their prime concern. For Akademgorodok's first decade the secretaries generally fit the former profile.

The cases of E. K. Ligachev and F. S. Goriachev illustrate this point. Egor Kuzmich Ligachev, later a member of the Politburo and a critic of Gorbachev's perestroika, occupied the position of first party secretary of the Sovet raikom when construction of Akademgorodok commenced. He was responsible for ensuring that Sibakademstroi came together as a competent construction organization and avoided meddling in scientific affairs. Ligachev, a native Siberian, began his long service to the Communist Party in Siberian Komsomol organizations in the late 1940s. A typical bureaucrat, he sought to facilitate Soviet economic development, believing that large-scale technologies connected with electrification and chemicalization were central to the construction of socialism. He argued that modern management techniques whose essence involved the centralization of decision-making powers in the hands of party officials were the key to economic growth. For Ligachev, "modern management" meant that the government should continue to fix what should be produced, in what batches, and at what pace. This also meant that enterprise managers would have significantly less flexibility than their western counterparts in meeting output targets. For workers, it meant repeated calls to raise "labor discipline."[3] Ligachev's efforts in Akademgorodok focused on forcing the pace of construction and in building a strong district party organization. He also gave some attention to improving bus services, which surely had a greater effect than did his speeches in improving the mood of the Siberian laborers.[4]

Ligachev's first duty to Akademgorodok was to turn the Ob Hydropower Station construction organization into the huge Sibakademstroi trust. It will be recalled that this was no easy task because of inexperienced workers and inadequate infrastructure. Ligachev criticized the slow pace and poor quality of Sibakademstroi work and strove to keep workers focused on the task at hand. He valued Lavrentev's decision to enlist scientists in lectures for Sibakademstroi employees as part of party-directed technological, political, and cultural education to raise the workers' "construction culture."[5] After a brief tenure in Akademgorodok, he was briefly secretary of the Novosibirsk obkom, then secretary of the neighboring Tomsk regional party committee from 1965 to 1983 from which he embraced the Brezhnevite "Siberia" plan and called for administrative measures to tame Siberia's resources. From that position he was promoted to Moscow by Iurii Andropov, later KGB head and general secretary of the party.

Under F. S. Goriachev, first secretary of the Novosibirsk obkom in the 1970s, relationships between the scientists and party apparatchiks remained professional. Goriachev was not content merely to pass on resolutions from above but strove to facilitate research. On the other hand, Goriachev and Lavrentev "did not get along well," according to Georgii Migirenko, the first head of the Akademgorodok partkom. "They were antipodes and did not respect each other." The same was not true for Marchuk and Goriachev. They were on the same wave length, especially with regard to the role of science in production. Together they wrote a series of joint articles on the subject.[6] On the whole, however, party apparatchiks too often interfered Akademgorodok's financial and political life. They were administrators, specialists in the history of the party, or Marxist philosophers who "just gathered papers, read them, and informed the obkom secretary of what they had learned. They knew nothing about science," Migirenko complained. Considering the worsening state of party-science relations during the Brezhnev years, Lavrentev told Migirenko, "We succeeded in finishing Akademgorodok just in time."

The Communist Party succeeded in infiltrating the scientific enterprise somewhat later than other areas of Soviet society. In the late 1920s and 1930s it tried to force scientists to join the party. It subjugated scientific institutes and professional associations to its organs of control. Pressure from below,

TABLE 7.1

Party Membership among Leading Scientists at Akademgorodok, 1964

	Total	Party Members	Percentage
Scientific Workers	666	297	44.5
Academicians	10	6	60.0
Corresponding Members	26	14	53.8
Doctors of Science	69	40	58.0
Candidates of Science	561	237	42.3

Source: Moletotov, "Problema kadrov sibirskogo nauchnogo tsentra i ee reshenie (1957–1964 gg.)," in Shchereshevskii, *Voprosy istorii sovetskoi sibiri*, pp. 346–347.

from increasingly class-conscious young communists, accelerated the pace of infiltration. Try as it might, the party was unsuccessful in attracting natural scientists for membership. For instance, Leningrad, home to the Academy of Sciences and other leading research institutes, was the center of scientific activity, but in 1929, of more than 5,000 scientific workers in the city, only 39 were party members; of 25,286 scientific workers in the entire USSR surveyed in 1930, only 2,007, or 7.9 percent, claimed party membership, and of these only 8 percent worked in the exact sciences.[7]

In the war and postwar years leading scientists were encouraged to join the party, and many did. The exceptions—Petr Kapitsa, Igor Tamm, and others—prove the rule. In natural scientific institutes roughly 13–18 percent of all scientists were party members. In social science institutes the percentage was somewhat higher. As many as 60–70 percent of laboratory heads belonged to the party and almost 100 percent of institute directors. Obligatory party membership for institute directors was largely a feature of the post-Stalin era. Party membership was important for career advancement and for such perquisites as foreign travel, conference invitations, and access to better goods and services.

In Akademgorodok, Communist Party membership among scientists follows a similar pattern. The party boasted membership of roughly 20 percent of the scientists depending on the institute, with 60 percent of the supervisory personnel (including most candidates and doctors of science). Many of these individuals were devoted communists; others joined to serve their careers. From the founding of Akademgorodok, party leaders pressed the cells in the institutes to take in new communists in order to ensure its control over the fledgling city of science. Between 1957 and November 1963 the partkom grew rapidly from 250 to 1,655 members (and 77 candidates), with 85 percent of all institute directors and deputy directors for science party members (see Table 7.1.).[8]

The party organization of the Sovet district differed from most districts in the USSR in its preponderance of white-collar professionals—over 60 percent. This is not surprising since the largest employer was the Siberian division of

TABLE 7.2
General Profile of the Sovet Raikom, 1966–1982

	1966	1970	1976	1979	1982
Women	969	1,323	1,698	—	25.6%
Workers	1,222	—	—	25.2%	—
by social origin	—	2,125	2,463	—	—
by occupation	—	28.3%	1,723	24.6%	—
White Collar	2,710	—	—	61.5%	—
by social origin	—	2,593	4,171	—	—
by occupation	—	4,407	—	63.2%	—
of whom scientists and engineers	823	39.4%	1,776	15.7%	14.9%
of whom "leading personnel"	773	1,074	—	—	—
Peasants	152	229	223	—	—
Total	3,790	5,554	6,825*	7,624	—

Source: PANO, f. 269, op. 1, ed. khr. 209, 11. 1–2; op. 4, ed. khr. 24, 11. 72–74; op. 7, ed. khr. 45, 11. 1–1 ob., 10–12, 16, 26; op. 12, ed. khr. 44, 11. 1, 94–95; and op. 18, ed. khr. 2, 11. 36–38. Not all data are provided in archival materials.

* This is an average figure as the statistical account is unclear on this figure, with the total given variously as 6,774 and 6,857.

the Academy of Sciences. The working-class contingent in the Sovet raikom from Sibakademstroi grew rapidly but was half the national average. Within a year of its organization in 1959 the Sibakademstroi organization had grown 3.5 times, with 19 party cells and 485 communists (and 76 candidates). Two years later there were more than 1,000 communists. The Sibakademstroi party organization gained the rights of a raikom in 1968 (see Table 7.2.).[9]

The membership of the Novosibirsk regional party organization closely reflected national trends, balancing the white-collar membership of the Sovet party organization. Throughout the 1950s about one-third of the members of the Novosibirsk obkom were of working-class origin. In the 1960s their numbers increased to 45 percent and grew only slightly in the 1970s and 1980s to around 47 percent. There was a rapid decline in the percentage of members of peasant social background from about 20 percent in the early 1950s to 15 percent in the early 1960s to 12 percent in the early 1970s to around 9 percent in the 1980s.[10]

As for virtually all other areas of Soviet life, so in party matters women received equal rights and responsibilities in slogans only. The Novosibirsk obkom and Sovet raikom were no exceptions. The percentage of women in the Novosibirsk party organization reached a high of 32 percent during World War II when so many men were at the front. By 1960 a little less than one-fifth of party members were women, and in 1980 a little more than a quarter. For the Sovet raikom, at times the percentages of women are slightly higher than for party organizations throughout the nation.[11] In terms of the nationalities of

TABLE 7.3

Nationality of Members of the Sovet Raikom and the Novosibirsk
Obkom Party Organizations, 1965–1976

	Sovet Raikon			Novosibirsk Obkom	
	1966	*1970*	*1976*	*1965*	*1976*
Russian	3,301	4,842	6,052	101,928	119,495
Ukrainian	233	343	364	5,743	6,029
Belorussian	46	61	68	1,289	1,284
Jewish	98	133	131	1,884	1,816
Tatar	15	48	943	1,130	
Chuvash	10	19	393	424	
Azeri		6	11		
Armenian		9	15		
German		25	43		
Georgian		16	7		
Total	3,790	5,554	6,825	115,397	134,213

Source: PANO, f. 269, op. 1, ed. khr. 209, 11. 1–2; op. 7, ed. khr. 45, 11. 1–1
ob., 10–12, 16, 26; and op. 12, ed. khr. 44, 11. 1, 94–95. Data for the Novosibirsk
region come from Lubenikov, *Novosibirskaia organizatsiia KPSS*, pp. 50–51.

party members there was one striking trend. The percentage of Jewish members declined steadily throughout the 1960s and 1970s, from around 2.6 percent to under 2 percent. This seems to reflect a conscious policy of the central
apparatus to limit their numbers in a period increasingly marked by anti-Semitism (see Table 7.3.)

There were no doctors or candidates of science in the Novosibirsk obkom
until 1950, and their numbers grew very slowly until the founding of Akademgorodok. The percentage of party members who were doctors of science grew
from 0.02 percent in 1950 to 0.04 percent in 1960 to 0.12 percent in 1970 to
0.22 percent in 1980. The percentage of party members who were candidates
of science grew from 0.1 percent in 1950 to 0.4 percent in 1960 to 0.9 percent
in 1970 to 1.4 percent in 1980. Of course the vast majority of doctors of science
and a third of the candidates of science lived in Akademgorodok. Not surprisingly, the founding of Akademgorodok contributed to an increase in the number of party members with specialized education. In 1966, 13.6 percent of the
Novosibirsk regional party organization had a higher education, whereas 37
percent of the members of the Sovet raikom had a higher education. In 1976,
21.9 percent of the Novosibirsk party organization had a higher education,
whereas 41 percent of those in the Sovet raikom had a higher education, both
far exceeding average levels throughout the rest of the country (see Table
7.4).[12] In general, the percentage of communists of the Novosibirsk party organization connected with science grew from 1.2 percent in 1958 to 2.0 percent
in 1960 to 4.3 percent in 1970 to 5.3 percent in 1980 (see Table 7.5).[13] Archival
documents indicate that workers were more likely to hear the party's political,
ideological, and social messages than were the scientists.

TABLE 7.4

Educational Level of the Members of the Sovet Raikom, 1966–1976

	1966	1970	1976
Doctors of Science	61	107	180
Candidates of Science	355	429	592
Higher Education	1,386*	1,921	2,766
Unfinished Higher Education	76	97	158
Middle School Education	1,118	1,689	2,229
Total	3,790	5,554	5,825

Source: PANO, f. 269, op. 1, ed. khr. 209, 11. 1–2; op. 7, ed. khr. 45, 11. 1–1 ob., 10–12, 16, 26; and op. 12, ed. khr. 44, 11. 1, 94–95.

* In 1966 of those with higher education, 595 were engineers and architects, 60 agricultural specialists, 88 doctors, 70 economists, 253 educational specialists.

TABLE 7.5

Absolute Number of Communist Party Members Engaged in "Science," Novosibirsk Obkom, 1958–1980

Year	No. of Party Members	Year	No. of Party Members
1958	881	1970	4,645
1959	1,171	1971	4,948
1960	1,583	1972	5,081
1961	2,003	1973	5,330
1962	2,543	1974	5,446
1963	2,984	1975	5,569
1964	3,436	1976	5,761
1965	3,534	1977	5,919
1966	3,960	1978	5,945
1967	4,195	1979	6,093
1968	4,517	1980	6,291
1969	4,645		

Source: Lubenikov, Novosibirskaia organizatsiia KPSS, pp. 60–61.

THE PARTKOM OF AKADEMGORODOK

In the 1920s the central party apparatus gradually exerted full control over its Siberian counterparts. Yet Akademgorodok was initially permitted to have its own party organization in keeping with its special mission and unique status, and this provided relative independence in dealing with higher party organizations. The members of the Akademgorodok partkom included the secretary of the university party organization, the main engineer for capital construction, the secretaries of the primary party organizations of all the institutes (each institute party organization had between ten and thirty members at this early stage), the director of the housing and "cultural" fund, and of course representatives from the presidium of the Siberian division, eighteen positions in all. The partkom's own staff handled all bookkeeping, financial matters, political education, and membership tasks. The partkom implemented policy through ten commissions designed to respond rapidly to directives from Central Committee plenary sessions which included commissions to control industrial and agricultural assimilation (*vnedrenie*), lecture propaganda, political education (*politprosvet*), and competition for communist labor in science. The commissions should have worked well, drawing on the authority of a number of leading scientists who were among its members,[14] but they maintained a rather low profile.

The partkom was intended "to guarantee the effective work" of the apparatus of each department under its jurisdiction—personnel, foreign matters (usually involving a KGB representative), chancellery, planning and finance, and so on—and to assist the presidium of the Siberian division in considering measures to improve the state of scientific research through directives on planning, funding, and innovation. Yet the usual problems of construction delays, inadequate supply of machinery, equipment, instruments, and chemicals, and other bottlenecks always waylaid its efforts. Any criticism the partkom received for these problems it simply passed on to the institutes under its jurisdiction. It did not matter that the difficulties had their origin in the heavily bureaucratized and overly centralized Soviet system of management and supply.[15]

Georgii Sergeevich Migirenko (b. 1920) was first secretary of the partkom. He trained under Leonid Ivanovich Sidov, a pure mathematician and a student of Lavrentev's, and then worked in the Institute of Mathematics in the department of the future president of the Academy of Sciences, M. V. Keldysh. A specialist in hydrodynamics, Migirenko was particularly interested in ship and submarine building. While serving in Akademgorodok, he was made an admiral, leaving his military colleagues in Leningrad bemused at how he had managed to be promoted by the navy while living in the most land-locked city in the USSR. A handsome, fit man, appearing much younger than his years, Migirenko viewed himself as an enlightened party official, thrust into the position of secretary of the partkom at Lavrentev's request, and content to return to active scientific research when the partkom's work was done. Yet he was a firm, unyielding administrator, instrumental in calming workers' dissatisfaction over having to build cottages for the academicians when they themselves lived in spartan barracks with only the promise of communal facilities and shared toilets. He admonished his scientific colleagues for failing to follow through on worker education, although several hundred had volunteered for *politprosvet*, or political education. At all times he embraced party jargon and never failed to thank Lenin for his clairvoyant guidance in considering any issue.[16]

As Lavrentev's right-hand man and as first secretary of the partkom, Migirenko was in the thick of things. He had to deal with the creation of Akademgorodok's party apparatus from scratch, with construction problems and Sibakademstroi, with attracting scientists and students to Akademgorodok, and with managing relations between all sorts of party officials. Most notorious from the point of view of the Akademgorodok party apparatus in general, and Migirenko in particular, were the various "verification commissions" that visited Akademgorodok to ascertain the state of construction, research, or education. Some were high-level, headed by Khrushchev himself, Central Committee personnel, or representatives of the Academy of Sciences in Moscow; for example, Academy president Anatolii Petrovich Aleksandrov visited Akademgorodok to push automated management systems, and then the "Siberia" program, accompanied by V. I. Vorotnikov, who represented the Politburo, and twenty-five ministers and deputy ministers.[17] They were concerned with Akademgorodok's general performance. Others had specific tasks, such as examining the rumored resurgence of "Weissmannism-Morganism" in the Institute of Cytology and Genetics to the consternation of the Lysenkoists and Khrushchev. No doubt with Lavrentev's blessings, Migirenko protected the biologists whenever possible. When several scientists at the Botanical Garden (under the jurisdiction of the Institute of Biology) expressed the view that they favored Lysenko's science, Migirenko threatened to interfere with their work unless they kept their views to themselves.

The relatively quiet relationship between the Akademgorodok party apparatus and the regional party apparatus went from manageable to worse with Khrushchev's ouster. According to Migirenko, obkom party secretaries who saw in their responsibilities common goals with the scientists gave way to those who resented Lavrentev's rapport with the Central Committee. Even in the early days they believed that Akademgorodok usurped their economic and political rights, and drained scarce resources needed by industry and agriculture. Occasionally some local party official confiscated Akademgorodok construction supplies for projects in his region, claiming: "I took the materials and [railway] wagons properly. If you need more, order more." Lavrentev and the others had to fight to get back the materials that were rightfully theirs.[18] More critical was a national change in the philosophy of scientific management that accompanied Khrushchev's departure. The secretary of the Central Committee responsible for science under Brezhnev, Trapeznikov, was accustomed to using the word *administer* (*rukovodit'*) when referring to science rather than *help* or *support*. "The party didn't understand," Migirenko said, "To them science was just one aspect of various slogans." Referring to a work by Tommaso Campanella, a philosopher and poet who was incarcerated during the Spanish Inquisition, in which he describes a utopian society without private property and where reason and community reign, Migirenko added, "If only they had read Campanella's *City of the Sun*."

The decision in 1965 to disband the partkom of the Siberian division and incorporate what remained of its bureaucracy into the Sovet raikom reflected both changes in attitudes about the best way to administer science and quantitative changes. The city of science had grown to thirty thousand people and twenty-two institutes, and administered social, medical, and cultural services, housing, and food stores. The partkom could no longer manage all these activities. It was better equipped to deal with questions of scientific policy: the direction of research, funding, hiring, educational programs, and so on. Once the main construction of Akademgorodok was completed, the regional and central party apparatus determined that the partkom was redundant, especially since the rights of party committees had been given to PPOs in the larger institutes and the university. More crucial, the existence of the partkom created difficulties in questions of authority since the Novosibirsk obkom thought it should control the *nomenklatura*. A department of science with a salaried secretary was thereupon created in the Sovet raikom to assume the partkom's responsibilities, and the raikom took over supervision of the activities and membership of the PPOs.[19]

As a rule, the raikom was seldom satisfied with the work of the PPO in any institute. At each meeting where the performance of PPOs was considered, an official mechanically read a report for a given institute into the record that presented data on attendance at meetings, progress in attracting scientists to join the Communist Party, efforts to publicize plans and ensure that targets were fulfilled, and the number of articles published and talks and seminars given. These data were inevitably applauded and then just as inevitably followed by the "however (*odnako*) comment," a criticism of various aspects of institute life—from scientific to social to party matters. With regard to the Institute of Catalysis, for example, party secretary E. G. Grazhdannikov pointed out that only one of sixteen laboratories was actively engaged in "socialist competition," that the PPO was ignoring its responsibilities regarding the "communist upbringing" of impressionable workers, providing guidance for the Komsomol, or seeing to it that scientific results found their way rapidly into the production process.[20] The *odnako* comment was part of the Soviet tradition of engaging in "self-criticism."

In another case the raikom criticized the Institute of Nuclear Physics for rarely planning cell meetings beforehand, not keeping minutes of meetings (which could be used by higher party organizations as evidence of failure), and never engaging in "self-criticism." Worse, the institute's partkom had failed "to ascribe any significance" to these tendencies. The worst transgression still was the fact that these mistakes "had a chronic character." Perhaps because the institute's scientific reputation was secure the directors felt immune to this criticism. Political education was indicated as a tonic. The raikom instructed the PPO to improve its organizational work and to keep detailed accounts of its activities.[21] The nuclear physics institute continued to get poor grades for failings in scientific-organizational and ideological work throughout its history.

The PPOs dealt with social issues in concert with the local trade union organization, or mestkom. The social concerns included labor discipline and drunkenness, housing, schooling, transportation issues, and organizing workers and scientists to participate in "voluntary" group activities such as "*subbotniki*" (using a *free* Saturday to do labor-intensive spring or fall cleaning in an institute), "*kartoshki*" (manual labor at a nearby state farm to help in the harvest of potatoes), and socialist obligations and competitions.

With regard to science policy, from the PPOs on up responsibilities included making sure that directives from higher levels were implemented and that annual and five-year plans for research, innovation, and assimilation were met. The PPOs paid attention to manpower issues in so far as they concerned party membership, usually leaving the final decision about the qualifications of personnel and hiring in the hands of institute directors. Still, the party committee could block the hiring or promotion and force the firing of persons of whom it did not approve for whatever reason.

Once Akademgorodok was fully operating and research proceeded as a matter of course, the party apparatus became more concerned with how well scientists conformed to the ideological precepts of the Brezhnev era. These precepts raised the specter of "proletarian science" as distinct from capitalist science, and fed on the growing tension between the United States and the USSR. The party sought to regulate foreign contacts and literature ever more carefully, turning inward and abandoning Khrushchevean internationalism. The ideological precepts gave fodder to those officials throughout the nation who wanted to use party mechanisms to exert greater control over scientific research. Specifically, the precepts stressed the so-called science-production tie and increasing concern about the failure of industry to be innovative. They involved the exhortation to focus research efforts on Siberian development and BAM. Clearly, in the eyes of party officials, the development of Tiumen oil and gas, Kuznetsk coal, and Siberian forests were far more important than fundamental research. In order to understand all the implications of these policy changes for Akademgorodok, we need to turn briefly to an overview of the politics of science in the USSR under Stalin.

SCIENCE AND POLITICS IN SOVIET RUSSIA

Most of the policies toward science that stressed applied research, autarkic scientific relations, and strict control over the ideological component of science had roots in policies instituted under Stalin in the early 1930s.[22] At that time, in an effort to harness the scientist to the machine of industrialization, the party centralized administration of science policy, in part by transferring most major physics and chemistry institutes to the jurisdiction of the Commissariat of Heavy Industry (known as "Narkomtiazhprom"). It subjugated all professional associations to its organs; its personnel infiltrated research institutes. The party's aim was to control the entire scientific enterprise, from the individual researcher to the commissariat by means of planning on an all-union scale.

Scientists and their institutes were forced to submit annual and five-year plans for their research, spelling out targets and research products. In doing so they had to avoid giving the impression that their research lacked application, was divorced from the needs of "socialist reconstruction," or resembled "ivory tower reasoning" so common under capitalism. It may indeed be that the party and scientists shared certain goals during the institution of the five-year plans. For example, both wished to expand the research enterprise and to introduce scientific achievements into production more quickly. Both recognized that government support was vital for increasingly expensive, large-scale research programs in modern scientific institutions. But scientists worried about the state's encroachment on their enterprise and feared premature identification of targets, arguing that it was often impossible to predict fruitful areas of research. The government's behavior was motivated in part by the fear of duplicating research efforts, but western observers have argued that the concentration of resources to avoid duplication in fact prevented competition between scientific centers and made Soviet science less dynamic in many fields than western science, while the emphasis on applications interfered with the effectiveness of basic research.

The party also pursued autarky in science so that regular scientific contacts with the West ceased. After such leading specialists as the biologist Theodosus Dobzhansky and the physicist George Gamov failed to return from western sojourns, the party ordered visas issued to other scientists only with exception. There was even danger associated with sending reprints to foreign colleagues; collaboration with the enemy might be alleged. Heightened ideological scrutiny, coercion, arrest, and execution were used to ensure compliance with state goals. From this point until the late 1950s, as a rule only "party scientists" were allowed to travel abroad.

In the sphere of philosophy, an increasingly stormy debate among Marxist philosophers over the relationship between Soviet philosophy—dialectical materialism—and science spilled over into the scientific community at large. The participants disagreed over how to apply the teachings of Marx, Engels, Lenin, and ultimately Stalin to the form and content of modern science. The resolution of the debate was that Stalinist ideologues acquired the authority to inform scientists which approaches were acceptable in the proletarian USSR. In genetics, the result was the rejection of the gene. In physics, various fields of mathematical physics, quantum mechanics, and relativity theory fell beyond the pale. As discussed elsewhere in this book, the results for other disciplines were nearly as severe.

Under Khrushchev, scientists succeeded in somewhat loosening the constraints of party control. As in other countries, the state increased funding for research and development in areas important to national security, international prestige, and economic growth, especially in such areas as nuclear physics and space research. Scientific and technological expertise became crucial to the political process of resolving disputes between officials or in advancing new programs. Scientists reestablished autonomy in several areas of their activity. They pushed for increased control over fundamental research and, in 1961, secured the transfer of technical sciences from the purview of the Academy of Sciences to the industrial ministries, ending a thirty-year period during which technology had been foisted on the Academy by Stalinist planners. Khrushchev signaled the end of autarky by calling for peaceful competition with the West and encouraging Soviet scientists to participate more regularly in international forums, for example, the Geneva conferences on atomic energy. On the philosophical front, scientists regained control from ideologues over discussions of politics, ethics, and epistemology. Finally, scientists helped resurrect constructivist visions of the future and were expected to play a leading role in the creation of the so-called material-technological bases of communism. Akademgorodok played a key role in this rejuvenation of science.

For Lavrentev and Akademgorodok, the November 1962 plenary session of the Communist Party Central Committee confirmed the basic notion of academic freedom central to the reform of Soviet science policy. At the session Khrushchev spoke of the "perestroika" of both the party leadership and the political and economic life of the country. Lavrentev took this message to his Siberian colleagues at a general meeting of the partkom, telling them that science was vital to this perestroika of society and that it would modernize industry. In an apparent reference to Lysenkoism, he argued that a further aspect of perestroika in science was the rejection of "false science." He believed that the 1961 reorganization of the Academy of Sciences required scientists to prove themselves by conducting research of central importance to Soviet society. Although this included big science and applied research, it also meant pursuing new and promising fundamental projects.[23]

Under Brezhnev this perestroika ran out of steam. Officials attempted to improve the performance of science and technology by administrative fiat. Military research came to dominate national programs. A conservative group of elder statesmen of science whose primary goals were to preserve their authority and pet research programs and to advance their students into positions of responsibility dominated policy making. Such leading scientists as N. G. Basov, a founder of quantum electronics, and A. P. Aleksandrov, president of the Academy of Sciences, pushed through half-hearted measures of reform. These measures—the formation of national bureaucracies like the State Committee on Science and Technology to coordinate research and development, the creation of hybrid scientific-production associations, and incessant calls to embrace the advantages of the developed socialist state during the ongoing scientific-technological revolution—amounted more to reform on paper than in reality. Party officials and scientific bureaucrats believed that central planning of research and development and command economic mechanisms should remain in place. The hope was that long-range planning would permit greater flexibility in research and development by tying it less closely to short-term economic needs.[24] The fascination with economies of scale contributed to the slowing pace of reform as it became a simple matter to fund nuclear fusion or space research but not more innovative, small-scale projects in such sunrise fields as computers and biotechnology.

From an ideological standpoint, insidious controls persisted. Pressure to focus on the solution of economic problems increased. Strict regulation of western scientific literature and contacts remained. Researchers were rewarded with travel abroad more on the basis of party affiliation than scientific quality.

Overt discrimination toward Jews and other minorities grew more pronounced as entry to universities and institutes was restricted. Under Brezhnev, the effort to innovate found response in party pronouncements about the need to perfect existing mechanisms and in the convocation of party-economic "aktivy," where party officials, economic planners, and scientists discussed what might be done. But party meetings at the regional or district level rarely addressed genuine reform, resorting instead to slogans. The result was the continued deterioration of Soviet science. A system capable of pioneering efforts in space (Sputnik), nuclear fusion (Tokamaks) and fission, and elementary particle and theoretical physics could not maintain a lead nor catch up in areas where it lagged.

Two major Central Committee resolutions that embody the essence of science policy under Brezhnev focused on Siberian science. On January 27, 1977, Kremlin leaders passed a resolution "on the activity of the Siberian division of the Academy of Sciences for the development of fundamental and applied scientific research, the raising of their effectiveness, the assimilation of scientific achievements in the economy, and the training of cadres." While acknowledging that Akademgorodok had become crucial to the nation's scientific and technological potential, the international authority of Soviet science, and the creation of new forms of cooperation between science and industry, the resolution called for redoubled efforts in all these areas to secure the rapid development of Siberian resources, agriculture, and industry. The resolution was followed by a general meeting of the Siberian division chaired by Marchuk, and then by party-economic aktivy throughout the region (Novosibirsk, Krasnoiarsk, Tomsk, Iakutsk, Irkutsk, and Ulan-Ude) over the next two months.[25] At its November 1979 plenum, the Central Committee seriously criticized the efforts of the Siberian scientists to meet its January 1977 instructions. It instructed Akademgorodok to identify paths of Soviet economic development on "intensive rails" and suggested the convocation of "Days of Siberia" to bring scientists and engineers together from branch industries, factories, and higher educational institutions to work together toward these ends.[26]

The development of industry and agriculture was the sine qua non of Soviet science policy. Large-scale, centralized projects directed from above such as "Siberia" and "BAM" were seen as the most efficient way to accomplish that end. When targets were missed, the party apparatus often fixed blame on the Academy of Sciences research institute, not the branch ministry, for the failure to innovate. In the case of Akademgorodok, the constant call for assimilation (*vnedrenie*) led to a significant change in the research atmosphere. Institute directors were concerned about securing financial support for fundamental research and yet had to make provisions in annual and five-year plans for "discoveries," "intended applications," and "successful assimilation" in industry. The "Siberia" economic program and BAM required Akademgorodok scientists to earn a growing share of their contract income from these programs rather than line item funding from the government's budget. Contract funding was essential to fundamental research, but scientists would have preferred direct support to contract sources that carried with them increased accountability to short-term economic goals. Instead seismologists, mining specialists, economists, sociologists, specialists in animal husbandry and plant breeding, and materials science people were expected to devote a lion's share of their resources to gain a better understanding of the problem of "Siberia." By 1972 one-sixth of the Siberian division's budget came from contract research, up four times over earlier years.[27]

THE INNOVATION BELTWAY

The effort to make government-funded research more accountable to state goals, so prominent under Stalin, Khrushchev, and Brezhnev, had much in common with the policies of most western governments in the twentieth century. The increasingly large cost of research, and the creation of massive institutes, in the Soviet case with hundreds of employees, made government funding crucial. In exchange for financial support, policy makers naturally expected scientists to be more accountable by conducting research with short-time horizons for applications.

Three factors distinguished the Soviet case. The first was a unidirectional devotion to applications that shook the historically strong tradition of reverence for fundamental research. This was nurtured by a state-sponsored ideology, Soviet Marxism, which at one time equated basic science with "empty bourgeois theorizing" or "ivory tower reasoning." The second factor was the disproportionate share of resources allocated to military research and development. This distorted the face of research and led to the liberal use of a "secrecy" label that slowed the dissemination of information not only to the West but throughout the country and across ministerial and academic lines. Bureaucratic barriers between science, education, and industry were bad enough. The absence of moderating forces, for example, scientists willing to speak against breakneck development or for science in the public interest, further distorted the picture. Finally, fundamental research in the USSR was subordinated to big science projects, such as BAM and "Siberia." Scientists in the West experienced many of these pressures for accountability, secrecy, and applicability but not to the extent that Soviet scientists did and not with such an overt ideological component. Further, western scientists had other sources of funding—foundations, industry, universities—through which they secured autonomy from state goals.

The science-production desideratum had an impact on all branches of science in Akademgorodok. Soviet scientists were expected to do the work that in the West is conducted in the private sector; they had to be discoverers, creators, innovators, entrepreneurs, and salesmen all wrapped into one. Scientists were also expected to take their ideas directly to the enterprise for adoption by the factory manager. But only the rare individual had the time, the interest, or the ability to go to the factory laboratory or collective farm in order to promote industrial applications from scientific discoveries. Enterprise managers, for their part, rarely approached the scientists at Akademgorodok for help in introducing a new technology. They hesitated to interfere with production knowing the costs of the failure to fulfill plans. Managers tended to look up the ladder for directives and not down the ladder to underlings on the shop floor. They viewed inventions as something one could put in a briefcase and carry anywhere to become part of the production process. Few of the human or institutional factors that inhibit innovation were considered.

Yet Akademgorodok's founders frequently stated their intention to serve state goals directly through the so-called innovation beltway. Planned in the earliest position papers of its founders, the innovation beltway was to be an intermediary between fundamental research and production. This was not the first time that Soviet scientists had taken the initiative to create a tie between science and production. The Leningrad Physical Technical Institute (LFTI) had set the trend by regularly sending its physicists to such leading factories as the Krasnyi Putilovets (Red Pathfinder), the Bolshevik, the Izhorsk Steel Works, and Krasnyi Treugolnik (Red Triangle) in Leningrad, and as far away as the Kharkov Electromechanical Works, to consult on questions of metallurgy, motors and transformers, and production problems. The institute also trained factory specialists to work in its laboratories, a method that met with some success. P. P. Kobeko and I. V. Kurchatov (later head of the Soviet atomic bomb project) reported that the workers of Red Triangle did not understand the insulating properties of ebony although they were engaged in its production. After factory workers studied ebony at LFTI, production miraculously improved,[28] although that may have been the result of the attention given to the workers.

Lavrentev had the LFTI experience in mind when he set out to find new forms for the organization of industrial research in Akademgorodok. His goal was to speed the transformation of scientific results into applications. The founders of Akademgorodok had succeeded in attracting a large group of promising young scientists. Could they now attract engineers and inventors to special design institutes? Could they provide an outlet for those scientists who wished to work in settings closer to industry? In the leading government newspaper, *Izvestiia*, Lavrentev proposed creating a network of design bureaus and experimental factories around such science cities as Dubna and Akademgorodok to promote scientific innovation, and he recommended the regular exchange of engineers and scientists across academic and ministerial jurisdictions.[29] The special design bureaus ("*spetsovki*"), or SKBs, were the "innovation beltway," the third of Lavrentev's grand organizational schemes for Akademgorodok (the other two being cadres and an interdisciplinary approach).

Lavrentev had reason to believe that the innovation beltway would work. He had the example of the Fakel (Torch) experiment (see chapter 4), in which young komsomol scientists sold computer programs, shaking the national authorities with their impropriety of profits, market mechanisms, and the accumulation of wealth outside appropriate party channels. Lavrentev wrote:

In my view the experience of the young people's scientific production association, "Fakel," which exists in Akademgorodok under the jurisdiction of the raikom of the Komsomol, is quite interesting. There are about 800 people in this self-financing organization, 250 of whom are students . . . Since its founding in June 1966 the association has fulfilled more than a hundred contracts for a sum of more than 2 million rubles. The income from the firms is used by the Komsomol raikom for socially appropriate matters. For example, they are building a sports complex, [working] with pioneers, and have suggested building a Youth Club.[30]

For Lavrentev, Fakel was an indication of what the *spetsovki* might achieve.[31]

The chairman of the partkom, Migirenko, commenced the effort to attract young innovators to the Akademgorodok design bureaus with an article in the Communist Youth League newspaper, *Komsomol'skaia pravda*. He pointed out how scientists at Lavrentev's Institute of Hydrodynamics had already developed water cannons for use in mining, but they failed to get Kuzbas and Donbas mining organizations to use them, in spite of the promise of huge increases in productivity. Migirenko wondered if it were not easier for a scientist to make a discovery than to find a place for that discovery in Soviet industry. Indeed, in the first five years of the 1960s the Siberian division had produced more than 350 innovations, but many languished or were transferred to industry only on paper. Like Lavrentev, Migirenko called for Akademgorodok Komsomol organizations to establish standing committees in concert with design bureau and factory engineers to ensure the dissemination of up-to-date scientific information and make the innovation beltway a reality.[32]

The leadership of Akademgorodok—Lavrentev, Trofimuk, Liapunov, Migirenko, and others—all believed that the Council of Young Scientists, which operated within the Komsomol under Liapunov's supervision, was the crucial link between science and production. The council had created Fakel. Mirroring the social activities of Pod Integralom, the council and Komsomol organized a series of university clubs for dancing, politics, and the sciences, for example, "Kvant," the physics club. The council and Komsomol were dominated by younger scientists who tended to be free-thinking and progressive by Soviet standards. It was likely that the impetus to implement reform and innovation would originate with them. The council supervised the olympiads, summer school workshops, and the boarding school, and could identify promising young engineers. It sponsored conferences on all sorts of organizational, scientific, and other questions that were so popular that scores volunteered to work on program committees and local events and thousands of aspiring Siberian scientists came from as far away as Tomsk, Irkutsk, Omsk, Barnaul, Krasnoiarsk, and South Sakhalinsk to attend Akademgorodok events.[33] Unfortunately for Siberian industry, these young innovators never received financial or moral support to carry through beyond Fakel.

Lavrentev spent his last years selling the idea of the innovation beltway to ministries interested enough to pay for the creation of *spetsovki* near Akademgorodok. Ten initially signed on but within a few years the number had

TABLE 7.6
Special Design Bureaus (SKBs) of Akademgorodok

SKB	Year Founded
Hydroimpulse Technology	1964
Scientific Instrument Building	1971
Monocrystals	1978
Applied Geophysics	1979
Special Electronics and Analytical Instrument Building	1980
Computer Technology	1981

Source: N. A. Pritvits, and V. L. Makarov, Khronika. 1958–1982 gg. Akademiia nauk SSSR. Siberskoe otdelenie (Novosibirsk: Nauka SO, 1982), p. 296.

dropped to eight, and only six were ultimately formed (see Table 7.6). In the first years of the SKBs, Akademgorodok provided both the physical plants and the staff. In theory, the design bureaus were subordinate to both the Siberian division and a ministry. The scientists intended this arrangement to be flexible, but the ministries understood only too well that it was they who were putting up most of the finances and that they wanted to reduce significantly the time it took for an innovation to be assimilated. Not surprisingly, therefore, they viewed Akademgorodok's responsibilities quite narrowly.

To succeed, the beltway had to be accompanied by managerial reform, greater reliance on market mechanisms, and new incentives for innovation. Instead, the few industrial ministries that followed through on the promise to create the new design bureaus viewed them as the handmaiden of industrial production. A general lack of material and equipment and the absence of flexibility in funding were a great hindrance. Most of the investment in science and technology under Khrushchev and Brezhnev was in personnel and salaries, not equipment, leaving Soviet science in a perpetual state of physical poverty.[34]

Gurii Marchuk, who succeeded Lavrentev as chairman of the Siberian division, spent his tenure trying to get the innovation beltway to function more efficiently. He fully embraced Lavrentev's notion that there needed to be an intermediary between the Akademgorodok institute and the branch factory. The impact of Lavrentev's thinking on Marchuk about research, cadres, even his love of North Siberia and the Far East (which grew after he accompanied Lavrentev on trips to Chukotka and Magadan in the early days of Akademgorodok) was clearly illustrated in his book, Molodym o nauke (For young people—about science). He read a passage from the book to express to me his belief that the promise of the future lay in young scientists' hands and that the key to the success of a scientific community was its ability to attract talented, open-minded young people. Older scientists were more resistant to change and were often burdened by administrative responsibilities that steered them away from science.[35] Yet when it came to the innovation beltway, Marchuk's views turned out to be rather mechanical, and he did not see young cadres as being crucial to innovation.

Marchuk referred to the problem of creating a tie between science and production with the slogan "Vykhod na otrasl'," Roughly translated, it means how best to ensure the "export" of scientific ideas to branch industry through the innovation beltway. He had doubts about how responsible Akademgorodok should be in that process. The USSR had roughly a hundred thousand enterprises of different sizes that represented more than two hundred branches of industry. When specialists in an academy institute or university succeeded in exporting a scientific idea to one enterprise, it was the result of great energy and capital investment. At that point, Marchuk argued, the institute's work was over. It was inappropriate for fundamental researchers to supervise the same innovation in another one or two, let alone two hundred, factories, as the branch ministries expected.[36] The fault lay with the ministries' limited view of the importance of experimental science in the design bureaus. They should have been trying to turn those very factories into the "head enterprises of the branch and the basis for technological progress."[37]

Marchuk acknowledged that in a number of fields researchers had contributed to *uskorenie*, the speeding up of scientific technological progress. The physicists built synchrotron radiation sources with broad applications; the biologists tackled plant and animal husbandry with vigor; the mathematicians, computer specialists, and economists were engaged in modeling and management systems. Yet the Akademgorodok Komsomol alone seemed capable of overcoming bureaucratic barriers to innovation. The Komsomol organization in the Mining Institute developed a computer system for the Gornaia Shoriia Mine that calculated how much explosives were needed and what coal faces could be mined most efficiently. The result of this effort to introduce "the newest word of culture into production," Marchuk tells us, was a twofold increase in labor productivity and the likelihood that this experience would find its way into other mines. Young Akademgorodok scientists erected "the scaffolding of new construction"—BAM, the Enisei—Sianov-Shushensk Hydroelectric Power Complex, the oil industry of Tiumen, and the Ust-Illimsk Cellulose Factory—that spanned the entire republic "from the Ural Mountains to the Pacific Ocean." He believed, however, that further innovation required more capital investment, not more science.[38]

Akademgorodok signed agreements with five, then a dozen, and ultimately more than twenty ministries to build special design bureaus with experimental production facilities as the link between academic science and branch production. Marchuk did not want to see the number of *spetsovki* increase until all sides were clear as to what their roles would be. In 1966 Akademgorodok scientists and the Novosibirsk obkom gained reconfirmation of government support to build the innovation beltway, largely in the person of the chairman of the Council of Ministers, Aleksei Kosygin. Brezhnev and his followers, pointing to the USSR's rapid economic growth in the 1950s, called instead for measures intended to stabilize and perfect existing economic mechanisms, not its continued transformation. Ultimately only six *spetsovki* were formed in the settlements of Pravye and Levyi Chemy, Nizhniaia Eltsovka and Berdsk, all within 15 km of downtown Akademgorodok.[39]

Marchuk was critical of the conservative administrators who effectively killed economic reforms and essentially deadened the impetus in the innovation beltway. Yet he, too, took every opportunity to call for the "further perfection" of organizations responsible for transferring scientific advances into production and the development of automated management and scientific information systems, rather than true reform.[40] He was handicapped by three factors in his efforts. The first was his overwhelming faith that the computer was a panacea. He was unable or unwilling to see that the problems facing Siberia were social and complex, not machinelike and orderly. Many of his slogans, "Vykhod na otrasl'," for example, ring similarly of empty, mechanistic promises. Second, he did not have Lavrentev's authority, either with his scientific colleagues or among Communist Party officialdom, and he could not tame their reticence to push beltway innovations. Third, he could never win in the struggle to adopt new approaches against entrenched Soviet management techniques, economic administrators, and party bureaucrats. The current Siberian division chairman, chemist Valentin Koptiug, suffered from many of the same handicaps in his attempts at innovation.[41]

The effort to create a stable tie between science and production proved successful in one case. In the late 1960s Akademgorodok scientists embarked on a program of long-term cooperation with the Novosibirsk factory "Sibselmash" (Siberia Agricultural Machinery), a large enterprise that was a national leader in the production of farm equipment. Sibselmash was central to Siberian industrialization and collectivization efforts in the 1930s. It was converted to meet military needs during World War II and then returned to agriculture. In the late 1960s, like much of Siberia's industry dating to the prewar years, Sibselmash required extensive capital investment and reconstruction, a challenging task given that investment capital was growing tight because of the military's larger share. At the initiative of plant director F. Ia. Kotov, later a deputy minister, Akademgorodok scientists joined in the modernization of Sibselmash. Their tasks often concerned narrow, industrial issues like the introduction of robotics and automated assembly lines, not those one would associate with modern science.

The joint effort between Sibselmash, Akademgorodok institutes, and the Special Design Bureau of Hydroimpulse Technology (SKBGT) commenced in 1971 with the decision to develop high-pressured and high-speed technologies, automated management systems for machine tools, and industrial detergents, and quickly expanded into dozens of areas of study. By the end of 1975, 350 scientists, metallurgists, economists, and other specialists in twenty-four Akademgorodok laboratories were engaged in research on forty-four themes in consultation with Sibselmash managers in such areas as machine tools, a new detergent, "Termos," and a mechanized accounting system for analysis and control of personnel. When nineteen numerically controlled machine-tools were introduced at Sibselmash in 1974 most workers reacted with bewilderment. Soon, we are told, they overcame their "psychological barrier." This must have reminded officials of the early 1930s when inexperienced workers who had been forced into higher positions by virtue of class origin, not qualifications, destroyed equipment through mistreatment or misuse.[42]

Like the experiment involving the Leningrad Physical Technical Institute and Red Triangle, Akademgorodok-Sibselmash cooperation involved a special program to train upper-level students from the university and the Novosibirsk Electrical Technology Institute jointly at Sibselmash and in an Akademgorodok laboratory, after which the student would write his senior thesis based on the experience. By all indications, Sibselmash staff relished the opportunity to work with scientists. One worker said that "cooperation with scientists helped us to train our specialists to carry out experimental work in production conditions and to get them [the scientists] interested in the problem of the technological development of production."[43]

Brezhnev himself singled out Sibselmash for its achievements as measured by the early fulfillment of the ninth five-year plan (1971–75). He drew attention to the Sibselmash experiment with a letter in *Pravda* published on January 9, 1975, that referred to the "broad assimilation of the achievements of science and technology in close cooperation with the scientists of the Siberian division of the Academy of Sciences." For Brezhnev, these accomplishments confirmed that the system worked as intended. At the twenty-fifth party congress Brezhnev underlined the fact that a revolution in science and technology demanded "perfection of planning and economic stimulus in order to create conditions that facilitated the rapid penetration of new ideas into all links from invention to mass production."[44] He failed to understand, however, that in the Soviet system high-level attention alone was the guarantor of success.

Another example of successful *vnedrenie*, or assimilation, also concerned the Special Design Bureau for Hydroimpulse Technology, a spin-off of Lavrentev's Institute of Hydrodynamics. The effort centered on the modernization of the powder metallurgy industry in Novosibirsk through the assimilation of robots and manipulators, and the utilization of impact welding (the use of explosions for the welding and "stamping" of metal) at the Chkalov Aviation Factory, Siblitmash, the Kuzmin Metallurgical Factory, and Sibelektrotiazhmash. Powder metallurgy was founded on the effort to use filings and burrs for economy of materials.[45] Andrei Deribas, who moved to Siberia with Lavrentev in 1958 and is now the director of the SKBGT, spent a lifetime trying to understand better how to utilize explosions in the production process. Although SKBGT came into existence in 1965, it took until the late 1970s before any of its long-term accomplishments, for example, wear-resistant electrodes for spot welding, was successfully assimilated into industry, for which Deribas blames personnel as much as the ministries.[46]

Scientists in the Institute of Cytology and Genetics also participated in the effort to raise production through innovations in the agricultural sector. These activities (discussed in chapter 3) included soil science, plant selection, animal husbandry, and new melioration techniques. R. G. Ianovskii, first secretary of the Sovet raikom and F. S. Goriachev, the obkom secretary, joined forces with institute director D. K. Beliaev to focus biologists' efforts on the Brezhnev food program. Six Siberian division institutes worked on improving the soil and agrochemistry at the Iskitim State Farm, about 35 km south of Akademgorodok. The plant breeders produced the new Novosibirsk-67 grain, the Botanical Garden provided early potatoes, and the computer specialists developed a small management package for the farm. The physicists offered suggestions on how to improve the energy efficiency of the farm, and even the car park provided its expertise in areas of repair. The results were so successful that even during the terrible drought of 1977 in western Siberia the grain harvest was within 75 percent of normal yields.[47] But this assistance led to the expectation among collective farm managers and party secretaries responsible for agriculture that high-level scientific assistance was standard fare. The problem was that in cases like Iskitim, success was assured only because the Ministry of Agriculture, Minselkhoz, focused adequate attention and resources on the farm. At other times scientists themselves were apparently supposed to make up for inadequate resources and poor production. Indeed, as in the case of most successful "experiments" in the Soviet Union, the Sibselmash, Chkalov factory, and Iskitim sovkhoz efforts succeeded owing as much to the high level of attention they received as to any inherent qualitative changes.[48]

The problem was that other than the Special Design Bureau of Hydroimpulse Technology, the genetics institute, Siblitmash, Sibselmash, and the Chkalov Aviation Factory the search for a more effective tie between science and production rarely went beyond mathematical modeling, automated management systems, or various forms of reorganization. Cooperation between science and production was not systematic; it was one of a kind. Soviet industry backed away from any long-term programs for innovation when short-term plans were king. The party was hesitant to relinquish control to managers who might have wished to adopt new technologies. Scientists, for their part, were expected to toe the industrial line.

Hence the innovation beltway was doomed to failure. The ministries zealously guarded their priority in decision making. Since they paid most of the bills, they called the tune. Dual subordination of the *spetsovki* to the Academy and the ministries existed only on paper. The ministries' "narrow bureaucratic interests" limited the effectiveness of the *spetsovki*. Ministerial officials often blamed the scientists for the failure to force the pace of innovation. Their solution was always to resort to greater centralization of policy priorities in central bodies that strived to push innovation by proclamation.[49] Only occasionally would a bold communist assert that "centralized planning to assimilate scientific achievements is the long way around. No less important is initiative from below . . . from individual enterprises to the branch level."[50]

Finally, the obkom secretaries, who, in the last analysis, could use their contacts to overcome severe supply, equipment, and manpower bottlenecks, were concerned primarily with meeting industrial and agricultural targets imposed on them from above rather than striking out in uncertain territory. Some of the more sympathetic party officials realized that Akademgorodok's first order of business should be fundamental research. However, most viewed science and Akademgorodok merely as tools to increase industrial and agricultural production.[51] Obkom secretaries could write about the "fruitful cooperation" between Akademgorodok institutes and other sectors of the economy, but their real interests were in seeing outputs increase.[52] Perhaps part of the difficulty was that the conservative Brezhnevite administrator had not yet gotten over the scare of Fakel so that every innovation proposed in Akademgorodok, whether originating in Aganbegian's Institute of Economics or Lavrentev's Institute of Hydrodynamics, looked suspiciously like a flirtation with market mechanisms. In the ideological sphere, too, Akademgorodok experimented all too often with subversive ideas.

SCIENCE AND IDEOLOGY IN AKADEMGORODOK

Ideological controls also shaped the face of Soviet science. For the scientists, the most important tool to inspire agitprop (agitation and propaganda) were the philosophical-methodological seminars that were held in all Akademgorodok institutes. The Communist Party began systematically to infiltrate scientific research institutes in the late 1920s through the forerunners of these seminars—Marxist study circles.[53] At that time, when the sciences were dominated by older scholars who had received their training in the tsarist years, the study circles were the main vehicle for the proletarianization of science. Proletarianization would occur through the penetration of Marxist methodology and working-class cadres into research settings. This ensured that scientific workers were not only "natural" (stikhiinyi) materialists in their understanding of modern science but conscious dialectical materialists in all their work. Readings usually involved the classics of Engels, Marx, Lenin, and increasingly Stalin, and their relevance for physics, chemistry, and biology. Circles spread from Moscow and Leningrad to the rest of the nation. Eventually the party established study circles in all research institutes. Their purposes were to organize and plan scientific research, fight alleged bourgeois idealism in the sciences, train cadres with the proper methodological guidance, and accelerate the conversion of so-called bourgeois specialists to the Marxist worldview.

After Stalin's death, the importance of philosophical control in natural and exact scientific institutes abated a great deal. No longer would phenomena like the Zhdanovshchina, a postwar period of cultural reaction, threaten the very core of the scientific enterprise. Led by the physicists, whose successes in nuclear physics had given them great authority, scientists began to write much more freely about the philosophical content of their disciplines. On November 30, 1955, at a special session of the Academy's physical and mathematical sciences division devoted to the fiftieth anniversary of relativity theory, such leading physicists as I. E. Tamm, L. D. Landau, V. L. Ginzburg, V. A. Fock, and A. D. Aleksandrov, the latter a future scientist-philosopher in Akademgorodok, spoke about the philosophical implications of Einstein's work in positive terms. Academy scientists then suggested that a conference be held to establish the proper relationship between science and philosophy, essentially to criticize the Stalinist philosophical legacy. In February 1957, on the heels of the twentieth party congress, the academic secretary of the Academy of Sciences, A. V. Topchiev, reiterated the call for a conference. The Central Committee endorsed the notion, and the conference was held in October 1958 in Moscow. There leading scientists from around the nation set the philosophical house in order with a series of papers arguing that they, and not Stalinist ideologues or policy makers, should resolve issues in the philosophy of science. Only biology remained beyond the pale until 1965.

After the 1958 Moscow conference, philosophical seminars served as a forum for much more far-ranging discussions of epistemology and science, and occasionally domestic politics and foreign relations, than in the Stalin years. Scientists began to pay lip service to the pronouncements of Marxism-Leninism. Their discussions reflected recognition of a new world order, one in which peaceful competition between socialism and capitalism, not the inevitability of war, predominated. Soviet scientists were now expected to compete with western scientists in all fields and to win scientific preeminence. But for Akademgorodok, a place where academic freedom reigned, the philosophical seminars paradoxically served to extend party control. This is because in a city filled with so many impressionable young scholars party officials wanted to be certain of ideological hegemony. Hence the seminars were expected to provide "more than philosophical generalizations of the newest achievements of science, but also to educate the young scientific workers in the spirit of militant materialism and to help them master the Marxist dialectical method and know how to utilize it in their research."[54]

Within the apparatus of the partkom, and later of the Sovet raikom, a special office supervised "party education" within scientific research institutes. Its main activities involved the organization of agitprop through meetings, lectures, and the media. The media included both radio reports and newspapers, both official government newspapers and those of the institutes. Because of censorship restrictions, newspapers were permitted only in organizations with more than five thousand employees, so institutes normally employed "wall newspapers," that is, bulletin boards that were changed eight or ten times a year. Organizations as large as Sibakademstroi had their own weeklies. The party apparatus paid close attention to subscription levels of leading Communist Party publications, especially in periods of heightened international tensions, for example, after the invasion of Czechoslovakia, and to bulletin boards lest misinformation be spread.[55]

Agitprop meetings were organized for workers and scientists alike. They included special "evening schools" for party indoctrination and lectures by scientists on the newest scientific achievements that were intended to stir the masses to greater efforts in the construction of communism. This continued a tradition dating to the first days of the Russian Revolution when the Bolsheviks endeavored to reach recently demobilized soldiers, workers, and peasants, most of whom were illiterate, and who were streaming into the cities, through various mass educational forums. The authorities believed that this type of remediation was needed in Akademgorodok as well, since only a quarter of the workers had finished middle school and four percent had no schooling at all. The workers seemed starved for contact with scientists. On one occasion, after Budker gave a ninety-minute physics lecture, the audience questioned him for another two hours.[56]

Over the years scientists gave thousands of lectures as part of political indoctrination ("*politprosvet*") at enterprises, institutes, schools, collective farms, and higher educational institutions, and at special faculties for workers like the Novosibirsk Peoples' University and the hastily assembled evening University of Marxism-Leninism, where aspiring communists could study their glorious heritage and improve their qualifications.[57] Lavrentev, Sobolev, Trofimuk, Khristianovich, Prudenskii, Budker, Rumer, Migirenko, and many others lectured on such topics as "The Chemistry of Polymers," "The Use of Geothermal Water in the Economy," "Science Will Become to a Full Degree a Direct Productive Force," "Robots and Life." On one occasion the Stalinist nature-creator and forestry specialist G. V. Krylov lectured on "The Comparative Level of the Development of Forestry Science in the USSR and Abroad" from the point of view of the materialist dialectic.[58] It is unclear what the trees thought about this matter.

In 1959 the leadership of the Academy of Sciences established a central bureau to organize the philosophical-methodological seminars nationwide. The seminars were intended to lead to the "systematic expansion of the Marxist worldview among scientists toward raising the level of their scientific activity and attracting them to the ideological struggle on the front of natural science." Each institute was instructed to set up a seminar bureau, but usually an ad hoc body of party member scientists fulfilled its responsibilities. The Siberian division formed its own council to supervise the seminars. Early on, Aleksandr Danilovich Aleksandrov directed the natural science seminars. Members of his seminar committee included some of the most prominent Akademgorodok scientists, such as economist Abel Aganbegian, chemist G. K. Boreskov, future president of the Siberian division V. A. Koptiug, biologist R. I. Salganik, and future obkom secretary R. G. Ianovskii.[59] The mathematician Sergei Sobolev participated actively and headed the newly formed Akademgorodok Znanie (Knowledge) society for spreading social and scientific knowledge among the public.

Aleksandrov (b. 1912), a leading geometrician, former rector of Leningrad University, and devoted communist to this day, arrived in Akademgorodok in 1965. There are conflicting stories as to why he left Leningrad University. Aleksandrov says that Lavrentev invited him to transfer to Novosibirsk on several occasions, and he finally decided to accept the invitation when he had tired of his position as a university administrator. Others said that Khrushchev, after one too many run-ins with Lavrentev over Lysenko, planned to replace him with Aleksandrov. Another story held that Khrushchev intended to start up another huge science city near Moscow under Lavrentev, and Aleksandrov moved to Novosibirsk to take his place. But then Khrushchev was ousted, and Lavrentev remained chairman of the Siberian division until 1976. Still others maintain that Aleksandrov moved to Siberia to become a full academician, whereas in Leningrad his path was barred because of his outspoken defense of quantum mechanics, relativity, genetics, and cybernetics.

Aleksandrov, admired by communists for his devotion to orthodox dialectical materialism and for his hostility toward any perceived idealism, was a logical choice for chairman of the natural science methodological seminars. For scientists his orthodoxy served as a smokescreen for his defense of modern physics, genetics, and cybernetics. Aleksandrov argued that scientists should be permitted to do their work without interference from ideologues who did not understand science. At Leningrad University he was a friend of the dissident, biologist Raissa Berg, and sponsored some of the earliest postwar seminars on genetics and cybernetics. Later, in Akademgorodok, he took on the cause of Politaev who had written the first major popularization of cybernetics; Politaev had lost his laboratory because of the petty concerns of mathematics institute director (and notorious Stalinist) Sergei Sobolev. Yet when it came to the alleged ideological mistakes in the philosophy of physics committed by such colleagues as Mosei Markov, Sergei Khaikin, Evgenii Feinberg, and Iakov Frenkel, all of whom were Jews, he revealed deep-seated anti-Semitism that is barely hidden to this day. He accused those physicists of mistakes at a time when Kremlin physicians faced charges of attempting to poison the Communist Party leadership. The xenophobic attack on the physicians, many of whom were Jewish, also known as the "doctors' plot," triggered similar attacks on other prominent Jews throughout the country. The physicist Vladimir Fock admonished Aleksandrov for adding fuel to "uneducated" Stalinist ideologues in their polemics. Fock told Aleksandrov it was far more important to protect physics than to sharpen his knives. Later, Aleksandrov joined Sobolev in writing a letter to the Central Committee "urging stronger measures against dissident youth" after the podpisanty affair. Wishing to dampen the charge of anti-Semitism, he relates with relish how he, as secretary of the partkom of Novosibirsk University, saw to it that four student reactionaries, who, in a drunken stupor, beat up several Akademgorodok Jews, were expelled for their actions. He informed the raikom that they were all given the appropriate political education. "After all," he said, "this isn't Alabama."[60]

Aleksandrov recognizes that it was the Soviet system that made Lysenkoism flourish and mathematical economics meet with rejection. He blames bad communists for this, those who selfishly turned away from communist ethics toward careerism and from truth-seeking and self-criticism to dogma. To avoid this he stressed the essential need to recognize the moral and ideological content of science and to use the philosophical seminars to insist on mutual assistance, the open exchange of ideas, collectivism, communist idealism, and of course a tie between science and production.[61] Still, one wonders why Akademgorodok, a city dedicated to discovering the truth, required ideological supervision in order to do so.

In the early days of Akademgorodok participants in the philosophical seminars were expected to address the proposition that "Soviet science should occupy the leading position in world science in all regions of knowledge, [and] science must introduce the workers to scientific-technological creativity." Then increased contacts with western scientists, along with the desire to achieve a leading position in world science, encouraged the development of another theme: the effort to foster "communist labor" in science to engender feelings of responsibility among the scientists for the general welfare of the laboratory, the institute, and the entire country. Apparently these sentiments were in stark contrast to the selfish, profit-motivated nature of scientific research in the West.[62]

A major aspect of the drive for communist labor in science was the effort to develop "a new type of scientist," toward which every scientist was expected to aspire. In May 1962 the partkom even held a conference on the "temperament of the new type of scientist." More than thirteen hundred scientists gathered to discuss the qualities such an individual should have. They concluded that this new scientist would cooperate with capitalist scholars in a businesslike fashion but would largely compete for world leadership based on a new view of "partyness" (*partiinost'*) in science—conscious scientific activity conducted in the interests of the working class to raise the productivity of labor and improve the quality of life. The fight against idealism, mysticism, and religion, and the critique of capitalism, would continue. At the conference Trofimuk told the gathered throngs: "Future communist society is a society of scholars in the broadest sense of the word. At this very moment, here at this conference, not only professors and academics but also builders, engineers, agronomists, workers, and collective farmers participate in scientific creation."[63] Yet the May 1962 meeting turned out to be the last all-Akademgorodok theoretical conference on Marxist methodology. The presidium found that organizing these mass activities was onerous and largely put its trust in the institutes to take over *politprosvet*.

An adjunct of party control to the seminars was a series of specially created Soviet awards. Autarky curtailed the ability of Soviet scientists to attend conferences, publish in foreign journals, exchange reprints or letters, or compete for international prizes. The authorities therefore created Soviet counterparts to these activities. The scientists were forced to publish only in Russian-language journals that were largely inaccessible to their western counterparts in spite of the journals' quality. Soviet scientists competed for Stalin and Lenin prizes, for state prizes, for Orders of the Red Banner of Labor, Challenge Red Banners, and so on, and engaged in "socialist competitions," often after publicly stating their "socialist obligations." Needless to say, Akademgorodok scientists took their share of the prizes. Of course they would have preferred a

Nobel prize but only eight Soviet scientists won that prize, and this for work done primarily before World War II. Some argue that the small number is the result of the damage done to Soviet science by Communist Party policies. Whatever the case, the Soviet scientists had to be content with Soviet certificates, Soviet medals, Soviet rubles, and Soviet recognition.

Initially the award of a Lenin or state prize in Akademgorodok was an occasion for great celebration. In less than a quarter of a century Sibakademstroi workers garnered three honorary titles of "Meritorious Builders of the RSFSR" and one of "Meritorious Trade Worker of the RSFSR," one state prize, one Leninist Komsomol prize, five prizes of the Council of Ministers, seventeen Orders of Lenin, and eight Orders of the October Revolution. By 1982 such scientists of the Siberian division as Lavrentev, Marchuk, Sobolev, Budker, Deribas, Migirenko, Skrinskii, and Trofimuk had earned twenty Hero of Socialist Labor awards (the highest Soviet honor); forty-nine Lenin prizes; ninety state prizes, including two three-time winners—Sobolev and Khristianovich—and ten two-time winners; eighteen Council of Minister prizes; twenty-seven Leninist Komsomol prizes; and forty-six honorary titles of "Meritorious Scientist of the RSFSR."[64]

By the end of the Soviet period scientists were circumspect about these awards. Although they realized that the awards had been for doing good science, science at the world level, they also recognized that the awards did not carry the esteem of western prizes. In November 1989, for instance, physicists at the Institute of Nuclear Physics won a Lenin prize for their accurate measurements of elementary particles produced by electron-proton collisions. Still they sat around somberly in the VEPP-4 control room rehashing the results and told me that it really did not mean anything if they could not count on government support for future expansion.

In addition to awarding prizes, the state attempted to control the form and content of science through socialist competitions (*sotsialisticheskoe sorevnovanie*) and socialist obligations (*sotsialisticheskoe obiazatel'stvo*). Institutes participated in these friendly "socialist competitions," proclaimed their "socialist obligations," and sought such honorific titles as "Shock Worker of Communist Labor." Competition, however, was considered a characteristic of capitalist systems and hence inappropriate for a society in the final stage before communism, that of "developed socialism," a stage in fact that had been suggested by a quick-thinking ideologue under Brezhnev to postpone a foolhardy promise made at the twenty-second party congress that the USSR would achieve communism by 1980. In any event, the comradely socialist competitions, which dated back to the 1930s, were seen not as competitions characteristic of the West but as a way to overcome systemic barriers to innovation. They were essentially pro forma obligations to ensure economic, political, and social control in all Soviet institutions. To the scientists they were particularly meaningless and involved a veritable litany of competitions: those for the best scientific work among leading specialists, among young scientists, and among graduate students; "inspection competitions" for the title of "best laboratory"

and "best in a profession"; competitions within departments, between departments and laboratories, between institutes of the same profile; competitions for
the best plans and counterplans; and the dreaded "supplementary obligations"
which no doubt the institutes took on with some trepidation. Obligations were
especially popular, although not among the scientists, just before national
holidays or upcoming congresses, anniversaries, or birthdays of various lumi-
naries. In honor of the forty-sixth anniversary of the Great October Socialist
Revolution, for example, the Institute of Cytology and Genetics promised to
produce several hybrid corn strains ahead of schedule as well as several DNA-
steroid preparations.[65]

In quantitative terms the party's ideological controls on philosophy, economics, party history, and international relations were working well. By the
end of 1961 80 percent of approximately six thousand employees were involved in the *politprosvet* network of Akademgorodok institutes through
forty-two philosophical seminars and circles. By the 1966–67 academic year,
the system of party education was firmly established. Altogether seventy-eight
different seminars were offered that attracted party and nonparty people
alike—some thirty-four hundred in 1966. Forty-nine of the seminars focused
on methodological issues, four on Lenin and Leninism, and nine on international relations.[66]

In terms of quality, however, the seminars left much to be desired. The
scientists found them tedious. Iu. P. Ozhegov, deputy chairman of the partkom
council overseeing the seminars, acknowledged that they met irregularly, were
too crowded, ignored newer employees, rarely cooperated with one another,
and did little to improve the agitprop efforts of the institutes. Discussion
leaders were poorly trained, and most scientists with whom I spoke said the
seminars provided little of substance for their own work. Sixty percent of the
participants were social scientists; the natural scientists could not be bothered
to attend. Young Akademgorodok scientists lacked both the interest and philosophical background to apply Marxian concepts to their specific area of contemporary science.[67]

For its part, the local party apparatus was seldom satisfied with the work of
the methodological seminars. The Soviet raikom frequently drew attention to
"serious insufficiencies" in the seminars on Communist Party history. Nor did
the seminars on international relations adequately discuss issues concerning
"the necessity of peaceful competition between countries of different systems"
while attacking "false bourgeois theories."[68] It may be that just as interdisciplinary mathematical approaches were vital to the success of Akademgorodok, so
the imposition of a rigorous Marxian epistemology enriched fundamental research since it required scientists to debate in a common language. But the
seminars lost their importance once the construction of Akademgorodok was
largely completed and research shifted from basic science to applied science.
Further, far more important than the seminars to the Communist Party in
controlling the personality of research institutes was the right to hire and fire
and the power of the purse.

In the Brezhnev years the seminars became at once more political and more formal. The party took increasing interest in their activities by requiring detailed annual accounts of seminar results within the overall system of party education.[69] As with everything else in the Brezhnev era, the seminar participants spoke of the "perfection" of existing mechanisms and criticized bourgeois views rather than discussing fundamental change that might bring about improvements in those mechanisms. As Brezhnev grew on in years the seminars celebrated his faultless leadership, the correct direction of Soviet economic development programs, and the appropriateness of Soviet international behavior. The seminars were intended to communicate to scientists that Soviet involvement in Third World conflicts, the invasion of Afghanistan, and increasingly bellicose rhetoric toward the United States were the logical outcome of enlightened Soviet leadership. The seminars became forums for attacks on any perceived deviation from the norm, but their vitriolic tone represented failure. They remained poorly attended with little teaching of any substance, which led an official study to lament the "philosophical illiteracy" that continued to persist,[70] and in 1989 they ceased operation altogether.

IDEOLOGICAL SHOWDOWN

In the eyes of the increasingly conservative Brezhnevite officials who came to dominate Soviet politics after Khrushchev's ouster, all these ideological controls failed to create an atmosphere in Akademgorodok commensurate with that in Moscow and Leningrad. Openness reigned in the administration of institutions. At the social clubs scientists discussed forbidden topics in politics and the arts. The scientists' party organizations were insufficiently vigilant against perceived ideological deviation. Akademgorodok organizations allowed experiments that smacked of capitalism, for example, Budker's profitable sale of industrial accelerators and Fakel's multimillion-ruble software contracts. Profits were not even shared with Moscow. Even the incorporation of the Akademgorodok partkom into the district party organization had not measurably slowed the pace of these transgressions. Lavrentev assumed that the same rules that had prevailed under Khrushchev, namely, great leeway in behavior based on close personal ties, still held force. Such long-time, committed communists as Trofimuk had outrightly attacked policies on Siberian resource development. Loose cannons like Budker regularly spoke their minds. Clearly the novel research approaches of Akademgorodok scientists, especially the sociologists and economists, had ruffled feathers in Moscow— and all this in a city of young scholars, children of the twentieth party congress, a city frequented by international travelers and held up as a paragon of Soviet science.

In 1968 the personality of Akademgorodok was changed forever. Local, regional, and national party organizations extended their ideological scrutiny to all aspects of the city's scientific life in response to three events. The first, in

April, was the affair of the *podpisanty* (signatories), namely, the protest of forty-six Akademgorodok scientists against the regime's violations of Soviet law and human rights when it prosecuted four citizens for their legal protest. The second, in May, was the festival of bards—folk singers—whose protest songs critical of the Soviet regime were all too well received by the scientific community, provoking the wrath of party officials. The third, in August, was the invasion of Czechoslovakia by Warsaw Pact forces under the leadership of the USSR. There was a tradition in Russia of clamping down on perceived political liberalism when economic reforms were in the air. In this case the reforms were in Czechoslovakia, and they smelled of market mechanisms, bourgeois nationalism, and individual freedoms. Brezhnev, the party leadership, and the military repressed those forces with tanks. What followed was another Leninist tradition, that of dividing the world into two camps, good and bad, materialist and idealist, Soviet and hostile. Akademgorodok's liberalism fell into the latter camps.

For some time protest had been building against the Brezhnev government for its abandonment of de-Stalinization. A prominent event in the melding of dissent was the trial of the writers Andrei Sinyavsky and Yuli Daniel under Article 70 of the criminal code for "anti-Soviet activity." The dissenters called for an amnesty for all political prisoners and abolition of the death penalty. Aleksandr Ginzburg and Yuri Galanskov published the transcripts of the trial whose *samizdat* ("self-published" works in the USSR, often produced with typewriters and carbon paper) and *tamizdat* (works published abroad) were prose equivalents of protest songs. In England the transcripts appeared as *The White Book*. Then Ginzburg, Galanskov, Aleksei Dobrovolskii, and Vera Lashkova were also arrested and tried for protesting the treatment of Daniel and Sinyavsky.

In March 1968 forty-six Akademgorodok scientists, teachers, and students, perhaps emboldened by the openness of their environment, signed a letter of protest concerning Ginzburg, Galanskov, Dobrovolskii, and Lashkova. What was most troubling to the authorities was that a large number of young people had signed the letter, the party's efforts at political education notwithstanding. On March 27 the letter was read over Voice of America and then published in the *New York Times*. The publication abroad was clearly the KGB's doing, since the signatories sent the only seven extant copies by registered and certified mail to Leonid Brezhnev, the general procurator of the USSR, the Supreme Court of Russia, the chairman of the presidium of the Supreme Soviet, Nikolai Podgorny, and the editorial boards of *Komsomol'skaia pravda* and *Izvestiia*. Clearly a decision had been made to find a pretext for hitting Akademgorodok hard and getting rid of troublemakers. The obkom secretary, V. P. Mozhin, and the directors of the institutes were instructed to hold meetings to discuss the signatories' treason and mete out punishment.

This was critical in jelling the dissident movement in general and particularly crucial in the case of Andrei Sakharov. Sakharov learned of the incarceration of Ginzburg and the others in mid-1966 and became involved in several activities involving their defense. Sakharov and others say that the case of the four and the trial of Sinyavsky and Daniel "played a critical role in shaping public consciousness and in forging the human rights movement in our country." Sakharov wrote Brezhnev a letter on their behalf. Although the letter was not circulated, it "was a milestone" for Sakharov in that it was his "first intervention on behalf of specific dissidents." The authorities punished him. He lost his position as department head, and his salary was cut nearly in half. In 1967 Sakharov also learned of the destruction of Baikal through *Komsomol'skaia pravda* and *Literaturnaia gazeta* and actually telephoned Brezhnev directly to indicate his concern.[71]

On April 16, 1968, the bureau of the Sovet raikom took up the matter of the *podpisanty*. Trofimuk, Aganbegian, and other leading scientists joined Mozhin, the obkom secretary, to debate how to respond to the growing crisis. They were upset mostly that the signatories had shaken up the quiet life at Akademgorodok and that a storm was surely brewing, while others were clearly disturbed by the violation of Soviet social norms. Particularly disturbing was that a group of communists—Alekseev from the university; Kostitsyna, Rozhov, and Borisov from the economics institute; and others—had signed. Their actions were "politically harmful" and would be used by "organizations hostile to our country for ideological diversion." The bureau acknowledged that a few communists were against condemning the act but still called for censure since there were those "among the youth who do not understand how damaging this is."

They then turned to what they considered to be the source of the problem: undisciplined activity in the social clubs like "Under the Integral" and "Grenada" where "unprincipled ideological behavior, the pursuit of pleasure, and now and again amorality" were observed. Things often got quite interesting at Under the Integral. There were beauty contests, where the measurements of the young ladies were advertised; there were even prostitutes. The directors of these organizations even allowed intemperate political discussions and invited bards to wail their seditious songs. Worse still, some residents disseminated the lyrics! The Komsomol and the Council of Young Scientists were filled with "politically unhealthy" elements. It remained to get rid of these officials. Akademgorodok needed "to raise the level of communist conviction, feelings of Soviet patriotism, and proletarian internationalism." The Sovet raikom instructed institute directors and party officials to be more careful in the future in hiring employees and to pay more attention to instilling class and party consciousness in science. The rector and the partkom of the university were ordered to replace teachers in the humanities departments and the boarding school where so many of the *podpisanty* had worked. There was even talk of transferring the university's entire humanities department—long "considered the main breeding ground for sedition"—to Krasnoiarsk for punishment. In the end, such visible dissidents as biologist Raissa Berg were censured for "political irresponsibility," their allies fell under suspicion, and a number lost their jobs. Dozens received this treatment in Akademgorodok institutes. The party no longer tolerated outspoken political behavior.[72]

For party officials the *podpisanty* affair was the logical outcome of increasingly brazen activities in the Scholars' Club and the social clubs. Everything club members did seemed designed to calculate how far they could push the authorities and still get away with it. Art exhibits and folk concerts were at the top of the list. The founders of Akademgorodok seriously believed that the arts would flourish under mathematics. They permitted exhibits unheard of elsewhere in the empire, including the only Soviet show devoted exclusively to the works of Pavel Nikolaevich Filonov. Ianovskii, secretary of the Sovet raikom, approved this exhibition when Aleksandr Aleksandrov was the temporary chairman of the Scholars' Club. (Most of Filonov's paintings found a home in locked basement storerooms of the Russian Museum in Leningrad until the late 1980s.) Makarenko, the gallery director, organized shows of Falk, Neizvestny, Shemiakan, and Goya's exposé prints as well. In earlier years a raikom commission would come to take a look at the paintings, their hair would stand on end, and the matter would blow over. Filonov's paintings were hung in May 1968, but the exhibit was closed by the Novosibirsk obkom before most people got a chance to see it. Makarenko chose this time to plan an exhibition of Kandinsky or Chagall and then decided on Chagall. The authorities could stand it no longer. They arranged for two photos to be stolen from the club, then sent in local party big wigs and police. Makarenko threatened to send a telegram to the Minister of Culture that would have brought the KGB into the matter. Within an hour the police found the pictures (Makarenko had taken the precaution of putting the real ones in the safe). Within two weeks, however, Makarenko was removed as director, and he moved on to Moscow. Shortly thereafter, he was arrested for speculation in icons and other artistic treasures and received an eight-year prison term.[73] He died in emigration.

The festival of bards was bolder still. The letter of the forty-six had already been sent, yet the festival of bards went on as if nothing had happened. Almost thirty performers arrived, including Aleksandr Galich. It was nearly impossible to obtain tickets. People waited hours to hear Galich sing "Ode to Pasternak," "Ballad on Surplus Value," "Clouds," and other songs. Galich's performance in the Scholars' Club received a standing ovation. He was escorted back to the hotel after a banquet around midnight. But students from the university and the physics-mathematics school got word and insisted on hearing him perform as well. Galich gave another show at 2:00 A.M. in the movie theater where he again sang "Ode to Pasternak." This ballad refers to the announcement of the great author's death, not by the prestigious Union of Writers to which he belonged but by the Literary Fund (Litfond), a second-class official organization of writers. Sakharov later befriended Galich and became a regular visitor at his home, especially after Galich himself was expelled from the Writers' Union in December 1971 for anti-Soviet activity. Galich emigrated in the summer of 1974 and died three years later. Sakharov last spoke with him on October 9,

1975, the day Sakharov received the Nobel peace prize. On the day Galich was exiled to Norway, Berg was in Moscow for the funeral of Boris Lvovich Astaurov, an academician and leading defender of genetics through the difficult Lysenko years. Sakharov was one of those she met at the cemetery.[74] Here follows a translation of "Ode to Pasternak":

Memories of Pasternak

The administration of the Literary Fund of the USSR reports the death of the writer, a member of the Litfond, Boris Leonidovich Pasternak, which occurred on May 30 of this year, in the seventy-first year of his life, after a prolonged and serious illness, and expresses its sympathy to the family of the deceased.

 The only communication about the death of B. L. Pasternak to appear in newspapers, indeed in only one newspaper—*Literaturnaia gazeta*.[75]

> They've taken apart the funeral wreaths to make hearth brooms,
> For a half hour or so they mourned . . .
> How proud we contemporaries are
> That he died in his bed!

> And the hacks tormented Chopin's funeral march.
> And the farewell went solemnly . . .
> He didn't lather the noose in Elabuga[76]
> And didn't lose his mind in Suchan.[77]

> Even the Kievan "men of letters"
> Were on time for his wake! . . .
> How proud are we contemporaries
> That he died in his bed!

> Not just a bit over forty.
> But exactly at seventy—a proper age for death,
> And he was not simply some kind of pariah,
> But a member of the Litfond—a valued "deceased"!

> Oh, the fir boughs dropped their needles.
> His snowstorms have finished the death knell,
> How proud we swine are
> That he died in his bed!

> "It snowed and snowed throughout the world,
> All over . . .
> The candle burned on the table,
> The candle burned."[78]

> No, there was no candle at all,
> A chandelier burned!
> The eyeglasses on the mug of the hangman
> Sparkled brightly![79]

And the audience hall was yawning, and the audience was bored—
Meli, Yemelia![80]
After all, we're not sending him to jail on Suchan,
Nor to the "ultimate penalty"![81]

Nor to the crown of thorns[82]
Or to be broken on the wheel,
But like a stick smacked across the face—
Just to a vote.

And someone drunkenly inquired:
"What for? Who're they talking about here?"
And someone stuffed his face, and someone whinnied
At a joke . . .

We won't forget that laughter
Or that boredom!
We will remember precisely everyone by name
Who raised his hand!

    "The murmur died away. I came out onto the stage, leaning against
       the doorjamb . . ."[83]

And now the slander and arguments had fallen silent,
As if granted a holiday by eternity . . .
And over his tomb his persecutors have risen,
To mount a "guard of honor"! . . .
    Ka-ra-ul![84]

December 4, 1966

Yet even given the party's response to the *podpisanty* and the festival of bards, it was the Prague Spring and the August invasion of Czechoslovakia that had the greatest impact on Sakharov, other dissidents, and many of the free thinkers in Akademgorodok. Sakharov wrote, "What so many of us in the socialist countries had been dreaming of seemed to be finally coming to pass in Czechoslovakia: democracy, including freedom of expression and abolition of censorship; reform of the economic and social systems; curbs on the power of the security forces, limiting them to defense against external threats; and full disclosure of the crimes of the Stalin era." In this environment more than a thousand signatures were collected in defense of the group of four.[85] This frightened the KGB into action and a campaign was launched to end academic freedom in Akademgorodok. Galich responded to the invasion with a poem that included the lines "Citizens, the Fatherland is in danger. Our tanks are on alien soil!" This provoked a number of Akademgorodok's young people to put up banners and slogans throughout the city of science: "Freedom for Socialist Czechoslovakia!" "Invaders, Hands Off Czechoslovakia."
Even before the invasion of Czechoslovakia, the Akademgorodok clubs had

been closed down. Under the Integral opened again a few times when western scientists came to Akademgorodok for conferences, but it lacked the ambience of the earlier years. Indeed, Marchuk had directed Aganbegian to create a plan for the development of social services and cultural organizations at Akademgorodok. But planning for art and literature only further contributed to the stultification of Akademgorodok cultural life.[86] Attracting talented young scientists to Siberia was growing increasingly difficult. And now that the social clubs were closed, the excitement that had carried the older residents through the first hard years gave way to a feeling that Akademgorodok was no different from the rest of Russia.

An End to Academic Freedom

For the remainder of the Brezhnev years the party apparatus paid far greater attention to personnel matters, striving to keep young troublemakers out of Akademgorodok and regulating all contacts with the West. To Communist Party leaders Akademgorodok was a hotbed of anti-Soviet activity and a remnant of the misguided de-Stalinization thaw. The Central Committee instructed local party organizations to pay strict attention to Akademgorodok cultural, social, and scientific life.

Until Gorbachev's rise to power and the introduction of glasnost Akademgorodok increasingly came to resemble other towns and cities in the Soviet empire. V. A. Mindolin, who served as secretary of the Novosibirsk obkom in its final days in 1991, showed just how much things had changed under Brezhnev. Just after the invasion of Afghanistan, Mindolin, then second secretary of the university's party organization, spoke at the fourth plenary session of the Sovet raikom where he fully endorsed a recent Central Committee resolution drafted in response to growing international criticism of the invasion. The resolution called for heightened ideological vigilance and increased attention to political education. One had to be especially careful in a scientific town, he warned, because of the growing number of foreign contacts. An anti-Semite, Mindolin saw evidence that Peking and "Zionist organizations" were engaged in "libelous" anti-Soviet propaganda. He therefore called for renewed efforts to stress the peaceful aspirations of Soviet science and to counter those who falsely appealed to "human rights" in their criticism of Russia's policies.[87]

At the same time social problems endemic in Soviet society began to tell upon the productivity of Akademgorodok scientists. The major difficulty facing Akademgorodok since the day construction officially ended, and one that plagues it to this day, concerned inadequate housing and related social services. The local, district, and regional party organizations devoted significant effort to this problem, as they did throughout the USSR. But because Akademgorodok had only one "boss," namely, the Academy of Sciences, and not a slew of economic, social, educational, and other organizations as was characteristic of any other city, it fell precisely on the shoulders of the Siberian division to handle this problem. Akademgorodok did not even have the assistance of local factories, collective farms, and hospitals in this endeavor as these organizations had their own problems with workers and housing. Without Lavrentev's resolve and Khrushchev's protection, Akademgorodok had to engage in outright "socialist competition" with dozens of other organizations for scarce resources. The housing problem became acute in the mid-1970s. S. A. Arkhipov, president of the local union, pointed out that housing construction had slowed to such a degree that less than 100 square feet—less than the average space given to scientists in the European USSR—was being allocated for housing for each new inhabitant. Construction of summer rest homes and sanitoriums also lagged. The one thousand one-room "apartments" (8 by 10 feet, with efficiency and toilet) to be constructed during the tenth five-year plan would hardly do the trick.[88]

Was Akademgorodok losing its main attraction to young scientists in Leningrad, Moscow, and Ukraine, that is, the allure of better housing? Aganbegian complained how powerless Akademgorodok had become in dealing with the complexities of housing, libraries, movie theaters, food stores, clubs, and kindergartens. He suggested using computers to automate production and research in order "to limit seriously the number of workers in the institutes." The solution, he pointed out, was to slow the city's growth. In 1976 the institutes set out to cut the number of workers they employed, hence eliminating the need to provide them housing. Nearly four hundred workers left the economics institute. At the same time, however, local higher-educational institutions continued to graduate scores of students every year according to plan, and inevitably they were taken on for work. After all, unemployment did not exist under socialism. So the institutes continued to grow, and the pressure on available housing stocks remained.[89]

What happened to the glory years of Akademgorodok? Some observers believe that the decline of the city of science from its vibrant founding years was the natural result of aging. Its leaders were unable to continue bringing the best young people to Siberia. Akademgorodok had lost its attraction. Heading the list was the problem of housing. Moreover, under Marchuk and Koptiug, the leadership and administration had grown far more conservative. Formal channels—like elsewhere in the empire—had to be followed. Even the aesthetics of Akademgorodok gave way to the natural aging of its infrastructure. The town was becoming rundown as budgets provided far more for new construction than for repair. The creative impulse Lavrentev had given to Siberian science had also run its course. Scientists no longer needed to relocate to Siberia to develop new research programs. Colliding beam accelerators, genetics, and mathematical models had become standard fare. And perhaps the scientists had gotten fat since they knew their funding was fairly stable, if too low year after year.

Was it only a matter of aging? I think not. Party controls were crucial in changing the face of scientific and social life in Akademgorodok. Scientists had been able to ignore many of them. Discussions of foreign policy or the recent directives of the Central Committee rarely had a direct impact on research. The methodological seminars had lost the vigilance of the Stalin years. Scientists had sufficient autonomy to select research topics within the constraints of finance and manpower. Yet controls were insidious. The party apparatus had grown rapidly, with representatives in every group and laboratory. They were the ears and eyes of conformity to directives established in Moscow requiring that Akademgorodok science reflect Soviet norms of collectivism, conformity, and practice (the science-production tie). Akademgorodok had grown out of a Russian tradition of excellence in fundamental research. Now its finished laboratories were required to show at every step that science contributed to economic growth. Researchers lost their enthusiasm for work. Accountability to the state weakened the fundamental basis that had been the strength of Akademgorodok science. Paradoxically, the more the state required that Akademgorodok research devote itself to national development programs, the more it resembled that of institutes elsewhere in the USSR and the less likely it would have a significant impact on the production process.

Most significant were ideological controls over the social and cultural life of the scientific community. The *podpisanty* affair brought down the party's wrath on everything unique in Akademgorodok. The social clubs were closed. The bards voices were stilled. The paintings were taken down from the walls. The Council of Young Scientists, the social clubs, the round table in the Institute of Nuclear Physics, all symbols of Akademgorodok's vitality, freedom, and openness, remained just that, only symbols of the original intent and spirit of Akademgorodok.

THE ECONOMIC and political turmoil that envelops the countries of the former Soviet Union presents Akademgorodok physicists, chemists, biologists, and social scientists with a set of challenges they could not have imagined in the Soviet era. The political intrigues between President Boris Yeltsin and his detractors have diverted the Russian government's attention from science. The economic decay caused by high inflation and the uncertain transition to a market economy from a centrally planned economy threaten the very foundations of Russian science. Shortages of materials for research, inability to repair already antiquated equipment, and even dismissals, the closing of institutes, and increasing international isolation have resulted from these conditions.[1] For Russian science generally, and Akademgorodok in particular, scientists must weather political and economic uncertainty before they can begin to come to grips with the persistent legacy of Soviet science policy and its top-heavy, centralized and conservative administrative apparatus. This in turn requires seizing onto nascent trends of democratization and decentralization in Russian science.

Akademgorodok remains Russia's third most important scientific center after Moscow and St. Petersburg. Today there are more than four score corresponding and full members of the Siberian Division of the Academy of Sciences, more than 450 doctors of science, and around 4,000 candidates of science.[2] The city's institutes continue their tradition of scientific excellence. Akademgorodok is a magnet for international ventures in science and technology, and a gateway to Siberian resources for Western firms. It would seem that a market economy, political reform, and the outlawing of the Communist Party would create conditions in which the Siberian city of science could develop at long last, freed from the constraints of Soviet cultural, political, and economic institutions. Yet it should be clear by now that scientific institutions can never exist in the way Francis Bacon, Mikhail Lavrentev, and other visionaries intended, that is, as a utopia somehow removed from broader social, political, and cultural concerns. Indeed nascent market mechanisms and democratic institutions in Russia are far from a panacea.

Since the trends of openness and academic freedom were prominent in the organization and administration of Akademgorodok science from its early years, it would seem that the city of science is well poised to move into the 1990s. If government support in Akademgorodok's first decades was miserly, at least it was constant. If many scientists lamented the pressure to do research accountable to Siberian economic development programs, at least they had a vision of the future communist world and how their research might bring this world closer to reality. And if they resented the Communist Party's scrutiny of their public and private lives, many believed they were still relatively free to do research in an oasis of academic freedom. Akademgorodok's institutes and personalities were its strength. How will they cope with the circumstances of rapid change? As for Russian science, so for Siberian science: nothing is simple any more.

Akademgorodok must cope with a new Academy of Sciences whose leadership seems set on asserting firm control over its branches throughout the nation but whose funding from the government is uncertain. The once powerful Soviet Academy of Sciences, the policies of which shaped fundamental research for the entire realm, gave way to the Russian Academy of Sciences in 1991. The Russian Academy of Sciences absorbed virtually all the institutes and staffs of the Soviet Academy of Sciences, including the Siberian division and Akademgorodok. The Russian Academy of Sciences continues to be dominated by Moscow, in spite of the intention to elect new members territorially, in a word, to decentralize policy making from Moscow. A parallel effort to democratize the Russian Academy of Sciences, in part by giving corresponding members, and perhaps doctors of science, the voting rights and other privileges of full members, has stood still—making a distant memory of the early stages of perestroika when the rank-and-file scientists forced the presidium of the Academy to nominate Andrei Sakharov and other liberals for election to the Congress of the People's Deputies. In most cases the "scientific bosses" of the past still hold institutional and administrative power. Thus efforts to decentralize administration and funding, both vital to engender competition among Russian scientific centers, are difficult, as are efforts to encourage the development of new fields of research, especially in biotechnology, computers, lasers, and communications, and to stimulate economic growth. The membership of the Academy has in fact become more conservative and nationalistic.

Since Akademgorodok is a company town that is responsible for its entire physical plant—from institutes and apartments to roads and stores—all the tensions associated with reform and decentralization of the Academy are more pronounced. The chairman of the Siberian division of the Academy of Sciences, Valentin Koptiug, is striving to hold onto his power and that of Akademgorodok vis-à-vis the Academy presidium in Moscow and nascent political groups hoping to bring democracy to the city of science. A major sore point concerns not the future direction of research nor how it will be funded but rather the disposition of living space. Privatization of living space moves in fits and starts throughout Russia; it has caused great confusion in Akademgorodok. Generally speaking, possession equals ownership in Russia today. Rather than permit privatization of the apartments and houses that scientists, technicians, and custodial and municipal workers and their families already live in, Koptiug wanted the presidium of the Siberian division to control their disposition. Some scientists who share his views are resentful that family members who have nothing to do with science have gained ownership of desirable housing solely by virtue of blood ties, not by a direct contribution to science. Apartment managers who used to defray part of their rents through clean-up and maintenance activities now refuse to do their jobs since they believe they have assumed ownership of their apartments outright. The result is a visible decline in the physical stock of Akademgorodok housing and increasing filth and litter throughout the town. The town fathers seem resigned to relying on the Institute of Nuclear Physics, with its relatively flush capital construction budget, to expand apartment inventory.

The current housing crisis has merely exacerbated a problem dating back to the 1980s when Akademgorodok was already finding it difficult to attract the best young scientists as it had done in its glorious first decade. According to official surveys conducted by Akademgorodok sociologists, the scientific youth of Akademgorodok has long been frustrated with career options. In the last years of the Soviet regime fully three-quarters of young scientists believed that Akademgorodok failed to make good use of their scientific skills and offered little prospect for career advancement. A major reason for this dissatisfaction, and for the growing outmigration of thirty- to forty-year-old scientists, was that many had to wait ten to twelve years for their own apartments; they had to share flats with their immediate families or even with strangers. As the Soviet period drew to a close nearly a quarter of Akademgorodok residents were waiting for apartments, 13 percent for places in schools, and 9 percent for hospital beds. Indeed a quarter of those ill on any day were unable to see a doctor in a timely fashion. Mirroring the problems of early building under Sibakademstroi, the construction of stores, schools, day care facilities, hospitals, sports facilities, pools, and medical services also lagged significantly. In addition, personnel who had finished high school were in short supply, drastically limiting the number of available laboratory assistants, nurses, and kindergarten teachers, which further discouraged young scientists from staying in Akademgorodok.[3]

Conditions are no better for senior scientists. There is a growing tension between those who welcome the ongoing reforms in science for their promise of democratization in policy making and those who, with Koptiug, wish to keep the reigns of power firmly in their own hands. The conservative scientists resent the independence of such facilities as the Institute of Nuclear Physics owing to its long-term financial ties with the West. They believe that official contacts, especially lucrative ones, ought to be run through the presidium of Akademgorodok which can then take its share of the "profits." Although powerless to stop the attrition of young scientists from institutes, these scientists also oppose the establishment of independent scientific cooperatives outside the grasp of their administration. Many individuals see the hundreds of private ventures that are springing up as a promising alternative to exclusive state budget funding of research, and as an efficient means of facilitating innovation and raising productivity. But others are concerned that middlemen will profit most and that the government or Akademgorodok will end up underwriting initial expenses, not to mention training costs, through its facilities and funding without any form of reimbursement.[4]

In this environment some of the best Russian scientists in many fields are leaving Akademgorodok, which seems to be suffering more of this "brain drain" than either Moscow or St. Petersburg. Still the phenomenon has been exaggerated by observers in both Russia and the West. Although thousands of scientists have gone abroad, many Russian observers are well aware that there is a limit to the number of scientists who can be absorbed abroad and they think this limit has already been reached.[5] The gravest danger is that scientists will quit their research entirely for business activities within the country.

More important, many scientific research institutes have dismissed workers either temporarily or permanently; others have not paid salaries for months at a time. Some scientists do not even come to their laboratories. The best universities in Moscow, St. Petersburg, and Novosibirsk are having trouble finding adequate candidates for admission to their science departments. Young persons are attracted by the prospect of business careers, science having ceased to attract them, which may result at some future date in an insufficient number of scientists. Thus the real danger of this brain drain is that a generation gap is forming.

This unsettling picture of political uncertainty and economic crisis has had a direct impact on Akademgorodok science, its institutes, and its individuals. At the Institute of Nuclear Physics scientists have been buffeted by the unexpected cancellation of the Superconducting Supercollider in Texas. The institute's physicists were at the center of both the experimental and theoretical aspects of the project. They were supposed to supply expertise and equipment, including superconducting magnets. Now, like high energy physicists everywhere, they must hook into projects at CERN near Geneva, Switzerland, DESY in Hamburg, Germany, and at KEK in Japan, hoping to turn these contacts into an opportunity to keep the institute going. Because of budget shortfalls, only grudgingly will the φ-factory come on line, a project intended to produce φ-mesons that may reveal the reasons for the apparent imbalance between matter and antimatter in the universe, and once considered the central hope for the institute's future. Further hampering research is that many leading physicists, like plasma specialist Dmitrii Riutov, spend months at a time abroad. Still, the skills with which director Aleksandr Skrinskii and his colleagues have maintained financial contacts with foreign governments and programs make it likely that the Institute of Nuclear Physics will weather the 1990s, as it has past storms.

In spite of the importance of genetic engineering for agriculture and medicine, the Institute of Cytology and Genetics is on less certain ground. In some respects the institute never fully recovered from Lysenkoism, having had only a decade of relative academic freedom before being forced into more narrow applied research with equipment inadequate to the task. The institute director, Vladimir Shumnyi, and deputy director, Anatolii Ruvinskii, remain confident that they will find talented, young biologists to fill the positions vacated by retiring senior scientists. But in addition to many young people abandoning science for business, others have abandoned the institute to set up their own biological ventures that involve rudimentary genetic engineering problems. Worse still, since Siberia is no longer needed as a refuge from Lysenkoism, Moscow has solidified its grasp as the center of Russian biology.

The Computer Center stands poised to take advantage of a long-awaited computer revolution that has finally hit Russia, its major impetus the burgeoning market economy. Software and hardware firms have sprung up everywhere. Still, most of these cater to business offices and use Western equipment at costs prohibitive to most people. In March 1991, in response to a review I wrote that was published in *Science*, I received a letter from Russian schoolteachers and their students requesting information about possible funding opportunities to create a Center for Ecological Education and Information in Baikalsk near Lake Baikal. They wanted between fifteen thousand and thirty thousand dollars but said they would settle for old computer catalogs. For the Akademgorodok Computer Center, contacts with such former staff members as Vadim Kotov at Western computer firms provide welcome financial and moral support.

For the environmental sciences and the Siberian environment in general, the picture is more discouraging. The last programs to be funded during an economic downturn are those relevant to environmental hazards, protection, and cleanup. Granted, the plan to divert Siberia's rivers has been canceled, but the clean-up costs from seventy-five years of pollution under Soviet power are greater than anyone could have imagined. Indeed, the runaway spill of tens of millions of gallons of oil into fragile Siberian tundra near the Arctic Circle village of Usinsk in the fall of 1994 is a chilling reminder that Soviet technology remains in place, even if the USSR has broken up. Furthermore, a new danger is posed by Russia's new capitalists and their Western partners who hope to make a killing on Siberia's rich timber, ore, and fossil fuel resources before complex environmental protection legislation is in place. Grigorii Galazii and Aleksei Trofimuk, who withstood thirty years of assault on Siberia's rivers and lakes, now live their retirement years in fear of irrational resource development promoted by market mechanisms. BAM and the long-range "Siberia" economic development program limp on, threatening to tame Siberia's rich mineral, land, forest, water, oil, and gas resources without taking environmental issues into consideration.

At one time Akademgorodok scientists felt the pressure to build communism, a decidedly political goal, at the same time as they were trying to build an apolitical city of science, one somehow divorced from the social, economic, and cultural pressures of Soviet life. It proved to be too much to be responsible for developing Siberian resources and for seeing to it that the Marxian urban paradigm was applied to an oasis in Siberia. Certainly it was too much to expect that scientists could preserve their autonomy in the face of Soviet ideological and political constraints. Now Akademgorodok scientists feel the pressure to carry out their research, keep their institutes open, hire and fire employees, and attract talented young scientists in competition with Moscow, St. Petersburg, and Western scientific institutes, all the while dealing with rapidly changing political and economic institutions. We can no more expect the creation of a Russian New Atlantis in these circumstances than a Soviet one in decades recently past.

NOTES

A WORD ON THE SOURCES

Archives

In addition to such published sources as local and national newspapers, books, and pamphlets, I used the following archives in the preparation of this book, which are referred to by the following abbreviations in the notes:

A AN   Arkhiv Akademii Nauk SSSR, Archive of the Academy of Sciences, Moscow
A ITsiG   Archive of the Institute of Cytology and Genetics, Akademgorodok
A IEiOPP   Archive of the Institute of Economics and the Organization of Industrial Production, Akademgorodok
A KIAE   Archive of the Kurchatov Institute of Atomic Energy, Moscow
A IIaF   Archive of the Institute of Nuclear Physics, Akademgorodok
NASO   Nauchnyi arkhiv Sibirskogo otdeleniia, Scientific Archive of the Siberian division, Novosibirsk
PANO   Partiinyi arkhiv Novosibirskoi oblasti, Communist Party Archive of the Novisibirsk region, Novosibirsk
Personal papers of Andrei Ershov, Computer Center, Akademgorodok
Personal papers of Andrei Trofimuk, Institute of Geology and Geophysics, Akademgorodok

Interviews

I was able to conduct interviews with the following individuals:

Abel Aganbegian, economist, Moscow
Murad Akhundov, philosopher, Cambridge, Mass., Moscow
Aleksandr Aleksandrov, mathematician, St. Petersburg
Anatolii Alekseev, computer scientist, Akademgorodok
Spartak Beliaev, physicist, Moscow
Andrei Bers, computer scientist, Akademgorodok
Andrei Deribas, director, Institute of Hydroimpulse Technology, Akademgorodok
Vadim Dudnikov, physicist, Akademgorodok
Iurii Eidelman, physicist, Akademgorodok
Zamira Ibragimova, correspondent, Novosibirsk
Aleksandr Kerkis, biologist, Akademgorodok
Tatiana Khodzher, biologist, Irkutsk
Valentin Koptiug, chairman, Siberian division, Akademgorodok
Vadim Kotov, computer scientist, Palo Alto, California
Mikhail Lavrentev, mathematician, Akademgorodok
Boris Mordukhovskii, Sibakademstroi photographer, Akademgorodok
Gurii Marchuk, mathematician, Moscow
Georgii Migirenko, mathematician, Akademgorodok
Aleksandr Nariniani, computer scientist, Akademgorodok
Boris Orlov, economist, Akademgorodok
Igor Pottosin, computer scientist, Akademgorodok

Natalia Pritvits, presidium of the Siberian division of the Academy of Sciences, Akademgorodok
Vladimir Ratner, biologist, Akademgorodok
Iurii Reshetniak, mathematician, Akademgorodok
Dmitrii Riutov, physicist, Akademgorodok, Moscow
Margarita Riutova, physicist, Akademgorodok
Anatolii Ruvinskii, biologist, Akademgorodok
Rudolf Salganik, biologist, Akademgorodok
Petr Shemetov, economist, Novosibirsk
Petr Shkvarnikov, biologist, Kiev
Vladimir Shumnyi, biologist, Akademgorodok
Andrei Trofimuk, geophysicist, Akademgorodok
Iurii Voronov, economist, Akademgorodok
Nikolai Vorontsov, biologist, Moscow
Georgii Voropaev, hydrologist, Moscow
Tatiana Zaslavskaia, sociologist, Moscow

INTRODUCTION

1. Sir Francis Bacon, *Essays and New Atlantis*, ed. Gordon S. Haight (New York: Van Nostrand, 1942), p. 288.

CHAPTER ONE
FROM MOSCOW, LENINGRAD, AND UKRAINE TO THE GOLDEN VALLEY

1. Among the many brief biographical notes on Lavrentev, see V. Davydchenkov, "Akademik Mikhail Lavrent'ev," *Sibirskie ogni*, no. 11 (1970): 155–159; and G. S. Migirenko, "Mikhail Alekseevich Lavrent'ev," in *Nekotorye problemy matematiki i mekhaniki* (Novosibirsk: SO AN SSSR, 1961), pp. 5–23, which includes a lengthy bibliography of Lavrentev's works.

2. See N. S. Khrushchev, *Khrushchev Remembers*, trans. Strobe Talbott (Boston: Little, Brown, 1974), pp. 58–71, for a discussion of Khrushchev's relationship with the scientific intelligentsia.

3. On space culture, see Paul Josephson, "Rockets, Reactors, and Soviet Culture," in Loren Graham, ed., *Science and the Soviet Social Order* (Cambridge, Mass.: Harvard University Press, 1990), pp. 168–191.

4. Z. Ibragimova and N. Pritvits, *"Treugol'nik" Lavrent'eva* (Moscow: Sovetskaia Rossiia, 1989), pp. 9–25. This study is the best single history of Akademgorodok available in Russian.

5. On the Novosibirsk Hydroelectric Power Station, see *Novosibirskaia GES*, n.d., n.c., Vneshtorgizdat; V. Grebennik, "Novosibirskaia GES," *Akademstroevets*, no. 41 (2121), October 17, 1991, 3; and ibid., no. 44 (2124), November 12, 1991, 3.

6. On the history of the trans-Siberian railroad, see Steven G. Marks, *Road to Power: The Trans-Siberian Railroad and the Colonization of Asian Russia, 1850–1917* (Ithaca: Cornell University Press, 1991).

7. Paul Josephson, "Science Policy in the Soviet Union, 1917–1927," *Minerva* 26, no. 3 (fall 1988): 342–369.

8. *Kul'turnoe stroitel'stvo v sibiri, 1917–1941. Sbornik dokumentov* (Novosibirsk: Zapadno-sibirskoe knizhnoe izdatel'stvo, 1979), pp. 148–151, 268–291.

9. On the prehistory of Akademgorodok, see E. T. Artemov, *Formirovanie i razvitie seti nauchnykh uchrezhdenii AN SSSR v sibiri, 1944–1980 gg.* (Novosibirsk: Nauka, 1990).

10. Lavrentev, "Nauka i pisateli," *Sibirskie ogni*, no. 9 (1958): 138–140.

11. A. Kulikov, "Gorod bol'shoi nauki," *Sibirskie ogni*, no. 3 (1958): 163–164.

12. NASO, f. 10, op. 3, ed. khr. 241, ll. 20–22, 50–57; ibid., ed. khr. 157, ll. 84–86; and PANO, f. 384, op. 1, ed. khr. 2, l. 21.

13. "N. S. Khrushchev v gostiakh u Akademstroevtsev," *Akademstroevets*, no. 77 (99), October 13, 1959, 1.

14. "N. S. Khrushchev—na stroike," *Akademstroevets*, no. 21 (242), March 14, 1961, 2.

15. A. M. Veksman et al., *Stroitel'stvo goroda nauki* (Novosibirsk: Novosibirskoe knizhnoe izdatel'stvo, 1963), pp. 9–13.

16. I. Pribytko, "Otsenka: khorosho!" *Akademstroevets*, no. 86 (108), November 17, 1959, 1. See also N. Mishin, "Gigant stroitel'noi industrii," ibid., 1–2; and NASO, f. 10, op. 3, ed. khr. 157, l. 3.

17. A. Leskov, "Zadachi stroitel'noi industrii," *Akademstroevets*, no. 16 (237), February 24, 1961, 2; M. Chemodanov, "Postanovlenie . . . o ser'eznykh nedostatkakh i o merakh po uluchsheniiu ekspluatatsii zhilishchno-kommunal'nogo fonda nauchnogo gorodka," ibid., no. 17 (238), February 28, 1961, 2; and N. Ivanov, "Po puti industrializatsii," ibid., no. 61 (282), August 4, 1961, 2.

18. *Za nauku v sibiri*, no.33 (58), August 15, 1962, 4.

19. "Nashi mysli, ruki i serdtsa—tebe, partiia!" *Akademstroevets*, no. 9 (31), February 3, 1959, 1.

20. *Za nauku v sibiri*, no. 13 (38), March 27, 1962, 3.

21. G. Migirenko, "Rasskazy ob uchenykh," *Sibirskie ogni*, no. 1 (1965): 139.

22. Interview with Andrei Andreevich Deribas, Director, Institute of Hydroimpulse Technology, Akademgorodok, December 7, 1991, and A. Deribas, "Istoriia nachalas' tak . . . ," *Za nauku v sibiri*, no. 26 (51), June 26, 1962, 4.

23. Lavrentev, "Tak nachinalsia Akademgorodok . . . " *Komsomol'skaia pravda*, May 14, 1977, p. 4

24. M. Nozdrenko, "Domovye griby," *Za nauku v sibiri*, no. 17 (94), April 28, 1963, 4; and no. 32 (57), August 8, 1962, 3.

25. *Za nauku v sibiri*, no. 11 (36), March 13, 1962, 2–3; ibid., no. 15 (92), April 11, 1963, 1; ibid, no. 38 (63), September 19, 1962, 4; and ibid., no. 28 (105), July 22, 1963, 4.

26. Ibid., no. 12, September 26, 1961, 4.

27. Ibid., no. 14 (91), April 4, 1963, 4.

28. S. Makarov, "Zelenyi drug v opasnosti," *Za nauku v sibiri*, no. 25 (50), June 20, 1962, 4.

29. *Za nauku v sibiri*, no. 28 (53), July 11, 1962, 3; ibid., no. 2 (79), January 10, 1963, 4; ibid., no. 8 (85), February 21, 1963, 4; and ibid., no. 13 (90), March 28, 1963, 4.

30. NASO, f. 10, op. 3, ed. khr. 241, ll. 27–28; and *Za nauku v sibiri*, no. 4 (81), January 24, 1963, 4.

31. P. M. Sidorov, ed., *Razvitie meditsinskoi sluzhby Novosibirskogo nauchnogo tsentra* (Novosibirsk: Nauka, 1973), pp. 4–6, 11–12, 30–33.

32. "Dizinteriia," *Za nauku v sibiri*, no. 26 (51), June 26, 1962, 4.

33. "V presidiume SO AN SSSR," *Izvestiia SO AN SSSR*, no. 3 (1959): 132; and NASO, f. 10, op. 3, ed. khr. 241.

34. *Za nauku v sibiri*, no. 4 (29), January 23, 1962, 4; and ibid., no. 8 (33), February 27, 1962, 2.

35. R. Gostrem, "Dva vazhnykh zvena v snabzhenii," *Za nauku v sibiri*, no. 50 (75), December 12, 1962, 1; and V. Al'tergot, "Bylo by tselesoobrazno," ibid., no. 51 (76), December 17, 1962, 1.

36. *Za nauku v sibiri*, no. 17, October 14, 1961, 2–3; and ibid., no. 4, January 23, 1961, 1, 3.

37. Ibid., no. 50 (75), December 12, 1962, 1; and ibid., no. 37 (117), September 23, 1963, 1.

38. Lavrentev, "Tak nachinalsia Akademgorodok"; and A. N. Nesmianov, "Novyi nauchnyi tsentr na vistoke," *Pravda*, June 8, 1957, p. 3.

39. P. L. Kapitsa, "Osnovnuiu stavku delat' na molodezh'," *Tekhnika—molodezhi*, no. 2 (1958): 2–3; L. A. Artsimovich, "Siberskii nauchnyi tsentr budet odnim iz krupneishikh nauchykh tsentrov strany," ibid., no. 2 (1958): 3; "Novyi nauchnyi tsentr v sibiri," *Pravda*, November 3, 1957; and N. A. Pritvits and V. L. Makarov, eds., *Khronika. 1957–1982 gg. Akademiia Nauk SSSR. Sibirskoe otdelenie* (Novosibirsk: Nauka SO, 1982), pp. 14–18.

40. PANO, f. 10, op. 4, ed. khr. 849, l. 1.

41. NASO, f. 4, op. 1, ed. khr. 7, l. 6; N. A. Pritvits and V. L. Makarov, *Khronika*, 20; and *Za nauku v sibiri*, no. 41 (118), October 21, 1963, 3.

42. On the life of Iu. B. Rumer, see Margarita Kemoklidze Riutova, *Kvantovyi vozrast* (Moscow: Nauka, 1989).

43. Riutova, "Priezzhaite. Einshtein vas primet . . . Okanchanie," *Siberskie ogni*, no. 2 (1989): 114.

44. Riutova-Kemoklidze, "Okonchanie," 115. See also G. A. Ozerov, *Tupolevskaia sharaga* (Munich, 1977). For a discussion of the role of *sharashki* (camps) in defense research and development, see David Holloway, "Innovation in the Defense Sector," in R. Amann and J. M. Cooper, eds., *Industrial Innovation in the Soviet Union* (New Haven: Yale University Press, 1982), pp. 334–341.

45. *Za nauku v sibiri*, no. 20 (45), May 16, 1962, 1.

46. N. A. Dediushina and A. I. Shcherbakov, "O formirovanii nauchnykh kadrov sibirskogo otdeleniia AN SSSR," in A. P. Okladnikov, ed., *Voprosy istorii nauki i profes-sional'nogo obrazovaniia v sibiri*, vol. 1 (Novosobirsk: n.p., 1968), p. 236; and *Za nauku v sibiri*, no. 34 (59), August 22, 1962, 1.

47. I. A. Moletotov, "Problema kadrov sibirskogo nauchnogo tsentra i ee reshenie (1957–1964 gg.)," in B. M. Shchereshevskii, ed., *Voprosy istorii sovetskoi sibiri* (Novosibirsk: Izdatel'stvo novosibirskogo gosudarstvennogo universiteta, 1967), p. 345.

48. NASO, f. 10, op. 1, ed. khr. 84, ll. 1–4.

49. *Za nauku v sibiri*, no. 20 (45), May 16, 1962, 1; Moletotov, "Problema kadrov," pp. 340–341; and Lavrentev, "Net uchenykh bez uchenikov!" *Za nauku v sibiri*, no. 6 (31), February 6, 1962, 1.

50. *Nauka i prosveshchenie* (Novosibirsk, n.p., 1965), pp. 10–19; and Dediushina and Shcherbakov, "O formirovanii," p. 242.

51. Lavrentev, "Molodaia nauka Sibiri," *Literaturnaia gazeta*, December 18, 1958.

52. M. Lavrentev and S. Khristianovich, "Vazhnoe uslovie razvitiia nauki," *Pravda*, April 2, 1957, p. 3; N. I. Vekua, "Universitet novogo tipa," ibid., June 19, 1959, p. 4; S. Zalygin, "Novosibirsk gosudarstvennyi," *Izvestiia*, September 4, 1959, p. 3; Vladimir Vinogradov, "Sibirskii tsentr nauki," *Sibirskie ogni*, no. 5 (1967): 128; *Za nauku v sibiri*, no. 21, November 28, 1961, 3; ibid., no. 49 (74), December 3, 1962, 3; Dediushina and Shcherbakov, "O formirovanii," p. 239; and Moletotov, "Problema kadrov," p. 342.

53. NASO, f. 10, op. 4, ed. khr. 849, ll. 52–53; G. S. Migirenko, ed., *Novosibirskii nauchnyi tsentr* (Novosibirsk: Izdatel'stvo SO AN SSSR, 1962), pp. 197–202; *Khronika. AN SSSR. SO*, pp. 76, 92–93, 114–115, 135; Dediushina and Shcherbakov, "O formirovanii," pp. 232–235; Dediushina, "Iz istorii organizatsii i deiatel'nosti zapadno-sibirskogo filial AN SSSR," in I. M. Razgon, ed., *Kul'turnoe stroitel'stvo v sibiri v 1917– 1960 gg.* (Novosibirsk: Izdatel'stvo SO AN SSSR, 1962), pp. 143–153; and *Narodnoe khoziaistvo novosibirskoi oblasti. Statisticheskii sbornik* (Novosibirsk: Gosstatizdat TsSU SSSR, 1961), p. 292.

54. Lavrentev, "Nauka i pisateli," 140.

CHAPTER TWO
COLLIDING BEAMS AND OPEN TRAPS

1. On Carlo Rubbia and the CERN program, see Gary Taubes, *Nobel Dreams* (Redmond, Wash.: Tempus, 1986).

2. For a first-rate study of the Soviet atomic bomb project and Kurchatov, see David Holloway, *Stalin and the Bomb: The Soviet Union and Atomic Energy, 1939–1956* (New Haven: Yale University Press, 1994).

3. A KIAE, f. 20, op. 1, ed. khr. 70/51.

4. A KIAE, 7/S-NT-14, 12/S-NT-52, 11/S-NT-40, 11/S-T, 56; and A. B. Migdal, "Fizik milost'iu bozhiei," in A. N. Skrinskii, ed., *Akademik G. I. Budker. Ocherki, Vospominaniia* (Novosibirsk: Nauka, 1988), p. 50.

5. Alla Melik-Pashaeva, "Piat' iablok Andreia Budkera," *Sovetskaia kul'tura*, August 4, 1990, p. 15. My thanks to Iurii Eidelman for drawing my attention to this article.

6. Ia. B. Fainberg, "Neobyknovennaia nauchnaia fantaziia," in Skrinskii, *Budker*, p. 89.

7. L. B. Okun, "Vstrechi s A. M. Budkerom," in Skrinskii, *Budker*, p. 87.

8. Melik-Pashaeva, "Piat' iablok Andreia Budkera."

9. A AN, f. 1854, op. 1, ed. khr. 19, ll. 34–35.

10. "Slovo stroitelei instituta iadernoi fiziki," *Akademstroevets*, no. 9 (230), January 24, 1961, 1.

11. E. P. Krugliakov, "Osobyi stil' raboty," in Skrinskii, *Budker*, p. 129.

12. See, for example, Grigori Freiman, *It Seems I am a Jew*, trans. and ed. Melvyn Nathanson (Carbondale: Southern Illinois University Press, 1980).

13. Krugliakov, "Osobyi stil' raboty," p. 126.

14. G. I. Dimov, "Cherez obsuzhdeniia k vyboru," in Skrinskii, *Budker*, p. 107.

15. V. Nifontov, ". . . postaraites' menia poniat'," *Energiia-Impul's*, nos. 4–5 (July–August 1990): 4.

16. Okun, "Vstrechi s A. M. Budkerom," p. 88.

17. "Kto chem bolen?" *Za nauku v sibiri*, no. 46 (123), November 25, 1963, 3.

18. NASO, f. 10, op. 3, ed. khr. 366, ll. 81–82; and Skrinskii, "Iz vospominanii ob Andree Mikhailoviche i IIaFe," in Skrinskii, *Budker*, pp. 109–113.

19. Michael Riordan, *The Hunting of the Quark* (New York: Touchstone, 1987), pp. 246–247.

20. B. Konovalov, "Proryv v mikromir," *Izvestiia*, July 10, 1964, p. 4.

21. Skrinskii, "Iz vospominanii ob Andree Mikhailoviche," pp. 112–113.

22. Konovalov, "Proryv v mikromir"; V. E. Balakin, V. A. Sidorov, and A. N. Skrinskii, "Electron-positron collisions at Novosibirsk," typewritten report, n.d. (1989?), pp. 4–6; Riordan, *The Hunting of the Quark*, p. 249.

23. A IIaF, f. 1, op. 1, ed. khr. 35, ll. 31–33.

24. NASO, f. 10, op. 5, ed. khr. 209, ll. 1–5.

25. Balakin, Sidorov, and Skrinskii, "Electron-positron collisions," pp. 4–6.

26. A IIaF, f. 1, op. 1, ed. khr. 82, ll. 1–5; and ibid., ed. khr. 30, ll. 1–8.

27. NASO, f. 10, op. 5, ed. khr. 524, ll. 1–5; and A IIaF, f. 1, op. 1, ed. khr. 37, ll. 1–4, 8.

28. A IIaF, f. 1, op. 1, ed. khr. 8. ll. 1–5.

29. Ibid., ed. khr. 37, ll. 1–4, 8. See Riordan, *The Hunting of the Quark*, pp. 246–260, for a discussion of the operation of SPEAR in 1973 and 1974, the data produced, and the difficulties it created for elementary particle theory.

30. Ibid., ed. khr. 67, ll. 29–34; ibid., ed. khr. 66, ll. 2–3, 5; ibid., ed. khr. 123, ll. 2–4; and *Institute of Nuclear Physics* (Vneshtorgizdat, 1980) (in Russian and English), pp. 15–16.

31. A IIaF, f. 1, op. 1, ed. khr. 123, ll. 6–8; and R. F. Schwitters, "Beam Polarization in High Energy e+e- Storage Rings," in S. T. Beliaev, ed., *Problemy fiziki vysokikh energii i upravliaemogo termoiadernogo sinteza* (Moscow: Nauka, 1981), pp. 31–32, 36, 41–42, 47.

32. Skrinskii, "Iz vospominanii ob Andree Mikhailoviche," pp. 113–114.

33. *Institute of Nuclear Physics*, pp. 18–19; and n.a., "V demilitarizovannom obshchestve—demilitarizovannaia nauka," *Priroda*, no. 11 (1991): 6–7.

34. A IIaF, f. 1, op. 1, ed. khr. 8, ll. 6–9; and Okun, "Vstrechi s Budkerom," p. 87.

35. L. C. Teng, "Electron Cooling and Its Application to Antiproton-Proton Colliding Beams at Fermilab," in Beliaev, *Problemy fiziki vysokikh energii*, pp. 50–51, 64; A IIaF, f. 1, op. 1, ed. khr. 66, ll. 2–3, 5; Balakin, Sidorov, and Skrinskii, "Electron-positron collisions," pp. 7–9; Skrinskii, "Iz vospominanii ob Andree Mikhailoviche," pp. 113–114.

36. NASO, f. 10, op. 3, ed. khr. 366, l. 79.

37. PANO, f. 5420, op. 1, ed. khr. 22, ll. 50–52.

38. Interview with Spartak Timofeevich Beliaev, Kurchatov Institute, Moscow, January 9, 1992; and Beliaev, "Nedolgaia, no iarkaia zhizn'," in Skrinskii, *Budker*, pp. 61–67.

39. A IIaF, f. 1, op. 1, ed. khr. 29.

40. On NPOs—scientific-production associations—see Julian Cooper, "Innovation for Innovation in Soviet Industry," in Amann and Cooper, eds., *Industrial Innovation in the USSR*, pp. 456–470.

41. A IIaF, f. 1, op. 1, ed. khr. 123, ll. 8, 10, 16, 22; ibid., ed. khr. 122, ll. 4–7; and ibid., ed. khr. 3, ll. 4–7.

42. V. Konovalov, "Vygodno vsem: Novyi opyt finansirovaniia fundamental'nykh nauchnykh issledovanii," *Izvestiia*, February 22, 1969.

43. A IIaF, f. 1, op. 1, ed. khr. 82, ll. 7–12; ibid., ed. khr. 30, ll. 1–8; ibid., ed. khr. 66, ll. 4, 6–7, 23; ibid., ed. khr. 123, ll. 8, 10, 16, 22; ibid., ed. khr. 122, ll. 4–7; and ibid., ed. khr. 3, ll. 4–7.

44. Ibid., ed. khr. 123, ll. 8, 10, 16, 22; ibid., ed. khr. 122, ll. 4–7; ibid., ed. khr. 3, pp. 4–7; and NASO, f. 10, op. 5, ed. khr. 524, ll. 6–10.

45. A IIaF, f. 1, op. 1, ed. khr. 60, ll. 1–11.

46. Ibid., ed. khr. 56, ll. 1–3; and ibid., ed. khr. 14.

47. Ibid., ed. khr. 8, ll. 6–9.

48. A stellarator, like a tokamak, has a system of toroidal magnetic surfaces that act to confine a plasma. However, the system of surfaces is produced not by a current excited in the plasma but by an external multipole magnetic field that rotates with distance along the system. Its advantage over the tokamak is that the magnetic configuration is a steady-state configuration, so that the undesirable alternation of heating and

cooling of structural elements may possibly be avoided. A stellarator, however, contains groups of particles that are not confined in a closed volume. As a consequence, the fluxes of heat and particles across the magnetic field may prove too high in operation with infrequent collisions. See B. B. Kadomtsev, V. S. Mukhovatov, and V. D. Shafranov, "Magnetic Plasma Confinement Dedicated to the Eightieth Birthday of Academician M. A. Leontovich," *Soviet Journal of Plasma Physics* 9, no. 1 (January–February 1983): 5.

49. G. Budker, "Ukroshchenie plazmy," *Za nauku v sibiri*, no. 34 (161), September 14, 1964, 2.

50. V. S. Strelkov, "Twenty-five Years of Tokamak Research at the I. V. Kurchatov Institute," *Nuclear Fusion* 25, no. 9 (1985): 1189–1194; Kadomtsev et al., "Magnetic Confinement," 2–10; and E. P. Velikhov and K. B. Kartashev, "Osnovnye rezul'taty issledovanii po UTS i fizike plazmy v SSSR za period s avgusta 1984 g. po avgust 1985 g.," *Voprosy atomnoi nauki i tekhniki. Seriia: termoiadernyi sintez* 1 (1986): 3–24.

51. V. M. Tuchkevich and V. I. Frenkel', *Plasma Physics* (Leningrad: Nauka, 1978); and V. E. Golant, "Tokamak Experiments at the A. F. Ioffe Physico—Technical Institute," *Nuclear Fusion* 25, no. 9 (September 1985): 1183–1188. In 1964 Boris Konstantinov, then vice president of the Academy, visited IIaF. As Budker showed him around, Konstantinov asked if the institute had a holography section. Holography was not yet widespread and Budker did not know the term. "What? Holography?" he repeated, thinking for a second. Then he answered, "Yes, but only in the privacy my home" (see Krugliakov, "Osobyi stil' raboty," in Skrinskii, *Budker*, p. 127).

52. N. G. Basov, "Laser Controlled Fusion," *Soviet Journal of Plasma Physics* 1 (1983): 10–14; and "Lazer i energetika budushchego," *Krasnaia zvezda*, July 2, 1983, p. 3; and B. E. Ivanov, ed., *50 let khar'kovskomu fiziko-tekhnicheskomu institutu AN UkSSR* (Kiev: Naukova Dumka, 1978), pp. 162–201.

53. PANO, f. 5420, op. 1, ed. khr. 22, ll. 3–4.

54. Ruitov, "Dva fragmenta iz vospominanii ob A. M. Budkere," in Skrinskii, *Budker*, pp. 174–177.

55. A IIaF, f. 1, op. 1, ed. khr. 35, ll. 31–33.

56. A. Likhanov, "Sibir' sporit s solntsem," *Komsomol'skaia pravda*, September 6, 1964, p. 4

57. I. N. Golovin, "Razvitie idei A. M. Budkera v oblasti otkrytykh lovushek," in Beliaev, *Problemy fiziki vysokikh energii*, pp. 125–133; and idem., "V Kurchatovskom institute," in Skrinskii, *Budker*, pp. 46–47.

58. A IIaF, f. 1, op. 1, ed. khr. 82, l. 6; ibid., ed. khr. 30, ll. 1–8; ibid., ed. khr. 8. ll. 1–5; ibid., ed. khr. 37, ll. 5–7; ibid., ed. khr. 66, ll. 2–3, 5; ibid., ed. khr. 123, ll. 6–8; and NASO, f. 10, op. 5, ed. khr. 524, ll. 5–6.

59. M. S. Gorbachev, as cited in B. B. Kadomtsev, "Tokomak," *Soviet Life* (August 1986): 13.

60. Riutov, "Budushchee termoiadernoi energetiki (tezy doklada na nauchnoi sessii OFTPE AN SSSR)," typewritten report, December 1991(?).

61. NASO, f. 10, op. 5, ed. khr. 209, ll. 1–5; PANO, f. 5420, op. 1, ed. khr. 22, ll. 49–50; *Atomnaia energiia* 22, no. 3 (March 1967): 163–164; and *Proceedings of the Third International Conference on Plasma Physics and Controlled Nuclear Fusion Research* (Vienna: IAEA, 1969), 2 vols.

62. A IIaF, f. 1, op. 1, ed. khr. 82, ll. 7–12; ibid., ed. khr. 30, ll. 22–25, 51; and ibid., ed. khr. 37, ll. 14–20, 27–30, 39–40.

63. PANO, f. 269, op. 1, ed. khr. 88, ll. 236–37; ibid., ed. khr. 125, ll. 86–87; and ibid., f. 3384, op. 1., ed. khr. 13, ll. 33–39.

64. Ibid., f. 269, op. 1, ed. khr. 86, ll. 237–241.

65. Ibid., f. 5420, op. 1, ed. khr. 22, l. 9.

66. Ibid., f. 5420, op. 1, ed. khr. 21, ll. 1–3; and ibid., ed. khr. 22, l. 11.

67. Ibid., f. 5420, op. 1, ed. khr. 22, ll. 10–17, 39–42.

68. See, for example, "Mutantov—dvoe. A prichin?" *NTR*, no. 16 (1989): 7; Sergei Trusevich, "Zapretnaia zona dlia dozimetrov?" *Poisk*, no. 22 (September 1989): 3; Vasil' Iakovenko, "Zemlia atomnogo vspolokha: Belorussiia," *Literaturnaia rossiia*, no. 43 (October 27, 1989): 6–7; M. Iur'ev, "Rentgeny eniseia," *Nauka v sibiri*, no. 40 (October 13, 1989): 4–5; Iu. Kotov, "Grozit li bezrabotitsa stroiteliam AES?" *Sovetskaia kul'tura*, November 23, 1989, p. 3; and V. Pokrovskii, "Strasti vokrug AES," *NTR*, no. 7 (1989): 1.

69. Balakin, Sidorov, and Skrinskii, "Electron-positron collisions," pp. 11–14.

70. A. M. Petros'iants, "Sovsem nedavno," in B. B. Kadomtsev, ed., *Vospominaniia ob Akademike L. A. Artsimoviche* (Moscow: Nauka, 1981), p. 55.

71. V. V. Parkhomchuk, "Uvlechennost', kotoroi zarazhalis' vse," in Skrinskii, *Budker*, pp. 181–182.

CHAPTER THREE
SIBERIA—LAND OF ETERNALLY GREEN TOMATOES

1. For more on Lysenko and Lysenkoism, see David Joravsky, *The Lysenko Affair* (Cambridge, Mass.: Harvard University Press, 1970); Zhores Medvedev, *The Rise and Fall of T. D. Lysenko* (New York: Columbia University Press, 1969); Mark Adams, "Science, Ideology, and Structure: The Kol'tsov Institute, 1900–1970," in Linda Lubrano and Susan Solomon, eds., *The Social Context of Soviet Science* (Boulder, Colo.: Westview, 1980), pp. 173–204; and Loren Graham, *Science, Philosophy, and Human Behavior in the Soviet Union* (New York: Columbia University Press, 1987), pp. 102–156.

2. Mark Popovskii, "Selektsionery," *Novyi mir*, no. 8 (1961): 198.

3. P. M. Zhukovskii, D. K. Beliaev, and S. I. Alikhanian, "50 let otechestvennoi genetiki i selektsii rastenii, zhivotnykh i mikroorganizmov," *Genetika* 8, no. 12 (1972): 21.

4. Popovskii, "Selektsionery," p. 202.

5. V. Enkena, "Pamiati uchitel'ia," *Za nauku v sibiri*, no. 51 (76), December 12, 1962, 4; on the Moscow pre-Lysenkoist tradition, see Adams, "Science, Ideology, and Structure," pp. 173–204.

6. Popovskii, "Selektsionery," p. 203.

7. G. Panderin, "Doroga k prostote," *Sibirskie ogni*, no. 7 (1971): 151.

8. *Za nauku v sibiri*, no. 45 (122), November 18, 1963, 4; N. P. Dubinin, *Vechnoe dvizhenie* (Moscow: Izdatpolit, 1973), pp. 365–379; and V. Gubarev, "Dva poliusa zhizni," *Komsomol'skaia pravda*, February 27, 1965, p. 4.

9. *Za nauku v sibiri*, no. 17 (42), April 24, 1962, 3.

10. R. Salganik, "Kod geneticheskoi informatsii raschifrovan," *Za nauku v sibiri*, no. 17 (42), April 24, 1962, 3.

11. Daniil Granin, *The Bison: A Novel about the Scientist Who Defied Stalin*, trans. Antonina W. Bouis (New York: Doubleday, 1990), p. 160; originally published in *Novyi mir*, no. 1 (1989). The "archive" of Timofeeff-Ressovsky, located in Pushchino, the biology city just outside Moscow, includes twenty-five kilometers of tape recorded reminiscences (see N. Dubrovina and M. Radzishchevskaia, "'Zubr' vspominaet," *Izvestiia*, April 3, 1993, p. 10).

12. Granin, *Bison*, pp. 171–191.

13. Ibid., pp. 195–196.

14. Raissa L. Berg, *Acquired Traits: Memoirs of a Geneticist from the Soviet Union*, trans. David Lowe (New York: Viking Penguin, 1988), pp. 219–221, 316–317; and Granin, *Bison*, pp. 225–226. On the boarding schools, see A. A. Liapunov, "FMSh—5 let. Sovremennik i shkola," *Za nauku v sibiri*, no. 14 (342), April 2, 1968, 3.

15. Granin, *Bison*, pp. 206–208.

16. Berg, *Acquired Traits*, pp. 306–308.

17. Gubarev, "Dva poliusa zhizni."

18. A KIAE, f. 2, op. 1, ed. khr. 442.

19. The original found in A KIAE differs slightly from the version published in *Pravda*, January 13, 1989, p. 4. The latter, I assume, has been edited for reasons of space.

20. N. P. Dubinin, ed., *Voprosy sovetskoi nauki: fizicheskie i khimicheskie osnovyi nasledstvennosti* (Moscow: Akademii nauk SSSR, 1958), classified report.

21. A KIAE, f. 2, op. 1, ed. khr. 406.

22. Dubinin, *Vechnoe dvizhenie*, pp. 365–379.

23. PANO, f. 1, ed. khr. 1, l. 11.

24. Ibid., ed. khr. 2, l. 22.

25. Ibid., ed. khr. 3, ll. 1–17.

26. Ibid., ed. khr. 1, ll. 10–16.

27. NASO, f. 50, op. 1, ed. khr. 30, ll. 10–11.

28. Ibid., ed. khr. 31, ll. 1–4, 7–9; and PANO, f. 5434, op. 1, ed. khr. 1, ll. 1, 10–12, 55–64, 71, 90–101.

29. NASO, f. 50, op. 1, ed. khr. 31, ll. 14–15.

30. Interview with Vladimir Konstantinovich Shumnyi, June 19, 1991, Institute of Cytology and Genetics, Novosibirsk.

31. PANO, f. 5434, op. 1, ed. khr. 1, ll. 57–58.

32. Ibid., ll. 58–59.

33. Ibid., ll. 9–10.

34. S. V. Argutinskaia, "Dmitrii Konstantinovich Beliaev," in V. K. Shumnyi and A. O. Ruvinskii, eds., *Problemy genetiki i teorii evoliutsii* (Novosibirsk: Nauka SO, 1991), pp. 19–20; and A. O. Ruvinskii, "Lider," *Priroda*, no. 4 (1988): 86.

35. Iurii Iakovlevich Kerkis, *Avtobiografiia*, unpublished manuscript, Novosibirsk, 1976. This autobiography of Kerkis was made available to me by his son, Sasha (Aleksandr Iuliovich) Kerkis, along with correspondence between Dubinin and Kerkis. Portions of the autobiography have been published as Iu. Kerkis, "Iz vospominanii genetika," *Znanie-sila*, no. 8 (1988): 52–59; idem., "Svoimi rukami. Dela Tomatnye," *Priroda*, no. 5 (1988): 81–86; and no. 3 (1989): 97–102.

36. NASO, f. 50, op. 1, ed. khr. 72, ll. 1–15.

37. Kerkis, *Avtobiografiia*.

38. D. K. Beliaev, "Nauku o nasledstvennosti—na sluzhbu narodu," *Sibirskie ogni*, no. 2 (1965): 154; and G. Paderin, "Doroga k prostote," ibid., no. 7 (1971): 141–142.

39. Argutinskaia, "Beliaev," pp. 8–10.

40. Ibid.

41. Ibid., pp. 11–13.

42. Ibid., pp. 13–15.

43. Ibid., pp. 14–16.

44. Ibid., pp. 16–17.

45. Beliaev, "Na osnove zakonov genetiki," *Pravda*, November 22, 1964, p. 3.

46. V. Rotner and V. Shumnyi, "Nasha Zoia Sofran'eva," *Za nauku v sibiri*, no. 6 (887), February 8, 1979, 3.

47. PANO, f. 5434, op. 1, ed. khr. 4, ll. 66–69; and Berg, *Acquired Traits*, pp. 309–310.

48. Berg, *Acquired Traits*, pp. 333–335.

49. Argutinskaia, "Beliaev," pp. 21, 23.

50. M. G. Savchenko, "Petro Klimentiiovich Shkvarnikov (do 85-richchia vid dnia narodzhennia i 65-richchia naukovoi ta pedagogichnoi diial'nosti)," *Tsitologiia i genetika* 25, no. 3 (1991): 169–170.

51. PANO, f. 5434, op. 1, ed. khr. 2, l. 25; Stenogram protokola zasedaniia uchenogo soveta Instituta tsitologii i genetiki SO AN SSSR, April 7, 1966, p. 1; D. K. Beliaev, "Raskryta zagadka nasledstvennosti," *Za nauku v sibiri*, no. 4 (81), January 24, 1963, 2; and ibid., no. 47 (124), December 2, 1963, 3.

52. PANO, f. 5434, op. 1, ed. khr. 2, ll. 29–34; and ibid., ed. khr. 4, ll. 100–101. See also I. Kiknadze, "Biologu fizik—i drug, i brat," *Za nauku v sibiri*, no. 41 (66), October 10, 1962, 3; and P. Shkvarnikov, "Mutatsiia i selektsiia rastenii," ibid., no. 2 (129), January 13, 1964, 3, for a description of Shkvarnikov's laboratory and work; and N. Tarasenko, "Selektsii—sovremennye pribory," ibid., no. 10 (137), March 9, 1964, 2–3.

53. Kerkis, *Avtobiografiia*, p. 13.

54. Ibid., pp. 27–28; A. E. Gaisinovich, "Ternistyi put' sovetskoi genetiki," *Priroda*, no. 5 (1988): 75–76; and Kerkis and Liapunov, "O rasshcheplenii gibridov," *Doklady AN SSSR* 31, no. 1 (1941): 43–46.

55. Kerkis, *Avtobiografiia*, pp. 109–110.

56. PANO, f. 5434, op. 1, ed. khr. 3, ll. 34–39; T. Sebeleva and L. Stepan'ian, "Genetika v shkole," *Za nauku v sibiri*, no. 50 (127), December 23, 1963, 4; Kerkis, "O 'nemolekul'iarnoi' biologii," ibid., no. 52 (77), December 26, 1962, 1, 3; Ruvinskii, "Lider," p. 86; and Protokoly zasedanii uchenogo soveta instituta tsitologii i genetiki SO AN SSSR, April 7, 1966, pp. 1–2.

57. PANO, f. 5434, op. 1, ed. khr. 3, ll. 7, 20–23.

58. NASO, f. 50, op. 1, ed. khr. 72, ll. 1, 8; PANO, f. 5434, op. 1, ed. khr. 2, l. 19; ibid., ed. khr. 3, ll. 17–19; and ibid., ed. khr. 5, ll. 37–38, 92.

59. Protokoly zasedanii uchenogo soveta instituta tsitologii i genetiki SO AN SSSR, January 15, 1965.

60. PANO, f. 5434, op. 1, ed. khr. 4, ll. 1–28; and Protokoly zasedanii uchenogo soveta instituta tsitologii i genetiki SO AN SSSR, March 1, 1969.

61. PANO, f. 5434, op. 1, ed. khr. 5, ll. 1–6, 14–15, 37–43.

62. Protokoly zasedanii uchenogo soveta instituta tsitologii i genetiki SO AN SSSR, February 22, 1965, October 4, 1965, and March 29, 1966.

63. PANO, f. 5434, op. 1, ed. khr. 4, 105–112; and ibid., ed. khr. 2, ll. 43–52.

64. Argutinskaia, "Beliaev," pp. 24–25.

65. NASO, f. 50, op. 1, ed. khr. 88, ll. 3–33; and PANO, f. 5434, op. 1, ed. khr. 3, ll. 1–14.

66. NASO, f. 10, op. 5, ed. khr. 690, ll. 1–54, 70.

67. Beliaev, "Avanpost sovremennoi biologii," *Sovetskaia sibir'*, February 1, 1978.

68. N. A. Kolchanov and V. A. Ratner, eds., *Modeling and Computer Methods in Molecular Biology and Genetics. Abstracts of the International Conference* (Novosibirsk: SO AN SSSR, 1990); and A. B. Osadchuk and A. A. Rodina, comps., *Genetika-narodnomu khoziaistvu* (Novosibirsk: SO AN SSSR, ITsiG, 1990), p. 4.

69. Interview with V. A. Ratner, Institute of Cytology and Genetics, Novosibirsk, December 2, 1991.

70. Graham, *Science, Philosophy, and Human Behavior in the Soviet Union*, pp. 145–146.

71. Ibid.

72. Paderin, "Doroga k prostote," p. 140.

73. Beliaev, "Genetika i problemy zhivotnovodstva v sibiri," *Pravda buriatii*, September 6, 1981, p. 2.

74. A. Vostiagin, "Cherga smotrit v budushchee," *Zvezda altaia*, October 16, 1980, p. 3.

75. N.a., *Novosibirsk Science Centre* (in Russian and English) (n.c.: Vneshtorgizdat, 1987).

76. Vostiagin, "Cherga smotrit v budushchee."

77. *Institut tsitologii i genetiki* (in Russian and English) (Novosibirsk: SO AN SSSR, n.d. [1989?]), pp. 59–62.

78. Ibid., introduction.

79. Interview with Academician Vladimir Konstantinovich Shumnyi, November 20, 1991, Institute of Cytology and Genetics, Novosibirsk.

80. Ibid.

81. *Institut tsitologii i genetiki*, pp. 6–15.

82. Ibid., pp. 16–23.

83. Ibid., pp. 24–37.

84. Osadchuk and Rodina, *Genetika-narodnomu khoziaistvu*, pp. 36–37.

85. *Institut tsitologii i genetiki*, pp. 38–49.

86. Interview with Academician Vladimir Konstantinovich Shumnyi, November 20, 1991, Institute of Cytology and Genetics, Novosibirsk.

87. Beliaev, "Tainy kletki," *Trud*, August 25, 1964.

CHAPTER FOUR
MACHINES CAN THINK, BUT CAN HUMANS?

1. Many of the references in this chapter are to materials held in the personal archive of Academician Andrei Petrovich Ershov (hereafter, the Ershov papers), maintained as he left them in precise and logical order in the Ershov Memorial Library of the Computer Center of the Siberian Division of the USSR Academy of Sciences. For encouraging me to use this archive I extend my thanks to Igor Vasilievich Pottosin, deputy director and acting director of the Institute of Informatics Systems in Akademgorodok.

2. For a thorough discussion of the recent history of computerization in the USSR, see Richard W. Judy and Virginia L. Clough, "Implications of the Information Revolution for Soviet Society: A Preliminary Inquiry," Hudson Institute Report, HI-4091-P, Indianapolis, January 9, 1989. See also Richard Staar, ed., *The Future Information Revolution in the USSR* (New York: Crane Russak, 1988); S. E. Goodman and W. K. McHenry, "Computing in the USSR: Recent Progress and Policies," *Soviet Economy* 2, no. 4 (1986): 327–354; and Loren Graham, "Science and Computers in Soviet Society," in Erik Hoffmann, ed., *The Soviet Union in the 1980s* (New York: Academy of Political Science, 1984), pp. 124–134.

3. A. P. Ershov and M. R. Shura-Bura, *Stanovlenie programmirovaniia v SSSR (nachal'noe razvitie)*, preprint 12 (Novosibirsk: VTs SO AN SSSR, 1976), p. 33.

4. A. P. Ershov and M. R. Shura-Bura, "Puti razvitiia programmirovaniia v SSSR," *Kibernetika*, no. 6 (1976): 152–153.

5. I. A. Poletaev, *Signal (o nekotorykh poniatiiakh kibernetiki)* (Moscow: Sovetskoe radio, 1958), pp. 275–330, 361, 383–384, 395–399.

6. Natal'ia Pritvits, "Akademik Mikhail Alekseevich Lavrentev," in G. N. Paderin, ed., *Sibirskaia postup'* (Moscow: Sovremennik, 1982), p. 28.

7. Pritvits, "Lavrentev," p. 27.

8. Interview with Iurii Grigorevich Reshetniak, corresponding member of the Academy of Sciences, the Institute of Mathematics, Akademgorodok, November 28, 1991.

9. A AN, f. 1854, op. 1, ed. khr. 19, ll. 1–8.

10. Among Sobolev's popular science and philosophy articles on computers and cybernetics, see "Mashina reshaet zadachi," *Iunost'*, no. 6 (1955): 92–94; "Drevnie rukopisi chitaet mashina," *Pravda*, January 25, 1961; "Raskrytaia taina. S pomoshch'iu EVM sibirskie uchenye prochitali drevnie pis'mena maiia," *Sovetskii soiuz*, no. 4 (1961): 24–25; "Poeziia matematiki," *Literaturnaia gazeta*, December 14, 1961; and, with A. A. Liapunov, *Kibernetika i estestvoznanie* (Moscow: AN SSSR, 1957) and "Matematicheskie problemy sovremennoi kibernetiki," *Izvestiia SO AN SSSR*, no. 5 (1962): pp. 3–13.

11. Sobolev and Liapunov, "Kibernetika i estestvoznanie," in P. N. Fedoseev, ed., *Filosofskie problemy sovremennogo estestvoznaniia* (Moscow: AN SSSR, 1960), pp. 237–267.

12. Sobolev and Liapunov, "Kibernetika i estestvoznanie," pp. 237–267.

13. Sobolev, "Poeziia matematiki."

14. Interview with I. G. Reshetniak, Akademgorodok, November 28, 1991.

15. On the so-called Luzin affair, see Aleksey E. Levin, "Anatomy of a Public Campaign: 'Academician Luzin's Case' in Soviet Political History," *Slavic Review* 49, no. 1 (spring 1990): 90–108; and A. P. Iushkevich, "Delo Akademika N. N. Luzina," in E. G. Iaroshevskii, ed., *Repressirovannaia nauka* (Leningrad: Nauka, 1991), pp. 377–394.

16. "Aleksei Andreevich Liapunov," *Za nauku v sibiri*, no. 26 (607), July 4, 1973, 2.

17. See L. A. Liusternik, A. A. Abramov, V. I. Shestakov, and M. R. Shura-Bura, *Reshenie matematicheskikh zadach na avtomaticheskikh tsifrovykh mashinakh: programmirovanie dlia BESM* (Moscow: Academy of Sciences, 1952).

18. Ershov and Shura-Bura, "Puti razvitiia programmirovaniia," pp. 152–153. The manual was entitled *Programmirovanie dlia bystrodeistvuiushchikh elektronnykh schetnykh mashin* (Moscow: Academy of Sciences, 1952).

19. *Problemy kibernetiki* (Moscow: Gosizdatfizmatlit, 1958), 1:265–266.

20. Interview with Murad Akhundov, Cambridge, Mass., June 18, 1992. Dr. Akhundov was the deputy of the cybernetics seminar in the Institute of Philosophy in Moscow.

21. Interview with I. G. Reshetniak, Akademgorodok, November 28, 1991.

22. This discussion of the history of computer science and technology in the USSR is based on Ershov and Shura-Bura, *Stanovlenie programmirovaniia* preprint 12, and *Stanovlenie programmirovaniia v SSSR (perexod ko vtoromu pokoleniiu iazykov i mashin)* preprint 13 (Novosibirsk: VTs SO AN SSSR, 1976).

23. Resolution No. 404 of the Presidium of the Academy of Sciences, April 10, 1975, and Prikaz 456 of the Ministry of Higher and Middle Specialized Education, May 27, 1974, in the Ershov papers.

24. Interview with Igor Vasilievich Pottosin, deputy director, Institute of Systems Information, Akademgorodok, November 22, 1991.

25. A. Kronrod, "Pamiatnaia zapiska" with "Avtomaticheskaia sovetskaia sistema AS-144," June 16, 1970, in the Ershov papers.

26. A AN SSSR, f. 1729, op. 1, ed. khr. 19, l. 3. At a special meeting of the Central Committee on January 1976, for example, Brezhnev learned that computer production in the USSR was one-twenty-second that of the United States, and power one-sixty-fifth, with a lag of seven to eight years.

27. "Proekt problemnogo doklada 'Matematichiskie i fizicheskie osnovy razvitiia elektronno-vychislitel'nykh mashin i problemy obshcheniia cheloveka s nimi,'" Academy of Sciences, Moscow, 1973, restricted use, in the Ershov papers, pp. 4–9, 56–77.

28. Letter from A. A. Logunov and B. N. Petrov, January 27, 1978, to A. P. Aleksandrov, in the Ershov papers.

29. Resolution of the Meeting of the Aktiv of the Academy of Sciences of the USSR, Devoted to Problems of Acceleration of Scientific Technological Progress, No. 49, June 26, 1985, Moscow, in the Ershov papers; "Novye grani davnei problemy," NTR, no. 8 (1989): 3; and V. Abramov, "Kontseptsiia est'—komp'iuterov net," Sovetskaia rossiia, March 20, 1990, p. 2. Regarding the failure to reach the schools, see M. Basina and T. Kamchatova, "Tekhnika na grani . . . primitiva," Leninskie iskry, December 2, 1989, p. 1.

30. Much of this discussion is based on Judy and Clough, "Implications of the Information Revolution for Soviet Society." See also Sergei Panasenko, "Mirazhi informatizatsii," Sotsialisticheskaia industriia, December 19, 1989, p. 2; and D. Pospelov, "Iskusstvennyi intellekt: nashi bedy i trudnosti," NTR, no. 7 (1989): 7.

31. Letter from A. P. Ershov to M. F. Zhukov, Scientific Secretary, Siberian Division, January 16, 1978, in the Ershov papers.

32. "Material k dokladu ak. E. P. Velikhova na Obshchem Sobranii AN SSSR," March 2, 1983, pp. 1–5, in the Ershov papers.

33. A. P. Ershov, Stikhi (Akademgorodok: SibPKTI, 1991), p. 28.

34. In October 1958 V. A. Ditkin, the director of the Moscow-based Academy Computer Center, wrote to the State Physics-Mathematics Publishing House with a request to publish translations of U.S. works on computer languages, the first such "publications in Russian that provide an impression of the latest achievements in the United States" in FORTRAN, JT, SOAR-2, and UNICOD.

35. A. P. Ershov, G. I. Kozhukhin, Iu. M. Voloshin, Vkhodnoi iazyk systemy avtomaticheskogo programirovaniia (predvaritel'nye soobshchenie) (Moscow: VTs AN SSSR, 1961). See also A. P. Ershov and K. V. Kim, PPS—programmiruiushchaia programma dlia vychislitel'noi mashiny STRELA-3 (Moscow: VTs AN SSSR, 1960); and A. P. Ershov, ed., FORTRAN. Rukovodstvo dlia pol'zovaniia. Iz sbornika perevodov (Moscow: Fizmatgiz, 1960).

36. A. P. Ershov, ed., Avtomatizatsiia programmirovanii (Moscow: Gosizdatfizmatlit, 1961), pp. 3–9.

37. Norbert Wiener, Cybernetics (New York: Wiley, 1948), p. 6, passim.

38. A. P. Ershov, "Put' k prizvaniiu (k 50-letiiu A. A. Bers)," April 20, 1984, p. 1, in the Ershov papers.

39. Ibid., pp. 2–3.

40. Ershov, The British Lectures (London: Heyden, 1980). See also A. P. Ershov, "A History of Computing in the USSR," Datamation (September 1975): 80–88.

41. Letters from M. A. Lavrentev to N. P. Zorin, deputy director, Main Administration for Preservation of Government Secrets in Print, October 10, 1967, and from G. I. Marchuk to E. I. Eiler, n.d. (sometime in 1967), in the Ershov papers.

42. Zamira Ibragimova, *Na vas nadezhda, muzhiki* (Novosibirsk: Novosibirskoe knizhnoe izdatel'stvo, 1989), pp. 77–102.

43. *Pionirskaia pravda*, September 1, 1981, p. 3; *Uchitel'skaia gazeta*, July 2, 1985, p. 2; and *Molodost' sibiri*, January 31, 1984.

44. A. P. Ershov, "Komp'iuternyi vseobuch: dvigat'sia vpered, povyshat' kachestvo," *Narodnoe obrazovanie*, no. 11 (1986): 24–26; and Vazhneishie nauchnye dostizheniia, VTs SO AN SSSR, typewritten report (Akademgorodok, 1984?), from VTs SO AN SSSR.

45. Ibragimova, *Na vas nadezhda, muzhiki*, pp. 77–102; and Ershov, "Komp'iuternyi vseobuch."

46. Ershov, "O razvitiii [*sic*] issledovanii v oblasti arkhitektury i matematicheskogo obespecheniia vychislitel'nykh sistem v vychisltel'nom tsentre sibirskogo otdeleniia akademii nauk sssr," n.d. (most likely 1981).

47. Some information in this section is taken from "Stengazety" (Bulletin Board Newspapers) from 1980, 1984, 1985, and 1986, and articles by L. B. Efros, I. V. Pottosin, V. V. Penenko, B. A. Kargin, A. N. Kremlev, V. P. Piatkin. A. S. Alekseev, M. K. Fage, Iu. N. Pervin, N. V. Kuliukov, V. A. Debelov, V. K. Gusiakov, and I. M. Bobko. I am grateful to S. V. Kutznetsov, scientific secretary of the center, and Anatolii Alekseev, its director, for giving me these *stengazety*. Unfortunately I received the articles from the *stengazety* after they were cut from the six- by fifteen-foot board to which they were glued and cannot be precise about dates and authors for all the articles.

48. Interview with I. G. Reshetniak, Akademgorodok, November 28, 1991.

49. NASO, f. 10, op. 3, ed. khr. 255, ll. 434–437; "Nam dorog kazhdyi den'—stroitelei instituta matematiki," *Akademstroevets*, no. 38 (363), May 19, 1962, 1; and I. Bobko and O. Moskalev, "VTs deistvuet," *Za nauku v sibiri*, no. 39 (116), October 7, 1963, 3.

50. NASO, f. 31, op. 1, ed. khr. 101, ll. 1–3.

51. Interview with Aleksandr Semenovich Nariniani, Akademgorodok, December 14, 1991.

52. Ibid.

53. NASO, f. 53, op. 1, ed. khr. 45, ll. 21, 24.

54. Ibid., ll. 20–21.

55. Letter from M. A. Lavrentev to M. E. Rakovskii, Deputy Chairman, Gosplan, October 16, 1967, in the Ershov papers.

56. N. V. Kuliukov, "20 let: liudi i EVM," Stengazeta VTs (1984?): 1–3.

57. NASO, f. 53, op. 1, ed. khr. 6.

58. Ibid., ed. khr. 45, ll. 3–8; and Stengazeta VTs (1984?).

59. NASO, f. 53, op. 1, ed. khr. 45, ll. 8–15; ibid., f. 10, op. 5, ed. khr. 358, ll. 2–6; A. P. Ershov, "Sozdaetsia sistema BETA," *Za nauku v sibiri*, no. 2 (633), January 9, 1974, 7; and Ershov, "Bers," pp. 3–5.

60. Personal communication to the author.

61. Ershov, "Sozdaetsia sistema BETA," p. 7.

62. NASO, f. 10, op. 5, ed. khr. 398, l. 36; ibid., ed. khr. 797, l. 13; and ibid., ed. khr. 358, ll. 7–8, 13–17.

63. Ibid., ed. khr. 338, ll. 26–27; and ibid., ed. khr. 797, l. 31.

64. Ibid., ed. khr. 797, ll. 2–7; *Za nauku v sibiri*, no. 2 (633), January 9, 1974, 1–2, 6–7; and Ershov, "Sozdaetsia sistema BETA."

65. NASO, f. 53, op. 1, ed. khr. 45, ll. 3–8; and V. K. Gusiakov, "Otdel matematicheskikh zadach geologii i geofiziki," Stengazeta VTs (1984?).

66. N.a., *Computing Center. Siberian Division of the USSR Academy of Sciences* (Vneshtorgizdat, 1980), in Russian and English.

67. NASO, f. 10, op. 5, ed. khr. 398, ll. 18, 36; ibid., ed. khr. 358, ll. 7–8, 13–17; ibid., ed. khr. 797, l. 22; ibid., f. 53, op. 1, ed. khr. 45, l. 19; Report of Research Activities of the Computer Center, 1976(?), folder "Vychislitel'nyi tsentr," n.a., n.d., pp. 1–7, in the Ershov papers; *Za nauku v sibiri*, no. 2 (633), January 9, 1974, 1–2, 6; and *Computing Center*.

68. Ibid.

69. "Spravka o nauchno-proizvodstvennoi deiatel'nosti vychislitel'nogo tsentra SO AN SSSR za 1982–1987 gg.," typewritten report.

70. On ASUs, see William McHenry, "MIS in USSR Industrial Enterprises: The Limits to Reform from Above," *Communications of the ACM* 29, no. 11 (November 1986).

71. NASO, F. 53, op. 1, ed. khr. 45, ll. 9–15.

72. R. K. Notman, "Otvet s pomoshch'iu EVM," *Sovetskaia sibir'*, August 10, 1979. See also I. M. Bobko, ed., *Ierarkhicheskie sistemy upravleniia i ikh adaptatsiia: sbornik nauchnykh trudov* (Novosibirsk: Computer Center, 1984); and I. M. Bobko, ed., *Integrirovannye ASU predpriiatiiami: tezisy dokladov vsesoiuznoi konferentsii* (Novosobirsk: Computer Center, 1985).

73. Vazhneishie nauchnye dostizheniia, VTs SO AN SSSR, typewritten report, Akademgorodok (1984?), from VTs SO AN SSSR.

74. Natasha Pritvits, "Prognoz 'AISTA,'" *Komsomol'skaia pravda*, July 27, 1971.

75. M. Glazyrin, A. Novikov, "ASU Raiona," *Sovetskaia sibir'*, April 16, 1980.

76. G. I. Marchuk, "XX vek. Problemy 'elektronnoi tsivilizatsii,'" *Literaturnaia gazeta*, March 24, 1971, p. 10; and Pritvits, "Prognoz 'AISTA.'"

77. *Computing Center*.

78. A. P. Ershov, "Chelovek i mashina," *Za nauku v sibiri*, no. 6 (133), February 10, 1964, 1–3. On the challenging semantic, semiotic, and logical problems, see, for example, John Haugeland, *Artificial Intelligence: The Very Idea* (Cambridge, Mass.: MIT Press, 1990).

79. Interview with A. S. Nariniani, Akademgorodok, December 14, 1991.

80. A. S. Nariniani, "Otchet o rabote laboratorii iskusstvennogo intellekta VTs SO AN SSSR," October 10, 1978; and A. P. Ershov, "Spravka o kh/dogovornykh rabotakh laboratorii iskusstvennogo intellekta," December 29, 1979.

81. A. S. Nariniani, "Vybor ideologii formal'noi modeli iazyka (neskol'ko zamechanii)," typewritten report, Computer Center, Akademgorodok (1974?), in the Ershov papers.

82. "Reshenie rasshirennogo zasedaniia sektsii semiotiki nauchnogo soveta po kibernetike AN SSSR, posviashchennogo aktual'nykh semioticheskim problemam obshcheniia cheloveka s mashinoi," September 1, 1975; and Informational Letter No. 9, September 1977, Scientific Council of Interdisciplinary Problem "Cybernetics," in the Ershov papers.

83. Interview with A. S. Nariniani, Akademgorodok, December 14, 1991.

84. NASO, f. 10, op. 5, ed. khr. 797, l. 15.

85. V. E. Kotov, "Laboratory of Theoretical Programming," in *Institute of Informatics Systems* (Novosibirsk: SibPKTI, 1991), pp. 5–8, in English.

86. Kotov, "Teoriia parallel'nogo programmirovaniia. Prikladnye aspekty," *Kibernetika*, no. 1 (1974): 1–16; no. 2 (1974): 1–18; idem., "Na starte 'START,'" *NTR*, no. 4 (July 2–15, 1985): 6; and A. S. Nariniani, "Teoriia parallel'nogo programmirovaniia. Formal'nye modeli," *NTR*, no. 3 (1974): 1–15; no. 5 (1974): 1–14.

87. V. E. Kotov, "Project Start: Concurrency + Modularity + Programmability = Mars," *Communications of the ACM* 34, no. 6 (June 1991): 32–45.

88. A. P. Ershov, "Kak prikhodit uspekh, pokazyvaet istoriia sozdaniia novykh moshchnykh EVM," *Izvestiia,* May 3, 1987, p. 2. The Gorbachev speech was reported in *Izvestiia,* April 16, 1987.

89. G. S. Pospelov, "Reshenie nauchnogo soveta AN SSSR po probleme 'Iskusstvennyi intellekt' ot 2 iiuliia 1987 goda," July 7, 1987, pp. 4–9.

90. E. T. Orlov, assistant director, VTs, "Polozhenie o propusknom i vnytriinstitutskom rezhemi v VTs i GP VTs SO AN SSSR," January 25, 1982.

91. A. S. Nariniani, "Vozmozhnost' russkogo chuda industriia intellekta" (1990) and "Poiasnitel'naia zapiska k proektu svobodnaiai tvorcheskaia territoriia," (1991?).

92. *Institute of Informatics Systems, Siberian Division of the Academy of Sciences of the USSR* (Novosibirsk: SO AN, SSSR, 1991) and *Current Topics in Informatics Systems Research* (Novosibirsk: IIS SO AN, 1991).

93. *Computer Center, Siberian Division of the Academy of Sciences* (Novosibirsk: SO AN, n.d.), in Russian and English.

CHAPTER FIVE
SIBERIAN SCIENTISTS AND THE ENGINEERS OF NATURE

1. Anatolii Zlobin, "Na sibirskoi magistrali," *Novyi mir,* no. 1 (1959): 127–128.

2. *XX S"ezd KPSS. Otchetnyi doklad* (Moscow, 1956), pp. 52–53.

3. Zamira Ibragimova, " 'Slavnoe more' na vesakh chesti i ekonomiki," *Literaturnaia gazeta,* February 19, 1986, p. 11.

4. Evgenii Bando, "Meridian Baikala," *Angara,* no. 1 (January–March 1961): 4–8.

5. Khr. D. Peev and A. K. Diunin, "Isskustvennoe regulirovanie snezhnogo pokrova v gornykh raionakh kak sredstvo upravleniia stokom," *Izvestiia sibirskogo otdeleniia AN SSSR, seriia tekhnicheskikh nauk,* no. 2, vyp. 1 (1963): 20–26; A. A. Babyr', "Kholod na sluzhby narodnogo khoziaistva," *Za nauku v sibiri,* no. 10, September 12, 1961, 3; D. I. Abramovich, "Voprosy gidrologii," ibid., no. 12, September 26, 1961, 3; and S. Seliakov, "Preobrazovanie prirody zapadnoi Sibiri," ibid., no. 15, October 17, 1961, 4.

6. G. I. Galazii and K. K. Votintsev, eds., *Put' poznaniia baikala* (Nauka: Novosibirsk, 1987), pp 275–281. Angara electrical energy potential is more than 19 million horsepower.

7. S. Grebel'skii, "Plotina unichtozhaet gnus," *Angara,* no. 4 (October–December 1964): 95–99.

8. Leonid Shinkarev, *Sibir': Otkuda ona poshla i kuda ona idet,* 2d ed. (Moscow: Sovetskaia rossiia, 1978), pp. 308–309, 315–317.

9. I. P. Bardin, M. A. Lavrentev, V. S. Nemchinov, et al., *Razvitie proizvoditel'nykh sil vostochnoi sibiri,* 13 vols. (Moscow: Academy of Sciences, 1960).

10. V. S. Nemchinov, "Vazhneishie problemy narodnogo khoziaistva vostochnoi sibiri v svete itogov konferentsii," Bardin et al., *Razvitie proizvoditel'nykh sil,* 13:107, 130.

11. Ibid., 13:122–123.

12. A. Azizian, "U obskogo moria," *Pravda,* September 12, 1958, p. 4.

13. "Dat' strane 50,000 tsentnerov ryb. Na malom more. Beseda s directorom malomorskogo rybzavoda T. Zakharovym," *Vostochno-sibirskaia pravda,* May 15, 1948, p. 3, as cited in A. S. Sobennikov, *Professor M. M. Kozhov: Biograficheskii ocherk* (Irkutsk: NIIB IGU, 1990), p. 32.

14. GAIO, f. 2901, op. 2, d. 51, as cited in Sobennikov, *Kozhov,* pp. 33–34.

15. Zlobin, "Na sibirskoi magistrali," pp. 133–134.

16. Ibid., pp. 129–131, 135–137.

17. For the early history of Baikal research, and for answers to 907 questions concerning every aspect of Baikal history, hydrobiology, geochemistry, economics, environment, in a word, everything you would like to know in short form, see G. I. Galazii, *Baikal v voprosakh i otvetakh* (Irkutsk: Vostochno-sibirskoe knizhnoe izdatel'stvo, 1987).

18. T. V. Khodzher and V. A. Obolkin, "The Present State and Main Results of Atmosphere Monitoring in the Baikal Region," unpublished paper [in English]; and Galazii and Votintsev, *Put' poznaniia baikala*, p. 139.

19. B. G. Ioganzen, "Spornye voprosy sovremennoi ekologii," *Izvestiia SO AN SSSR*, no. 9 (1959): 76–86; and G. M. Krivoshchekov, "Obsuzhdenie voprosov ekologii v biologicheskom institute SO AN SSSR," ibid., no. 12 (1960): 137.

20. G. V. Krylov and B. G. Ioganzen, "Biologicheskaia nauka i kompleksnoe osvoenie prirodnykh resursov zapadnoi sibiri," *Izvestiia Sibirskogo otdeleniia, seriia biologomeditsinskikh nauk*, no. 12, vyp. 3 (1966): 3–9.

21. N. G. Salatova, "IV vsesoiuznoe soveshchanie po okhrane prirody," *Izvestiia sibirskogo otdeleniia akademii nauk SSSR*, no. 11 (1961): 151–152.

22. G. I. Galazii, *Baikal i problemy chistoi vody v sibiri* (Irkutsk: Limnologicheskii institut SO AN SSSR, 1968), pp. 10–11.

23. G. I. Galazii, "Baikal v opasnosti. Pis'mo v redaktsiiu," *Komsomol'skaia pravda*, December 26, 1961, as cited in B. F. Lapin, ed., *Slovo v zashchitu Baikala* (Irkutsk: Vostochno-sibirskoe knizhnoe izdatel'stvo, 1987), pp. 22–25.

24. Shinkarev, *Sibir'*, pp. 319–321.

25. Galazii, *Baikal i problemy chistoi vody*, pp. 12–13.

26. Personal and official correspondence on the Baikal question, collected by A. A. Trofimuk, Institute of Geology and Geophysics, Akademgorodok, Siberia (hereafter, the Trofimuk papers), vol. 5: "Baikal: stroitel'stvo BTsBK, 1960–1970," pp. 75, 82–83. Not all volumes of Trofimuk papers have numbers.

27. Letter from M. V. Keldysh, president of the Academy of Sciences, to M. D. Millionshchikov, chairman of the Supreme Economic Council, August 8, 1964, in the Trofimuk papers, 5:64–67.

28. Letter from V. A. Kirillin, February 19, 1965, to D. F. Ustinov, in the Trofimuk papers, the volume on Baikal regarding the elimination of industrial discharges of BTsBK, 1965–1974, pp. 61–74.

29. The Trofimuk papers, 5:42.

30. K. K. Votintsev and G. I. Galazii, "Predel'no dopustimye kontsentratsii veshchestva, sbrasyvaemykh s promyshlennymi stochnymi vodami v ozero baikal" (Listvenichnoe: LI SO AN SSSR, 1973), Trofimuk papers, 6:74–86.

31. M. M. Odintsov, report to the presidium of the Siberian Division, December 8, 1974, in the Trofimuk papers, the volume on Baikal regarding the elimination of industrial discharges of BTsBK, 1965–1974, pp. 1–10; and idem., "Institute zemnoi kory," *Za nauku v sibiri*, no. 29 (407), July 16, 1969, 1, 3.

32. Letter from B. A. Smirnov, in the Trofimuk papers, 5:45.

33. A. B. Zhukov, "Budet les—budet Baikal," *Za nauku v sibiri*, no. 17 (598), April 26, 1973, 7.

34. "Baikal: stroitel'stvo BTsBK, 1960–1970," the Trofimuk papers, 5:55; I. P. Gerasimov and A. A. Trofimuk, "General'naia skhema kompleksnogo ispol'zovaniia resursov ozera baikal i ego basseina," speech at the presidium of the Academy of Sciences,

May 21, 1965, ibid., 4:2–18; and decision of the presidium of the East Siberian Filial of the Academy of Sciences, January 8, 1965, ibid., 5:51.

35. Letter from A. A. Trofimuk to L. I. Brezhnev, April 9, 1965, in ibid., 4.

36. Trofimuk, "Tsena vedomstvennogo upriamstva. Otvet ministru SSSR G. M. Orlovu," *Literaturnaia gazeta*, April 15, 1965.

37. Letter from Trofimuk to L. I. Brezhnev, July 19, 1966, in the Trofimuk papers, 4.

38. Iurii Makartsev, "Tekut v Baikal reki i stoki," *Sobesednik*, no. 41 (October 1988): 12–13.

39. Ibid.

40. Galazii, *Baikal i problemy chistoi vody*, pp. 5–9.

41. Letter from Iu. A. Azreal', for the State Committee on Hydrometeorology and Control of the Environment (Goskomgidromet), to the Commission of the Presidium of the Council of Ministers on environmental protection and the rational use of natural resources, June 15, 1983, in the Trofimuk papers, in "Problem of Maximum Allowable Concentration, 1971–1985"; Makartsev, "Tekut v Baikal," pp. 12–13; and Galazii and Votintsev, eds., *Put' poznaniia baikala*, p. 291.

42. N. M. Zhavoronkov, ed., *Novyi materialy v tekhnike i nauke: proshloe, nastoi-ashchee, budushchee* (Moscow: Nauka, 1966).

43. The Trofimuk papers, 7:72–73.

44. The Trofimuk papers, in "Problem of Maximum Allowable Concentration."

45. A. Beim, "Uchenye Baikal'ska—xii piatiletke," *Leninskoe znamia*, 22 July 1986, p. 2.

46. Beim, "Uchenye Baikal'ska"; and Sobennikov, *Kozhov*, pp. 61–62.

47. O. M. Kozhova, "Baikal i Baikal'skii kombinat," *Baikal'skii tselliuloznik*, nos. 21–22 (June 4, 1986).

48. Ibid.

49. Khozova, "Reshat' v komplekse," *Baikal'skii tselliuloznik*, nos. 15–16 (April 30, 1986): 3.

50. Resolution of GEK Gosplan SSSR, June 18, 1966, no. 10, "On the Preservation of Lake Baikal from Pollution of Waste Waters of TsBP," in the Trofimuk papers, 5:2; and Report from the Protocol of the Meeting of the Presidium of the Council of Ministers, May 20, 1970, no. 20, in ibid., 5:1.

51. The Resolution of the Siberian Division of the Academy of Sciences, August 1970, in the personal papers of N. N. Vorontsov.

52. Galazii and Votintsev, eds., *Put' poznaniia baikala*, pp. 285–287; and idem., *Problemy baikala* (Novosibirsk: Nauka, 1978).

53. The Trofimuk papers, 11:84–132.

54. Ibid., 17:62–67.

55. Ibid., 10:5, and thirty-eight-page report following.

56. Ibid., 11:43–51; and Valentin Rasputin, "Baikal u nas odin," *Izvestiia*, February 17, 1986, p. 4.

57. My great thanks to Professor Michael Bressler of Furman University for his critical comments on this section. See Furman, "Agenda Setting and the the Development of Soviet Water Resources Policy, 1965–1990," Ph.D. dissertation, University of Michigan, 1992; and Robert Darst, Jr., "Environmentalism in the USSR: The Opposition to the River Diversion Projects," *Soviet Economy* 4, no. 3 (1988): 223–252.

58. See *Trudy noiabr'skoi sessii AN SSSR 1933 goda: problemy volgo-kaspiia* (Leningrad: Izdatel'stvo Akademii nauk SSSR, 1934).

59. A. A. Tushkin, ed., *Teoriia i metody upravleniia vodnymi resursami sushi*, part 1 (Moscow: ONK VASKhNIL, 1988), pp. 8–13, 21; and Vadim Leibovskii, "Bespovorotno," *Ogonëk*, no. 40 (1987): 25.

60. G. V. Voropaev and D. Ia. Ratkovich, *Problema territorial'nogo pereraspredeleniia vodnykh resursov* (Moscow: IVP AN SSSR, 1985), pp. 252–253.

61. Ibid., p. 251.

62. Interview with G. V. Voropaev, January 15, 1992, Institute of Water Problems, Moscow.

63. Voropaev and Ratkovich, *Problema territorial'nogo pereraspredeleniia*, pp. 6–8.

64. V. I. Korzun, "O publikatsii 's kogo spros?'" in Tushkin, *Teoriia i metody*, part 1, p. 53.

65. Interview with G. V. Voropaev, January 15, 1992; and Korzun, "O publikatsii 's kogo spros?'" p. 53. By the end of the century total river flow into the Caspian may be reduced by 55–65 km^3 annually.

66. Voropaev and Ratkovich, *Problema territorial'nogo pereraspredeleniia*, pp. 414–416.

67. Ibid., pp. 417, 419–420, 424–429, 454–458.

68. Ibid., pp. 459–463.

69. Ibid., pp. 469–471.

70. Ibid., pp. 468–469.

71. Sergei Zalygin, "Povorot," *Novyi mir*, no. 1 (1987): 12.

72. "Ot redaktsii," in Tushkin, *Teoriia i metody*, part 1, pp. 109–111.

73. Douglas Weiner, *Models of Nature* (Bloomington: Indiana University Press, 1988), pp. 168–169.

74. On the glories of the Angarastroi construction trust, see, for example, Evgenii Bando, "Meridian Baikala," *Angara*, no. 1 (January–March 1961): 2–8.

75. N. Anpilogov, "Buriatskii lesokhimicheskii, *Svet nad Baikalom*, no. 4 (1958): 137, as cited in Sobennikov, *Kozhov*, p. 52.

76. Aleksandr Tvardovskii, "Iz rabochikh tetradei (1953–1960)," *Znamia*, no. 9 (1989): 147–150.

77. *XXIII s"ezd KPSS. Stenograficheskii otchet* (Moscow: Izdatpolit, 1966), 1:359.

78. Frants Taurin, "Baikal dolzhen byt' zapovednikom," *Literaturnaia gazeta*, February 10, 1959, as cited in Lapin, *Slovo v zashchitu Baikala*, pp. 17–20.

79. Zalygin, "Povorot," pp. 4–5.

80. "Zemlia, ekologiia, perestroika," *Literaturnaia gazeta*, January 18, 1989, pp. 1–4. On ecological luddism, see my "Science and Technology as Panacea in Gorbachev's Russia," in James Scanlan, ed., *Technology, Culture, and Development: The Experience of the Soviet Model* (Armonk: M. E. Sharpe, 1992).

81. Valentin Rasputin, "Baikal u nas odin," *Izvestiia*, February 17, 1986, pp. 3–4; and idem., "Chto imeem . . . Baikal'skii prolog bez epiloga," *Vostochno-sibirskaia pravda*, March 12, 1987, p. 3.

82. Zalygin, "Gosudarstvo i ekologiia," *Pravda*, October 23, 1989, p. 4.

83. Zalygin, "Proekt: nauchnaia obosnovannost' i otvetvennost'," *Kommunist*, no. 13 (1985): 63–73.

84. Zalygin, "Povorot," p. 7.

85. Zalygin, "Professionaly ot gigantomanii," *Literaturnaia gazeta*, February 8, 1989, p. 11. See also L. Filipchenko, "Baikal'skii sindrom," *Izvestiia*, May 4, 1989, p. 2; and "Aral-88," *Literaturnaia gazeta*, April 19, 1989, p. 12.

86. Zalygin, "Povorot," p. 7.

87. See for example, "Po povodu stat'i 'povorot'," Zvezda vostoka, no. 6 (1987): 10–14; and L. Epshtein, "Na shto potracheny sily arala," ibid., no. 12 (1987): 143–149.

88. Sotsialisticheskaia industriia, June 14, 1987.

89. A. L. Ianshin and G. Golitsyn, "Zemlia—nashe glavnoe bogatstvo," Izvestiia, February 12, 1986; Epshtein, "Na chto potracheny sily Arala?"; Korzun, "O publikatsii 's kogo spros?'" pp. 55–56; "Perebroska, ekologiia, moral,'" Zvezda vostoka, no. 9 (1987): 122–126; S. P. Tatur, "S vodoi i bez vody," in Tushkin, Teoriia i metody, part 2, pp. 38–58; and Vadim Leibovskii, "Bespovorotno," Ogonëk, no. 40 (1987): 24–27.

90. A. G. Aganbegian, "Kompleksnoe razvitie," Nauka i zhizn', no. 2 (1981): 18.

91. Nash sovremennik, no. 1 (1987): 134–137; and B. N. Laskorin and V. A. Tikhonov, "Novye podkhody k resheniiu vodnykh problem strany," Kommunist, no. 4 (1988): 90–100.

92. Interview with G. V. Voropaev, January 15, 1992, Institute of Water Problems, Moscow.

93. For criticism of V. Leibovskii's "Bespovorotno" for its "emotionalism" and inaccuracy, see Tatur, "S vodoi i bez vody," pp. 57–58. See also Korzun, "O publikatsii 's kogo spros?'" p. 61; "Perebroska, ekologiia, moral,'" Zvezda vostoka, no. 9 (1987): 122–126; and A. E. Asarin, "'Bespovorotno,' no predvziato," in Tushkin, Teoriia i metody, part 2, pp. 58–60.

94. See in particular Korzun, "O publikatsii 's kogo spros?'" p. 68; and Tatur, "S vodoi i bez vody," pp. 38–58.

95. N. Volkov and V. Karnaukhov, "Prebyvanie B. N. El'tsina v Baikal'ske," Vostochno-sibirskaia pravda, August 8, 1985.

96. Tushkin, Teoriia i metody, part 1, pp. 7, 30.

97. "Soveshchanie v TsK KPSS," Izvestiia, December 28, 1986, p. 2; and Rasputin, "Chto imeem . . . Baikal'skii prolog."

98. V. A. Koptiug, "Soiuz nauki i proizvodstva," Trud, August 21, 1980, p. 2.

99. Ibid.; and Pritvits and Makarov, Khronika, pp. 287–291.

100. Tushkin, Teoriia i metody, part 1, pp. 5–6, 8–13, 21–22.

101. Makartsev, "Tekut v Baikal."

102. "Baikal: vedomstvennoe solo," Vostochno-sibirskaia pravda, January 17, 1988, p. 2; and "Ekologicheskii shchit Baikala," Izvestiia, June 10, 1987.

103. "BTsBK, Baikal'sk i baikal'chane . . . ," Baikal'skii tselliuloznik, November 23, 1988, p. 4.

104. Ibid.

105. Koptiug, "Nauka i ekologiia," Vechernii novosibirsk, February 3, 1989, p. 2.

106. Koptiug, "Pervoocherednye zadachi organizatsii issledovanii po ekologicheskim problemam," Speech at the General Meeting of the USSR Academy of Sciences, December 28, 1988; Koptiug, "Nauka i ekologiia," p. 2; and Koptiug, "S pozitsii ustoichivogo razvitiia," Sovetskaia sibir', no. 111 (June 8, 1991), p. 2.

107. Ibid.

108. Koptiug, "Nauka i ekologiia," p. 2. Koptiug elaborates on the notion of "ecological passports" in "Ekologiia: ot obespokoennosti—k deistvennoi politike," Kommunist, no. 7 (1988): 24–33.

109. Koptiug, "Nauka i ekologiia," p. 2.

110. Ibragimova, "'Slavnoe more,'" p. 11.

111. "Baikal na planete odin," *Vostochno-sibirskaia pravda*, November 28, 1989, p. 1.

112. On the struggle of Siberian scientists against the Lower Ob Hydropower Station, see, for example, "Perebroska," *Za nauku v sibiri*, no. 7 (1390), February 24, 1989, 1, 4–5; and for a chronology of decisions on Baikal, see V. Ermikov, N. Meshkova, and N. Pritvits "Bor'ba za baikal," ibid., no. 8 (1391), March 3, 1989, 2–3.

CHAPTER SIX
THE SIBERIAN ALGORITHM

1. Michael Ellman, *Planning Problems in the USSR* (Cambridge: Cambridge University Press, 1973), pp. 1–2.

2. A. Pal'm, "Sibirskii algoritm," *Komsomol'skaia pravda*, February 19, 1965, p. 2.

3. A. A. Belykh, "Ob otnoshenii ekonomistov k matematike v period s nachala 30–x do serediny 50–x godov," in V. A. Zhamin, ed., *Voprosy istorii narodnogo khoziaistva i ekonomicheskoi mysli* (Moscow: Ekonomika, 1990), 2:174–181.

4. Ibid., 2:181–185.

5. Mark R. Beissinger, *Scientific Management, Socialist Discipline, and Soviet Power* (Cambridge, Mass.: Harvard University Press, 1988), p. 168.

6. V. S. Nemchinov, ed., *Primenenie matematiki v ekonmicheskikh issledovaniiakh* (Moscow: Sotsekgiz, 1959); and V. S. Nemchinov, L. V. Kantorovich, I. A. Kulev, L. E. Mints, and V. V. Voroshilov, eds., *Obshchie voprosy primeneniia matematiki v ekonomike i planirovanii* (Moscow: Academy of Sciences, 1961).

7. Nemchinov et al., *Obshchie voprosy*; and Ellman, *Planning Problems*, pp. 4–6.

8. Nemchinov et al., *Obshchie voprosy*, pp. 24, 239–241.

9. Many of the comments in this section are based on an interview with Abel Gezevich Aganbegian, Academy of Economic Sciences, Moscow, conducted on December 26, 1992.

10. Aganbegian, "Kurs politicheskoi ekonomii sotsializma i logika 'Kapitala,'" *Voprosy ekonomiki*, no. 6 (1962): 67; and Pal'm, "Sibirskii algoritm," p. 2.

11. A. G. Aganbegian and N. M. Voluiskii, *Dlia blaga sovetskogo cheloveka-stroitelia kommunizma* (Moscow: Gosplanizdat, 1960).

12. A. G. Aganbegian, "The Lessons of History," in Michael Barratt Brown, ed., *The Economic Challenge: Economics of Perestroika*, trans. Pauline M. Tiffen, with an introduction by Alec Nove (London: Hutchinson, 1988), p. 51.

13. Nemchinov et al., *Obshchie voprosy*, pp. 196–197, 203–206.

14. P. V. Shemetov, *Ekonomicheskie issledovaniia v sibiri* (Novosibirsk: Nauka, 1983), pp. 22–49.

15. See G. A. Prudenskii, *Vnutriproizvodstvennye rezervy (rezervy rosta proizvoditel'nosti truda)* (Moscow: Gospolitizdat 1954); idem., *Vremia i truda* (Moscow: Mysl', 1964); and the posthumous idem., *Problemy rabochego i vnerabochego vremeni. Izbrannye proizvedeniia* (Moscow: Nauka, 1972).

16. PANO, f. 5435, op. 1, ed. khr. 11, ll. 14–16.

17. G. A. Prudenskii, "O razvitii ekonomicheskikh issledovanii v sibiri," *Izvestiia SO AN SSSR. Seriia obshchestvennykh nauk*, no. 1, vyp. 1 (1963): 8–9.

18. Interview with Boris Pavlovich Orlov, senior scientist, IEiOPP, Akademgorodok, December 19, 1992.

19. PANO, f. 5434, op. 1, ed. khr. 11, ll. 67–74.

20. Interview with Iurii Petrovich Voronov, senior scientist, IEiOPP, and deputy director, *EKO*, Novosibirsk, December 4, 1991; and Shemetov, *Ekonomicheskie issledovaniia*, pp. 55–60.

21. Interview with Boris Pavlovich Orlov, December 19, 1992; and PANO, f. 5435, op. 1, ed. khr. 11, ll. 16–25.

22. *Za nauku v sibiri*, 13 (90), March 28, 1963, 1.

23. "Otchet o rezul'tatakh NIR, osnovnykh rezul'tatkh ispol'zovaniia zakonchennykh razrabotakh v narodom khoziaistve i nauchno-organizatsionnoi deiatel'nosti za 1984," (Novosibirsk: IEiOPP SO, 1984), p. 84.

24. PANO, f. 5435, op. 1, ed. khr. 10, l. 123; ibid., ed. khr. 11, l. 17; NASO, f. 55, op. 1, ed. khr. 76, l. 15; Aganbegian, "V interesakh razvitiia," *Sovetskaia sibir'*, December 7, 1979, p. 2; and idem., "O razvitii ekonomicheskoi nauki v 1957–1975 gg. Institut ekonomiki i organizatsii promyshlennogo proizvodstva SO AN SSSR (Novosibirsk[?]: IEiOPP[?], 1975[?]), p. 23.

25. "Otchet o deiatel'nosti IEiOPP SO AN SSSR za 1971–1975 gg.," (Novosibirsk: IEiOPP SO AN SSSR, 1976), p. 30; PANO, f. 5435, op. 1, ed. khr. 10, l. 123; ibid., ed. khr. 9, ll. 72–84 ibid., ed. khr. 11, ll. 45–49; "Instituty otdeleniia ekonomiki AN SSSR," *EKO*, no. 2 (1975): 20–27; and F. A. Baturik, "Ekonomicheskaia nauka v sibiri za 1964 god," *Izvestiia SO AN SSSR. Seriia obshchestvennykh nauk*, no. 5, vyp. 2 (1965): 135–138.

26. Interview with Iurii Voronov, December 4, 1991; "Predvidet'!" *Literaturnaia gazeta*, August 5, 1970, p. 5, and "'EKO.' Zhurnal khoziaistvenika," ibid., November 17, 1971.

27. For further discussion on the work of these economic institutes, laboratories, and bureaus, see E. Lazutkin, "Raboty uchenykh-ekonomistov sibiri i dal'nego vostoka," *Voprosy ekonomiki*, no. 3 (1964): 153–157; F. Baturin and L. Malinovskii, "Ekonomicheskie issledovaniia v sibiri," ibid., no. 12 (1965): 150–152; F. Baturin and P. V. Shemetov, "Dela sibirskikh ekonomistov," ibid., no. 5 (1969): 151–155; and Shemetov, *Ekonomicheskie issledovaniia*, pp. 12–24.

28. A. B. Gorstko and N. V. Sysoletina, "Igra kak metoda obucheniia studentov-ekonomistov (iz opyta Novosibirskogo gosudarstvennogo universiteta)," *EKO*, no. 3 (1971): 130–147; A. G. Granberg, "Spetsial'nost', rozhdennaia nauchno-tekhnicheskoi revoliutsiei," ibid., no. 2 (1975): 48–59; Pal'm, "Sibirskii algoritm"; and Shemetov, *Ekonomicheskie issledovaniia*, pp. 66–67, 72–84.

29. PANO, f. 5435, op. 1, ed. khr. 10, l. 18; Prudenskii, "O razvitii ekonomicheskikh issledovanii v sibiri," p. 7; and Baturik, "Ekonomicheskaia nauka v sibiri za 1964 god," pp. 135–138.

30. PANO, f. 5435, op. 1, ed. khr. 10, ll. 123–125; NASO, f. 55, op. 1, ed. khr. 76, ll. 1–3; V. G. Fomin, "O razvitii ekonomicheskikh issledovanii v sibirskom otdelenii AN SSSR," *Izvestiia SO AN SSSR*, no. 10 (1961): 10–13; and Baturik, "Ekonomicheskaia nauka v sibiri za 1964 god," pp. 135–138.

31. On Taylorism, see Hugh G. J. Aitken, *Scientific Management in Action: Taylorism at Watertown Arsenal, 1908–1915* (Princeton, N.J.: Princeton University Press, 1985).

32. Kendall E. Bailes, "Alexei Gastev and the Soviet Controversy over Taylorism, 1918–1924," *Soviet Studies* 29, no. 3 (July 1977): 373–394.

33. Bailes, "Alexei Gastev"; Beissinger, *Scientific Management*, pp. 159–186; Pal'm, "Sibirskii algoritm."

34. "Vnedrenie NOT—vazhnaia zadacha proizvodstvennykh kollektivov," *Izvestiia SO AN SSSR. Seriia obshchestvennykh nauk*, no. 9, vyp. 3 (1965): 3; Prudenskii, "O

razvitii ekonomicheskikh issledovanii v sibiri," pp. 3–6; and Fomin, "O razvitii ekonomicheskikh issledovanii v sibirskom otdelenii," pp. 10–13.

35. See "Chitateli—o 'budniakh not,'" *EKO*, no. 1 (1978): 159–168; E. I. Kissel', "NOT v nauchnykh i proektno-konstruktorskikh organizatsiiakh," ibid., no. 6 (1979): 156–159; and D. A. Bedrii, E. B. Koritskii, R. Kh. Borisov, and A. I. Kravchenko, "A. K. Gastev i nauka o trude," ibid., no. 6 (1983): 99–112. See also Beissinger, *Scientific Management*, pp. 159–186, 234–237.

36. Aganbegian, "O razvitii ekonomicheskoi nauki v 1957–1975 gg.," p. 5.

37. Nemchinov et al., *Obshchie voprosy*, pp. 70–71.

38. Belykh, "Ob otnoshenii ekonomistov k matematike," p. 183; Nemchinov et al., *Obshchie voprosy*, pp. 70–71; G. Rubinshtein, ""Pervootrkryvatel,'" *Za nauku v sibiri*, no. 5 (132), February 3, 1964, 2; and "Leonid Vital'evich Kantorovich," ibid., no. 4 (29), January 23, 1962, 3.

39. G. Rubinshtein, "'Pervootkryvatel,'" p. 2; and "Leonid Vital'evich Kantorovich," p. 3.

40. P. Larin, "Matematika sluzhit ekonomike," *Sovetskaia sibir'*, December 27, 1974. The two works are *Metodiki optimizatsii odnomernogo parametricheskogo riada* and *Rekomendatsii po optimizatsii mnogomernogo parametricheskogo riada*.

41. Fomin, "O razvitii ekonomicheskikh issledovanii v sibirskom otdelenii AN SSSR," pp. 13–14, and Larin, "Matematika sluzhit ekonomike."

42. A. G. Aganbegian, "Primenenie matematichikh metodov v planirovanii," *Za nauku v sibiri*, no. 44 (69), October 31, 1962, 2; and "Sodruzhestvo ekonomistov, matematikov, inzhenerov," *Ekonomicheskaia gazeta*, no. 50 (71), December 8, 1962, 18.

43. Fomin, "O razvitii ekonomicheskikh issledovanii v sibirskom otdelenii AN SSSR," pp. 13–14; Baturik, "Ekonomicheskaia nauka v sibiri za 1964 god," pp. 135–138; and Pal'm, "Sibirskii algoritm."

44. NASO, f. 55, op. 1, ed. khr. 76, ll. 10–12.

45. A. G. Aganbegian, "Poisk optimal'nogo varianta," *Molodoi kommunist*, no. 12 (1974): 88.

46. Interview with Boris Pavlovich Orlov, December 19, 1992.

47. Otchet o deiatel'nosti IEiOPP SO AN SSSR za 1967 god (Novosibirsk: IEiOPP SO AN SSSR, 1967), pp. 6–13.

48. PANO, f. 5434, op. 1, ed. khr. 16, ll. 34–36; and A. G. Aganbegian, "Poisk optimal'nogo varianta," i "Ekonomicheskaia nauka i trebovaniia praktiki," *Ekonomicheskaia gazeta*, no. 47 (November 1972): 12; and idem., "Institut ekonomiki i organizatsii promyshlennogo proizvodstva," ibid., no. 11 (March 1969): 15. See *The Methodological State of Optimal Branch Planning in Industry* (Metodicheskie polozhenie po optimal'nomu otraslevomu planirovaniiu v promyshlennosti) (1972); K. K. Val'tukh, *Narodno-khoziaistvennye modeli. Tendentsii razvitiia ekonomiki SSSR* (Novosibirsk: Nauka, 1974); A. G. Granberg, ed., *Ekonomichesko-matematicheskii analiz razmeshcheniia proizvoditel'nykh sil SSSR* (Novosibirsk: Nauka, 1972); idem., *Mezhotraslevye balansy v analize territorial'nykh proportsii SSSR* (Novosibirsk: Nauka, 1975); A. G. Aganbegian and A. G. Granberg, *Ekonomiko-matematicheskii analiz mezhotraslevogo balansa SSSR* (Moscow: Mysl', 1968); A. G. Aganbegian and K. K. Valtukh, *Ispol'zovanie narodno-khoziaistvennykh modelei v planirovanii* (Moscow: Ekonomika, 1975); A. G. Granberg, ed., *Metody i modely territorial'nogo planirovaniia*, vol. 3 (Novosibirsk: IEiOPP, 1975); M. K. Bandman, ed., *Regional Development in the USSR. Modeling the Formation of Soviet Territorial-Production Complexes*, trans. F. E. Ian Hamilton and Boris

P. Kutyriev (Oxford: Pergamon, 1976); and A. G. Aganbegian, ed., *Regional Studies for Planning and Projecting* (Paris: Mouton, 1981).

49. "Otchet nauchno-issledovatel'skikh rabot instituta ekonomiki i organizatsii promyshlennogo proizvodstva na 1981 g.," pp. 19–27. There have been a number of monographs based on this work, including L. M. Ruvinskaia, *Modeling of the Dynamics of Complexes of Consumer Demand* (1981); K. K. Valtukh and I. A. Itskovich, eds., *Applied Economic Models* (1981); R. M. Shniper, ed., *Problems of Research of Regional Programs* (1981); K. K. Bandman, ed., *Territorial Production Complexes* (1981); and E. A. Burov, *Economic Problems of the Development of Productive Forces of the Region* (1981).

50. A. G. Aganbegian, "Institut ekonomiki," p. 15.

51. T. Gaponova, V. Pushkarev, and V. Bykov, "Pochemu uezzhaiut novosely," *Trud*, June 24, 1972, p. 2; and Zaslavskaia, "Tvorcheskaia aktivnost' mass: sotsial'nye rezervy rosta," *EKO*, no. 3 (1986): 17.

52. PANO, f. 5434, op. 1., ed. khr. 11, l. 71.

53. A. G. Aganbegian, "Ekonomicheskaia nauka i trebovaniia praktiki."

54. PANO, f. 5434, op. 1, ed. khr. 16, ll. 11–13.

55. Ibid., op. 1, ed. khr. 16, ll. 36–45.

56. PANO, f. 269, op. 18, ed. khr. 1, ll. 92–95.

57. K. V. Mokhortov, "Baikalo-amurskaia magistral' na rubezhe zavershaiushchego etapa stroitel'stva," in A. G. Aganbegian and A. A. Kin, eds., *BAM: Pervoe desiatiletie* (Novosibirsk: Nauka, 1985), pp. 29–33.

58. Zamira Ibragimova, "Doroga v zavtra," *EKO*, no. 2 (1976): 56.

59. V. A. Lamin, "Iarostnoe preodolenie prostranstva," in Aganbegian and Kin, *BAM*, pp. 14–28.

60. PANO, f. 269, op. 12, ed. khr. 36, ll. 5–32; Vladimir Sbitnev, "Nauka osvaivaet BAM," *Zabaikalskii rabochii*, July 8, 1975; A. G. Aganbegian, "V interesakh razvitiia," *Sovetskaia sibir'*, December 7, 1979, p. 2; and Mokhortov, "Baikalo-amurskaia magistral,'" pp. 33–35.

61. R. I. Shniper, "Programnaia prorabotka problemy khoziaistvennogo osvoeniia zony BAM," in Aganbegian and Kin, *BAM*, pp. 68–105.

62. Mokhortov, "Baikalo-amurskaia magistral,'" pp. 34–37; and B. A. Bessolov and A. G. Kleva, "Tonneli BAMa," in Aganbegian and Kin, *BAM*, pp. 38–44.

63. Iu. Medvedev, "Lishnei dorogi v sibiri ne byvaet," *Pravda*, January 29, 1992, p. 1.

64. T. I. Zaslavskaia, *The Second Socialist Revolution: An Alternative Soviet Strategy*, trans. Susan M. Davies with Jenny Warren, with a foreword by Teodor Shanin (London: Tauris, 1990), p. 2; and Archie Brown, "Tat'yana Zaslavskaya and Soviet Sociology: An Introduction," *Social Research* 55, nos. 1–2 (spring/summer 1988): 264–266.

65. T. I. Zaslavskaia and R. V. Ryvkina, *Sotsiologiia ekonomicheskoi zhizni: ocherki istorii* (Novosibirsk: Nauka SO, 1991).

66. G. Krovitskii and B. Revskii, *Nauchnye kadry VKP (b)*, (Moscow, 1930), pp. 16, 51–83.

67. On the history of Soviet psychology, see Raymond Bauer, *The New Man in Soviet Psychology* (Cambridge, Mass.: Harvard University Press, 1952); and Loren Graham, *Science, Philosophy, and Human Behavior* (New York: Columbia University Press, 1987), 157–219; as well as the writings of V. S. Vygotskii and A. R. Luria.

68. Interview with Tatiana Ivanovna Zaslavskaia, Moscow, December 29, 1992; and T. I. Zaslavskaia, *The Second Socialist Revolution*, p. 29.

69. Interview with Tatiana Ivanovna Zaslavskaia, Moscow, December 29, 1992.

70. Ibid.

71. Zaslavskaia, *The Second Socialist Revolution*, p. 11.

72. Interview with Tatiana Ivanovna Zaslavskaia, Moscow, December 29, 1992.

73. Ibid.

74. Ibid.

75. Zaslavskaia and Ryvkina, *Sotsiologiia ekonomicheskoi zhizni*, pp. 3–4.

76. Interview with Tatiana Ivanovna Zaslavskaia, Moscow, December 29, 1992; and Zaslavskaia, *The Second Socialist Revolution*, p. 33.

77. Interview with Tatiana Ivanovna Zaslavskaia, Moscow, December 29, 1992.

78. Interview with Iurii Voronov, Novosibirsk, December 4, 1991.

79. Zaslavskaia, *The Second Socialist Revolution*, p. 23; and idem., *Sovremennaia ekonomika kolkhozov* (Moscow: USSR Academy of Sciences, 1960).

80. T. I. Zaslavskaia, "Puti obshchestvennogo soizmereniia truda v kolkhozakh," *Voprosy ekonomiki*, no. 10 (1963): 64–69.

81. T. I. Zaslavskaia, *Raspredelenie po trudy v kolkhozakh* (Moscow: Ekonomika, 1966). See also T. I. Zaslavskaia, ed., *Migratsiia selskogo naseleniia* (Moscow: Mysl', 1970).

82. A. G. Aganbegian and V. N. Shubkin, "Sotsiologicheskie issledovaniia i kolichestvennye metody," in *Kolichestvennye metody v sotsiologii* (Moscow: Nauka, 1961), pp. 16–33; A. G. Aganbegian, G. V. Osipov, and V. N. Shubkin, eds., *Kolichestvennye metody v sotsiologii* (Moscow: Nauka, 1966); R. V. Ryvkina, "K voprosu o printsipakh klassifikatsii metodov, ispol'zuemykh v sotsiologii," in T. I. Zaslavskaia, V. A. Kalmyk, and R. V. Ryvkina, eds., *Nauchnyi seminar po primeneniiu kolichestvennykh metodov v sotsiologii* (Novosibirsk: LPSMME, 1966), 2:58–68; and V. E. Shliapentokh, "Chelovek kak istochnik sotsiologicheskoi informatsii," in ibid., 2:74–91. At the same seminar Shliapentokh, who became a professor of sociology at Michigan State University, laid out criteria for designing rigorous survey instruments.

83. A. G. Aganbegian et al., eds., *Modelirovanie sotsial'nykh protessessov* (Moscow: Nauka, 1970), pp. 5–7. See also T. I. Zaslavskaia, "Problemy issledovaniia i modelirovaniia mobil'nosti trudovykh resursov," in ibid., pp. 139–154; and T. I. Zaslavskaia and R. V. Ryvkina, *Metodologischeskie problemy sistemnogo izucheniia derevni* (Novosibirsk: Nauka, 1977).

84. T. I. Zaslavskaia, "Zadachi sotisal'no-ekonomicheskogo izucheniia derevni v svete sistemnogo podkhoda," in T. I. Zaslavskaia and R. V. Ryvkina, eds., *Problemy sistemnogo izucheniia derevni* (Novosibirsk: IEiOPP, 1975), pp. 5–15; and T. I. Zaslavskaia and R. V. Ryvkina, "Material'noe potreblenie kak podsistema derevni," in ibid., pp. 87–95.

85. T. I. Zaslavskaia, "Sotsiologicheskie issledovaniia trudovykh resursov," *EKO*, no. 1 (1972): 43–56.

86. Tatiana Zaslavskaia, "Sibirskoe selo," *Izvestiia*, November 4, 1971; and A. G. Aganbegian, "Poisk optimal'nogo varianta," *Molodoi kommunist*, no. 12 (1974): 89.

87. "Otchet nauchno-issledovatel'skikh rabot instituta ekonomiki i organizatsiia promyshlennogo proizvodstva na 1981 g.," pp. 5–10.

88. T. I. Zaslavskaia, *The Second Socialist Revolution*, p. 2.

89. T. I. Zaslavskaya, "The Novisibirsk Report," *Survey* 28, no. 1 (spring 1984): 88–89.

90. Ibid., p. 92.

91. Ibid., p. 98.

92. Ibid., pp. 98–99.

93. Ibid., pp. 102, 106–108.

94. Archie Brown, "Tat'yana Zaslavskaya and Soviet Sociology," pp. 262–276.

95. Ibid., pp. 267–276; Zaslavskaia and Ryvkina, *Sotsiologiia ekonomicheskoi zhizni*, pp. 32–33.

96. Zaslavskaia and Ryvkina, *Sotsiologiia ekonomicheskoi zhizni*, pp. 5, 13–15, 33. On the recognition of the Communist Party leadership of the growing importance of sociological research to achieve perestroika, see "V tsentral'nom komitete KPSS," *Pravda*, June 12, 1988, pp. 1–2.

97. Zaslavskaia and Ryvkina, *Sotsiologiia ekonomicheskoi zhizni*, pp. 8, 12.

98. For further discussion of Aganbegian's early views on perestroika, see his "Acceleration and Perestroika," in A. G. Aganbegian and Timor Timofeev, *The New State of Perestroika* (New York: Institute for East-West Security Studies, 1988), pp. 25–43.

99. "Otchet nauchno-issledovatel'skikh rabot instituta ekonomiki i organizatsiia promyshlennogo proizvodstva na 1986–1990 gg.," pp. 19–30.

100. Ibid., pp. 26–28, 71–76.

101. See, for example, Voronov, "Pochemu ukazy El'tsina ne vedut k rynochnoi ekonomike," *Sibirskaia birzhevaia gazeta*, no. 21 (November 1991): 2, 7.

102. T. I. Zaslavskaia, *The Second Socialist Revolution*, pp. 2–17. See also T. I. Zaslavskaia, *A Voice of Reform*, ed. Murray Yanowitch (Armonk, N.Y.: M. E. Sharpe: 1989); and Murray Yanowitch, ed., *New Directions in Soviet Social Thought* (Armonk, N.Y.: M. E. Sharpe, 1989).

CHAPTER SEVEN
CRACKDOWN: THE COMMUNIST PARTY AND ACADEMIC FREEDOM
IN AKADEMGORODOK

1. See, for example, the impact of the fall 1976 Central Committee resolution concerning "Work on the Selection and Training of Ideological Cadres in Party Organizations of Belorussia" on "labor discipline" (a case of drunkenness) in the Institute of Cytology and Genetics, in PANO, f. 269, op. 12, ed. khr. 3, ll. 114–116.

2. PANO, f. 269, op. 1, ed. khr. 86, ll. 237–241; ibid., ed. khr. 88, ll. 236–37; and ibid., ed. khr. 125, ll. 86–87.

3. Egor Kuz'mich Ligachev, *Ekonomika, politika, printsipy upravleniia* (Moscow: Sovetskaia rossiia, 1965), pp. 13–14, 19, 65–67.

4. PANO, f. 269, op. 1, ed. khr. 2, ll. 113–118. See also E. K. Ligachev, "Zakladyvaia fundament bol'shoi nauki v sibiri," in *Izbrannye rechi i stati* (Moscow: Izdatpolit, 1989), pp. 8–11.

5. E. K. Ligachev, "Za vysokuiu stroitel'nuiu kul'turu," *Akademstroevets*, no. 9 (31), February 3, 1959, 2; idem., "Zakladyvaia fundament bol'shoi nauki," pp. 8–11; and idem., "Taktika osvoeniia prirodnykh resursov," *EKO*, no. 1 (1981): 22–27.

6. F. S. Goriachev, "Soiuz nauki i proizvodstva," *Ekonomicheskaia gazeta*, no. 13 (March 1972): 10.

7. G. Krovitskii and B. Revskii, *Nauchnye kadry VKP (b)*, (Moscow, 1930), pp. 16, 51–83.

8. PANO, f. 384, op. 1., ed. khr. 13, l. 2; ibid., ed. khr. 24, l. 1; I. A. Moletotov, *Partiinaia rabota v nauchnom tsentr* (Novosibirsk: Novosibirskoe knizhnoe izdatel'stvo,

1963), p. 4; and idem., "Problema kadrov sibirskogo nauchnogo tsentra i ee reshenie (1957–1964 gg.)," in B. M. Shchereshevskii, ed., *Voprosy istorii sovetskoi sibiri* (Novosibirsk: Izdatel'stvo NGU, 1967), pp. 346–347.

9. G. D. Lykov et al., *Gordoe zvanie stroitel'* (Novosibirsk: Zapadno-sibirskoe knizhnoe izdatel'stvo, 1982), p. 48; and M. P. Chemodanov, "2,590 vedut za soboi," *Akademstroevets*, no. 93 (212), November 29, 1960, 1.

10. N. I. Lubenikov et al., *Novosibirskaia organizatsiia KPSS v tsifrakh (1920–1980 gg.)* (Novosibirsk: Zapadno-sibirskoe knizhnoe izdatel'stvo, 1981), pp. 44–46.

11. Ibid., pp. 62–63.

12. Ibid., pp. 47–49.

13. Ibid., pp. 60–61.

14. PANO, f. 269, op. l, ed. khr. 24, ll. 126–128; and Moletotov, *Partiinaia rabota*, pp. 42–43.

15. PANO, f. 269, op. 1, ed. khr. 118, ll. 31–34.

16. Interviews with G. S. Migirenko, December 8 and 12, 1991, Akademgorodok; "Po plechu!" *Akademstroevets*, no. 94 (213), December 2, 1960, 1; PANO, f. 384, op. 1, ed. khr. 17; and ibid., ed. khr. 25.

17. PANO, f. 4, op. 99, ed. khr. 120, l. 47.

18. Lavrentev, "Tak nachinalsia Akademgorodok . . ." *Komsomol'skaia pravda*, May 14, 1977, p. 4.

19. PANO, f. 269, op. 1, ed. khr. 193, ll. 73–74.

20. Ibid., ed. khr. 191, ll. 69–71.

21. Ibid., op. 7, ed. khr. 9, ll. 118–120; and ibid., ed. khr. 3, l. 77.

22. For the best single treatment of the issues and actors in Soviet science policy, see Stephen Fortescue, *Science Policy in the Soviet Union* (New York: Routledge, 1990). On industrial R and D and technology policy, see Robert Lewis, *Science and Industrialization in the USSR* (New York: Holmes and Meier, 1979), and Bruce Parrott, *Politics and Technology in the Soviet Union* (Cambridge, Mass.: MIT Press, 1983).

23. PANO, f. 384, op. 1., ed. khr. 13, ll. 97–98; and ibid., ed. khr. 17, ll. 36–37.

24. Louvan Nolting, *The Planning of Research, Development, and Administration in the USSR*, Foreign Economic Report No. 14, U.S. Department of Commerce, Census Bureau (Washington: U.S. GPO, 1978).

25. "V tsentral'nom komitete KPSS," *Pravda*, February 11, 1977, p. 1; *Za nauku v sibiri*, no. 16 (797), April 14, 1977, 1; and Marchuk, "Sibirskii potential," *Trud*, March 27, 1977, p. 2.

26. Several Akademgorodok scientists were stung by what they viewed as unfair criticism of their performance. See PANO, f. 269, op. 20, ed. khr. 9, ll. 4–23, 36.

27. V. Kadzhaia, "Ot zamysla do vnedreniia," *Ekonomicheskaia gazeta*, no. 25 (June 1972): 22; and I. Lavrov, "Sviaz' priamaia i obratnaia," *Sovetskaia sibir'*, March 19, 1980, p. 2.

28. *Fizika i proizvodstvo*, no. 1 (1930): 51–56; and ibid., no. 2–3 (1930): 61–65.

29. M. A. Lavrentev, *Izvestiia*, January 17, 1966; see also his "Uchenye sibiri—narodnomu khoziaistvu," *Literatura i zhizn'*, February 27, 1959; "Plody nauki—proizvodstvu," *Izvestiia*, January 24, 1967; "Uskorenie vremeni," *Komsomol'skaia pravda*, July 19, 1969; and "Ot idei do mashiny," *Literaturnaia gazeta*, January 1, 1970, p. 10.

30. M. A. Lavrentev, "Molodezh' i nauka," *Izvestiia*, March 8, 1968.

31. PANO, f. 384, op. 1, ed. khr. 17, ll. 36–37; and M. A. Lavrentev, "Ot idei do mashiny"; and idem., "Uskorenie vremeni."

32. G. Migirenko, "Nauka v rabochei spetsovke," *Komsomol'skaia pravda*, March 10, 1966.

33. V. Mindolin, "Spetsialist pridet v nauku," *Vechernii novosibirsk*, July 8, 1974; and Moletotov, *Partiinaia rabota*, pp. 45–46. For Trofimuk's endorsement of the special role of the Komsomol in the innovation beltway, see PANO, f. 269, op. ed. 20, khr. 9, l. 37.

34. PANO, f. 269, op. 12, ed. khr. 3, ll. 9–10; ibid., ed. khr. 27, ll. 25–30; and G. I. Marchuk, "Gliadia v griadushchee," *Literaturnaia gazeta*, April 23, 1980, p. 12.

35. Marchuk, *Molodym o nauke* (Moscow: Molodaia gvardiia, 1980), pp. 4–5, 208–211, 215.

36. Ibid., p. 55.

37. Marchuk, "'Poias vnedreniia,'" *Sotsialisticheskaia industriia*, May 12, 1974, p. 2.

38. Marchuk, "Sibirskii potential," *Trud*, March 27, 1977, p. 2; "Strategiia nauki—kompleksnost'," *Komsomol'skaia pravda*, May 16, 1978, p. 2; "Sibirskie traditsii," *Sotsialisticheskaia industriia*, October 25, 1977; and "Uchenye sibiri—piatiletke," *Sovetskaia sibir'*, May 7, 1976.

39. Interview with G. I. Marchuk, January 28, 1992, Moscow; M. Monusov, "Nauka i proizvodstvo: kurs na sotrudnichestvo," *Vechernii novosibirsk*, December 4, 1975; and Marchuk, *Molodym o nauke*, pp. 63–64.

40. PANO, f. 269, op. 12, ed. khr. 27, ll. 25–30.

41. "Dostizheniia nauki i tekhniki—na sluzhbu narodnomu khoziaistvu," *Sovetskaia sibir'*, April 26, 1981, pp. 1–3.

42. *Znamiia truda* (Sibselmash newspaper), November 17, 1971, January 15, 1975, and December 12, 1975.

43. *Znamiia truda*, October 22, 1975.

44. Kadzhaia, "Ot zamysla do vnedreniia"; R. Notman, "Novyi uroven' sotrudnichestva," *Sovetskaia sibir'*, February 25, 1978, p. 2; and "Nauka na perednem krae bor'by za kommunizm," *Sovetskaia sibir'*, April 14, 1979, pp. 1–2. Brezhnev noted a series of other successful cooperative ventures between science and technology: work with Sibakdemstroi on new apparatuses and mechanisms for construction using prefabricated concrete forms; an automated workplace (ARM) based on an M-6000 computer developed by the SKB of Scientific Instrument Building; and the nuclear physics institute's industrial accelerators that made it possible to use such soft woods as birch or linden in construction.

45. N. Solovykh, "Otvetstvennaia i vazhnaia rol'," *Vechernii novosibirsk*, June 10, 1981; A. Deribas, "Ekonomiiu daet vzryv," *Sovetskaia sibir'*, Janaury 23, 1980; and Marchuk, *Molodym o nauke*, p. 60.

46. For more on Deribas, see "Ego stikhiia—vzryz," *Za nauku v sibiri*, no. 24 (1005), June 11, 1981, 6. See also Z. Ibragimova, "Vzryv bez ekha," *Literaturnaia gazeta*, August 26, 1981, p. 10, and PANO, f. 269, op. 30, ed. khr. 3, ll. 59–60.

47. PANO, f. 269, op. 14, ed. khr. 2, ll. 39–42; and Marchuk, *Molodym o nauke*, pp. 66–67, 69.

48. Ianovskii, "Vosrastanie roli partiinogo rukovodstva nauchno-tekhnicheskim progressom," *Kommunist*, no. 18 (1977): 55–58; and Marchuk, "Rezervy effektivnosti," *Trud*, October 15, 1977.

49. PANO, f. 269, op. 30, ed. khr. 3, l. 62.

50. Ibid., ll. 59–60.

51. R. G. Ianovskii, "Vysokii dolg uchenykh," *Sovetskaia sibir'*, July 22, 1972, p. 2, and Goriachev, "Soiuz nauki i proizvodstva," *Ekonomicheskaia gazeta*, no. 13 (March 1972): 10.

52. See also the thoughts of A. P. Filatov, first secretary of the Novosibirsk Obkom in the late 1970s, as set forth in "Nauka na perednem krae bor'by za kommunizm," *Sovetskaia sibir'*, April 14, 1979, pp. 1–2; and "Dostizheniia nauki—v praktiku," *Ekonomicheskaia gazeta*, no. 25 (1981): 5.

53. On the Marxist study circles, see Josephson, *Physics and Politics in Revolutionary Russia* (Los Angeles: University of California Press, 1991), pp. 203–208.

54. Iu. P. Ozhegov and R. G. Ianovskii, "Krepit' soiuz filosofii i estestvoznaniia," *Voprosy filosofii*, no. 12 (1962): 153–155.

55. PANO, f. 5434, op. 1, ed. khr. 8, l. 54.

56. Ibid., f. 269, op. 1, ed. khr. 86, ll. 237–241; ibid., op. 7, ed. khr. 38, ll. 32–33; and Moletotov, *Partiinaia rabota*, pp. 9–11.

57. PANO, f. 269, op. 1, ed. khr. 86, ll. 237–241; ibid., ed. khr. 88, ll. 236–37; ibid., ed. khr. 125, ll. 86–87; and ibid., op. 26, ed. khr. 6, ll. 173–175.

58. Moletotov, *Partiinaia rabota*, pp. 7–8.

59. PANO, f. 269, op. 1, ed. khr. 225, ll. 57–58.

60. Interview with Aleksandr Danilovich Aleksandrov, January 5, 1993, St. Petersburg; Arkhiv Leningradskogo otdeleniia AN SSSR, f. 1034, op. 3, ed. khr. 3, letters from V. A. Fok to A. D. Aleksandrov, letter 1, ll. 1–3; and Mark Popovsky, *Manipulated Science*, trans. Paul Falla (Garden City, N.Y.: Doubleday, 1979), p. 162.

61. Aleksandrov, "Kommunist v nauke," *Pravda*, February 12, 1966; and idem., "Razmyshleniia ob ekonomike i etike," *EKO*, no. 12 (1986): 78–87. See also his collection of essays, *Problemy nauki i pozitsiia uchenogo* (Leningrad: Nauka, 1988).

62. Moletotov, *Partiinaia rabota*, pp. 5–7, 16–19.

63. Ibid., pp. 23–25.

64. Pritvits and Makarov, *Khronika*, pp. 315–322; and Lykov, *Gordoe zvanie stroitel'*, pp. 60–61.

65. PANO, f. 269, op. 12, ed. khr. 4, ll. 42–47; ibid., op. 20, ed. khr. 11, ll. 81–83; and NASO, f. 10, op. 3, ed. khr. 403, ll. 1, 16.

66. PANO, f. 269, op. 1, ed. khr. 225, ll. 32–33.

67. Ozhegov, "Novye zadachi—novye trebovaniia," *Za nauku v sibiri*, no. 9 (34), February 27, 1962, 1; G. Brediuk, "Partprosveshchenie—na uroven' novykh zadach," ibid., 35 (60), August 29, 1961, 1; "Partiinoe prosveshchenie v svete novykh zadach," ibid., 22 (99), June 6, 1963, 1; and V. F. Volin and A. E. Kolesnikov, "Metodologicheskie seminary—forma ideino-politicheskoi i professional'noi podgotovki nauchnykh kadrov," in A. L. Ianshin et al., eds., *Problemy razvitiia sovremennoi nauki* (Novosibirsk: Nauka, 1978), pp. 359–360.

68. PANO, f. 296, op. 1, ed. khr. 118, ll. 70–72.

69. V. I. Kovalev, "K voprosu o sisteme i osnovnykh printsipakh deiatel'nosti filosofskikh (metodologicheskikh) seminarov," in A. L. Ianshin et al., *Problemy razvitiia sovremennoi nauki*, pp. 349–354.

70. Ianovskii, *Ideino-politicheskoe vospitanie nauchno-tekhnicheskoi intelligentsii* (Moscow: Nauka, 1982), pp. 58–63.

71. Andrei Sakharov, *Memoirs*, trans. Richard Lourie (New York: Knopf: 1990), pp. 272–279.

72. PANO, f. 269, op. 7, ed. khr. 9, ll. 134–138; ibid., f. 5434, op. 1, ed. khr. 8, ll. 6–13, 27; and Berg, *Acquired Traits*, pp. 382, 457–462, 467.

73. Interview with A. S. Nariniani, December 14, 1991, Akademgorodok; Berg, *Acquired Traits*, pp. 219–221, 316–317; and Granin, *The Bison*, pp. 225–226.

74. Berg first heard Galich sing his poetry in Novosibirsk when a colleague brought a tape to her home. Not wanting to expose her daughters to the seditious singer, she asked them to leave the room before playing the tape. When the voice "sang out from the tape recorder" one of her daughters cried out from another room, "Mama, that's Galich!" (Berg, *Acquired Traits*, pp. 316, 377). See also Sakharov, *Memoirs*, p. 355.

75. Boris Pasternak, the poet and author who wrote the novel *Doctor Zhivago*, came under attack when he was awarded the Nobel prize in literature in 1958: Khrushchev threatened to expel him from the country; the Writers' Union ousted him from their midst; and mobs came from Moscow to his dacha at Peredelkino to hurl anti-Semitic slogans at him.

76. A reference to the poetess Maria Tsvetaeva, who was exiled in the Stalin years.

77. This refers to Osip Mandelshtam, the great poet who was imprisoned in Suchan where he died.

78. This is a quote from a Zhivago poem, one of Pasternak's best known.

79. This stanza about the "chandelier" refers to those in the Writers' Union, called "chandeliers," who expelled Pasternak from the Union after he received the Nobel prize in literature in 1958.

80. Meli-Yemelia was a traditional Russian fairy-tale hero.

81. The "ultimate penalty" was execution by bullet at the base of the skull.

82. An obvious reference to Christ's crucifixion.

83. This quote is from Pasternak's poem "Hamlet," a poem about a poet's isolation and responsibility.

84. Aleksandr Galich, *General'naia repetitsiia* (Moscow: Sovetskii pisatel, 1991), pp. 57–59. My deep thanks to Lydia Kesich who corrected my translation of this poem and provided its annotation, and to Zora Essman for polishing the final version.

85. Sakharov, *Memoirs*, pp. 280–281.

86. PANO, f. 269, op. 12, ed. khr. 27, ll. 30–32.

87. Ibid., ed. khr. 36, ll. 131–135.

88. Ibid., ed. khr. 3, ll. 9–10; and ibid., ed. khr. 27, ll. 25–32, 116–120.

89. Ibid., ed. khr. 27, ll. 98–101.

EPILOGUE

1. Some of the issues addressed in this epilogue I raise in "Russian Scientific Institutions: Internationalisation, Democracy, and Dispersion," *Minerva* 32, no. 1 (spring 1994): 1–24, and "Scientists, the Public, and the Party under Gorbachev," *Harriman Institute Forum* 3, no. 5 (May 1990).

2. NASO, f. 10, op. 4, ed. khr. 849, ll. 52–53; Migirenko, *Novosibirskii nauchnyi tsentr*, pp. 197–202; Pritvits and Makarov, *Khronika*, pp. 76, 92–93, 114–115, 135; and N. A. Dediushina and A. I. Shcherbakov, "O formirovanii nauchnykh kadrov sibirskogo otdeleniia AN SSSR," in A. P. Okladnikov, ed., *Voprosy istorii nauki i professional'nogo obrazovaniia v sibiri*, vol. 1 (Novosibirsk, 1968), pp. 232–235.

3. Iu. D. Tsvetkov, *Nauchno-organizatsionnaia deiatel'nost' sibirskogo otdeleniia AN SSSR v 1986 gody i zadachi na perspektivy* (Novosibirsk: SO AN SSSR, 1987), pp. 4–12, tables 4 and 5; a classified report.

4. V. Dadykova, "Kooperativy v nauke: zlo ili blago," *Nauka v sibiri*, no. 6 (February 17, 1989): 2.

5. See Vladimir D. Shkolnikov, *Scientific Bodies in Motion: The Domestic and International Consequences of the Current and Emergent Brain Drain from the Former USSR* (Santa Monica, Calif.: Rand Corporation, 1994). Although Shkolnikov's estimates of potential brain drain are speculative and I believe exaggerated, his work remains one of the few good studies that exist on the subject.

ABOUT THE AUTHOR

Paul R. Josephson is currently writing a cultural history of the Soviet nuclear era. His other books include *Physics and Politics in Revolutionary Russia* and *Totalitarian Science and Technology*.